Hydrogen Bonds

Hydrogen Bonds

Editor

Mirosław Jabłoński

MDPI • Basel • Beijing • Wuhan • Barcelona • Belgrade • Manchester • Tokyo • Cluj • Tianjin

Editor
Mirosław Jabłoński
Nicolaus Copernicus University in Toruń
Poland

Editorial Office
MDPI
St. Alban-Anlage 66
4052 Basel, Switzerland

This is a reprint of articles from the Topical Collection published online in the open access journal *Molecules* (ISSN 1420-3049) (available at: https://www.mdpi.com/journal/molecules/topical_collections/Hydrogen_Bonds).

For citation purposes, cite each article independently as indicated on the article page online and as indicated below:

LastName, A.A.; LastName, B.B.; LastName, C.C. Article Title. *Journal Name* **Year**, *Volume Number*, Page Range.

ISBN 978-3-0365-6770-9 (Hbk)
ISBN 978-3-0365-6771-6 (PDF)

© 2023 by the authors. Articles in this book are Open Access and distributed under the Creative Commons Attribution (CC BY) license, which allows users to download, copy and build upon published articles, as long as the author and publisher are properly credited, which ensures maximum dissemination and a wider impact of our publications.

The book as a whole is distributed by MDPI under the terms and conditions of the Creative Commons license CC BY-NC-ND.

Contents

About the Editor vii

Preface to "Hydrogen Bonds" ix

Mirosław Jabłoński
Hydrogen Bonds
Reprinted from: *Molecules* **2023**, *28*, 1616, doi:10.3390/molecules28041616 1

Frank Weinhold
High-Density "Windowpane" Coordination Patterns of Water Clusters and Their
NBO/NRT Characterization
Reprinted from: *Molecules* **2022**, *27*, 4218, doi:10.3390/molecules27134218 5

Miguel Gallegos, Daniel Barrena-Espés, José Manuel Guevara-Vela, Tomás Rocha-Rinza and Ángel Martín Pendás
A QCT View of the Interplay between Hydrogen Bonds and Aromaticity in Small
CHON Derivatives
Reprinted from: *Molecules* **2022**, *27*, 6039, doi:10.3390/molecules27186039 23

Wojciech Pietruś, Rafał Kafel, Andrzej J. Bojarski and Rafał Kurczab
Hydrogen Bonds with Fluorine in Ligand– Protein Complexes-the PDB Analysis and
Energy Calculations
Reprinted from: *Molecules* **2022**, *27*, 1005, doi:10.3390/molecules27031005 41

Poul Erik Hansen
Isotope Effects on Chemical Shifts in the Study of Hydrogen Bonds in Small Molecules
Reprinted from: *Molecules* **2022**, *27*, 2405, doi:10.3390/molecules27082405 55

Elena Yu. Tupikina, Mark V. Sigalov and Peter M. Tolstoy
Simultaneous Estimation of Two Coupled Hydrogen Bond Geometries from Pairs of Entangled
NMR Parameters: The Test Case of 4-Hydroxypyridine Anion
Reprinted from: *Molecules* **2022**, *27*, 3923, doi:10.3390/molecules27123923 75

Mirosław Jabłoński
On the Coexistence of the Carbene···H-D Hydrogen Bond and Other Accompanying
Interactions in Forty Dimers of N-Heterocyclic-Carbenes (I, IMe$_2$, I^{i}Pr$_2$, I^{t}Bu$_2$, IMes$_2$, IDipp$_2$,
IAd$_2$; I = imidazol-2-ylidene) and Some Fundamental Proton Donors (HF, HCN, H$_2$O,
MeOH, NH$_3$)
Reprinted from: *Molecules* **2022**, *27*, 5712, doi:10.3390/molecules27175712 89

Rafał Wysokiński, Wiktor Zierkiewicz, Mariusz Michalczyk, Thierry Maris and Steve Scheiner
The Role of Hydrogen Bonds in Interactions between [PdCl$_4$]$^{2-}$ Dianions in Crystal
Reprinted from: *Molecules* **2022**, *27*, 2144, doi:10.3390/molecules27072144 109

Paweł A. Wieczorkiewicz, Tadeusz M. Krygowski and Halina Szatylowicz
Intramolecular Interactions in Derivatives of Uracil Tautomers
Reprinted from: *Molecules* **2022**, *27*, 7240, doi:10.3390/molecules27217240 121

Sebastian Szymański and Irena Majerz
Theoretical Studies on the Structure and Intramolecular Interactions of Fagopyrins—Natural
Photosensitizers of *Fagopyrum*
Reprinted from: *Molecules* **2022**, *27*, 3689, doi:10.3390/molecules27123689 137

Emran Masoumifeshani, Michał Chojecki, Dorota Rutkowska-Zbik and Tatiana Korona
Association Complexes of Calix[6]arenes with Amino Acids Explained by
Energy-Partitioning Methods
Reprinted from: *Molecules* **2022**, *27*, 7938, doi:10.3390/molecules27227938 157

Kamil Wojtkowiak and Aneta Jezierska
Exploring the Dynamical Nature of Intermolecular Hydrogen Bonds in Benzamide, Quinoline
and Benzoic Acid Derivatives
Reprinted from: *Molecules* **2022**, *27*, 8847, doi:10.3390/molecules27248847 199

Nina Podjed and Barbara Modec
Hydrogen Bonding and Polymorphism of Amino Alcohol Salts with Quinaldinate:
Structural Study
Reprinted from: *Molecules* **2022**, *27*, 996, doi:10.3390/molecules27030996 215

About the Editor

Mirosław Jabłoński

Mirosław Jabłoński, Dr., is the Associate Professor in the Department of Quantum Chemistry and Atomic Spectroscopy in the Faculty of Chemistry at the Nicolaus Copernicus University in Toruń. He deals with computational chemistry in the field of molecular modeling and physical organic chemistry. The subject of his scientific research concerns various intra- and intermolecular interactions, especially hydrogen, hydride, halogen, triel and agostic bonds. He is also interested in methods of studying electron density distribution, endohedral chemistry, carbenes and methods of determining the energy of intramolecular interactions.

Preface to "Hydrogen Bonds"

The eponymous hydrogen bonds are generally considered to be the most important type of so-called non-covalent interactions. This position results from both their ubiquity and their enormous importance in practically all natural sciences. Undoubtedly, the main example is the presence of hydrogen bonds O-H...O in liquid and solid water, which gives it specific properties, e.g., an exceptionally high boiling point compared to its heavier counterpart, hydrogen sulfide, or the relatively low density of ice. Nevertheless, hydrogen bonds are just as important, if not more so, in macromolecules including amino acids and proteins. In this case, the template example is the N-H...O and N-H...N hydrogen bonds in the complementary nitrogen base pairs cytosine–guanine and adenine–thymine. Their presence allows the binding of two DNA strands, giving them a double-helical structure, which is of fundamental importance in the replication of genetic information. In addition to these instances of intermolecular hydrogen bonding, intramolecular hydrogen bonding is of similar importance. Particularly important in this case is the impact of their presence and strength on conformational preferences, which in turn may translate into a specific structure in the solid state determined by crystallographic measurements. Another area of the manifestation of hydrogen bonds is the movement and even transfer of the proton in X...H...Y hydrogen bridges.

Hydrogen bonds owe their extraordinary role and importance mainly to the fact that they are defined by a unique atom—the smallest hydrogen. On the one hand, its presence between the larger X and Y atoms in the X-H...Y bridge allows these atoms to approach a fairly short distance without significant steric effects. On the other hand, the partial positive charge on the hydrogen atom allows for relatively strong attractive interactions with a high-electron-density region, most often the electron lone pair on a strongly electronegative Y atom. All these specific properties make hydrogen bonds act as a glue in the rich world of various intermolecular interactions, binding single molecules into dimers and larger aggregates. At the same time, their relative weakness allows for the full dynamics of this process.

It might seem that after more than a hundred years of research into hydrogen bonding, this area has been completely exploited; there is nothing more to study on the subject of hydrogen bonds. However, this reprint is excellent proof that despite more than a hundred years of research into hydrogen bonds, they are still an interesting and fundamental topic in the natural sciences. The collected articles deal with various aspects of the existence of hydrogen bonds and show human ingenuity in the field of inventing and applying various techniques, both experimental and theoretical, in order to describe them more fully.

Mirosław Jabłoński
Editor

Editorial

Hydrogen Bonds

Mirosław Jabłoński

Faculty of Chemistry, Nicolaus Copernicus University in Toruń, Gagarina 7, 87-100 Toruń, Poland; teojab@chem.umk.pl; Tel.: +48-56-611-4695

The Topical Collection "Hydrogen Bonds" is a continuation of the previous Special Issue "Intramolecular Hydrogen Bonding 2021" [1]. Nevertheless, there is a crucial difference: by also including intermolecular hydrogen bonds, the topic is significantly generalized. This collection consists of 12 research articles, including one review article. Once more, the collected articles address various aspects of hydrogen bonding, which are analyzed by both experimental and various theoretical methods; these articles demonstrate that hydrogen bonding remains a major topic in chemistry.

Water is a quintessential example of the occurrence of hydrogen bonding; therefore, Weinhold's theoretical study on high-density water clusters [2] deserves special attention. The water clusters considered in this article are formed via the use of a cyclic $(H_2O)_4$ "windowpane" cluster, characterized by a quadrilateral coordination motif and then by limiting itself to the requirement of the maximal Grotthuss-type proton ordering and organization according to the Aufbau fashion. Importantly, Weinhold demonstrates the utility of the NRT-based bond order and emphasizes that the nature of hydrogen bonding can be completely explained in terms of resonance-covalency ("charge transfer").

By applying quantum chemical topology (QCT) tools (QTAIM, IQA) and electronic delocalisation indicators (FLU and MCI), Gallegos, Barrena-Espés, Guevara-Vela, Rocha-Rinza, and Pendás detail the influence of aromaticity and antiaromaticity on the characteristics of the so-called aromaticity and antiaromaticity-modulated hydrogen bonds (AMHB) [3]. Significant differences are observed, depending on whether a proton donor or proton acceptor atom is incorporated into the monomer ring. The results that the authors obtain show that aromaticity and antiaromaticity may be considered on a common scale.

Based on data from both the Protein Data Bank and theoretical calculations, Pietruś, Kafel, Bojarski, and Kurczab detail the results of their research on the abundance, structure, and strength of the hydrogen bond with the fluorine atom as an acceptor [4]. The authors confirm that fluorine is a weak proton acceptor and thus forms weak hydrogen bonds; these are rather forced by the presence of stronger ligand–receptor interactions. For this reason, X-H$\cdots$F hydrogen bonds are, typically, considerably more bent (120°–150°) than standard hydrogen bonds.

In a review article [5], Hansen shows that the influence of isotopes on NMR chemical shifts is an important tool in chemistry; isotopes can help describe hydrogen bonds, determine structural parameters, and define dimers, trimers, etc. It is particularly constructive to analyze the impact of isotopic effects upon chemical shifts in the study of common tautomeric systems. The fact that two-bond deuterium isotope effects (TBDIE) may be employed in order to estimate the energy of intramolecular hydrogen bonds is also accentuated.

The advantages of NMR spectroscopy are also exhibited in an insightful and original article by Tupikina, Sigalov, and Tolstoy [6]. Namely, the authors propose a method for determining the geometry of two coupled hydrogen bonds by defining a pair of NMR chemical shifts for various atoms. This method is based on the determination of two-dimensional maps, which present the dependence of two chemical shifts upon the position of hydrogen atoms in the coupled hydrogen bonds, and then the determination of the intersection point of two isolines. It is especially recommended that the chemical shifts

Citation: Jabłoński, M. Hydrogen Bonds. *Molecules* **2023**, *28*, 1616. https://doi.org/10.3390/molecules28041616

Received: 13 January 2023
Accepted: 6 February 2023
Published: 8 February 2023

Copyright: © 2023 by the authors. Licensee MDPI, Basel, Switzerland. This article is an open access article distributed under the terms and conditions of the Creative Commons Attribution (CC BY) license (https://creativecommons.org/licenses/by/4.0/).

of the carbons and protons in the CH groups of the linking system are used. While this method may be promising, its possible limitations are also identified.

Jabłoński presents the results of the first systematic theoretical study of the hydrogen bond between the carbene carbon atom of the commonly used imidazol-2-ylidene (I) derivatives (IMe$_2$, I^{i}Pr$_2$, I^{t}Bu$_2$, IPh$_2$, IMes$_2$, IDipp$_2$, IAd$_2$) and the fundamental proton donors (HF, HCN, H$_2$O, MeOH, NH$_3$) [7]. The author found that, for a given carbene, the dissociation energy values of the IR$_2\cdots$HD dimers increase in the following order: NH$_3$ < H$_2$O < HCN $\leq$ MeOH $\ll$ HF; in addition, for a given HD proton donor, IDipp$_2$ forms the strongest dimers. The roles of the various accompanying secondary interactions are also analyzed.

Wysokiński, Zierkiewicz, Michalczyk, Maris, and Scheiner show that the presence of the hydrogen bonds of the N-H$\cdots$Cl type, formed between the counterion (e.g., [NH$_3$-(CH$_2$)$_4$-NH$_3$]$^{2+}$) and the two anions [PdCl$_4$]$^{2-}$, facilitates a relatively short contact between these anions in the crystal [8]. Thus, these hydrogen bonds play the role of glue binding both anions.

Applying theoretical methods (cSAR, QTAIM, NCI), Wieczorkiewicz, Krygowski, and Szatylowicz investigate the influence of the substituent (-NH$_2$ or -NO$_2$ group in positions 5 and 6) and solvent (PCM model; $1 \leq \varepsilon < 109$) on the electronic structure and the presence of intramolecular hydrogen bonds in tautomeric forms of uracil [9]. Neither substitution nor solvent has been shown to affect tautomeric preferences significantly.

Szymański and Majerz introduce a theoretical study on the spatial structure and intramolecular hydrogen bonds of fagopyrins, which are characterized by a double-anthrone moiety and are natural photosensitizers of the *Fagopyrum* species [10]. Although these systems are characterized by the presence of a strong double hydrogen bridge O-H$\cdots$O(=C)$\cdots$H-O in the "peri" region, the presence of N-heterocyclic piperidine and pyrrolidine rings in these systems affects the possibility of breaking these bridges due to the formation of competitive hydrogen bonds O-H$\cdots$N.

An essential area of manifestation of the importance of intermolecular interactions, especially hydrogen bonds, are inclusion complexes based on the guest–host interaction. Masoumifeshani, Chojecki, Rutkowska-Zbik, and Korona use energy partitioning approaches (SAPT, F-SAPT, SSMF3 (i.e., the authors' modification of SMF)) to systematically characterize interactions in the complexes between amino acids as the potential guests and three calix[6]arene and hexa-*p*-*tert*-butylcalix[6]arene conformers as the hosts [11]. One of the many interesting findings of this study is that the most stable pinched-cone conformer is the one least prone to interact with amino acids. Methodologically, the SSMF3 procedure was revealed to be suited to reproducing the interaction energy of the complexes, particularly in regard to its dispersion component.

By applying Car–Parrinello and Path Integral molecular dynamics, Wojtkowiak and Jezierska investigate (also after taking into account nuclear quantum effects, NQEs) the dynamic nature of intermolecular hydrogen bonds in crystalline and gaseous phases of 2,6-difluorobenzamide, 5-hydroxyquinoline, and 4-hydroxybenzoic acid [12]. The authors reveal that the inclusion of NQEs engenders a reduction in the donor–acceptor distance and an increase in proton delocalization in hydrogen bridges (except for C-H$\cdots$O in the first system).

Podjed and Modec, by performing in-depth solid-state studies, characterize polymorphisms in the salts of three amino alcohols (3-amino-1-propanol, 2-amino-1-butanol, and 2-amino-2-methyl-1-propanol) with quinaldinic acid (i.e., quinoline-2-carboxylic acid) [13]. All of the investigated structures contain the NH$_3^+\cdots{}^-$OOC heterosynthon. Nevertheless, individual polymorphic forms also vary in their motifs of hydrogen bonds and $\pi\cdots\pi$ stacking interactions.

Funding: This research received no external funding.

Acknowledgments: I thank all the Authors for their valuable contributions to the Topical Collection "Hydrogen Bonds" and all the Reviewers for their responsible effort in evaluating the submitted manuscripts.

Conflicts of Interest: The author declares no conflict of interest.

References

1. Jabłoński, M. Intramolecular Hydrogen Bonding 2021. *Molecules* **2021**, *26*, 6319. [CrossRef] [PubMed]
2. Weinhold, F. High-Density "Windowpane" Coordination Patterns of Water Clusters and Their NBO/NRT Characterization. *Molecules* **2022**, *27*, 4218. [CrossRef] [PubMed]
3. Gallegos, M.; Barrena-Espés, D.; Guevara-Vela, J.M.; Rocha-Rinza, T.; Pendás, Á.M. A QCT View of the Interplay between Hydrogen Bonds and Aromaticity in Small CHON Derivatives. *Molecules* **2022**, *27*, 6039. [CrossRef] [PubMed]
4. Pietruś, W.; Kafel, R.; Bojarski, A.J.; Kurczab, R. Hydrogen Bonds with Fluorine in Ligand–Protein Complexes-the PDB Analysis and Energy Calculations. *Molecules* **2022**, *27*, 1005. [CrossRef] [PubMed]
5. Hansen, P.E. Isotope Effects on Chemical Shifts in the Study of Hydrogen Bonds in Small Molecules. *Molecules* **2022**, *27*, 2405. [CrossRef] [PubMed]
6. Tupikina, E.Y.; Sigalov, M.V.; Tolstoy, P.M. Simultaneous Estimation of Two Coupled Hydrogen Bond Geometries from Pairs of Entangled NMR Parameters: The Test Case of 4-Hydroxypyridine Anion. *Molecules* **2022**, *27*, 3923. [CrossRef] [PubMed]
7. Jabłoński, M. On the Coexistence of the Carbene···H-D Hydrogen Bond and Other Accompanying Interactions in Forty Dimers of N-Heterocyclic-Carbenes (I, IMe$_2$, I^{i}Pr$_2$, I^{t}Bu$_2$, IMes$_2$, IDipp$_2$, IAd$_2$; I = imidazol-2-ylidene) and Some Fundamental Proton Donors (HF, HCN, H$_2$O, MeOH, NH$_3$). *Molecules* **2022**, *27*, 5712. [CrossRef] [PubMed]
8. Wysokiński, R.; Zierkiewicz, W.; Michalczyk, M.; Maris, T.; Scheiner, S. The Role of Hydrogen Bonds in Interactions between [PdCl$_4$]$^{2-}$ Dianions in Crystal. *Molecules* **2022**, *27*, 2144. [CrossRef] [PubMed]
9. Wieczorkiewicz, P.A.; Krygowski, T.M.; Szatylowicz, H. Intramolecular Interactions in Derivatives of Uracil Tautomers. *Molecules* **2022**, *27*, 7240. [CrossRef]
10. Szymański, S.; Majerz, I. Theoretical Studies on the Structure and Intramolecular Interactions of Fagopyrins–Natural Photosensitizers of Fagopyrum. *Molecules* **2022**, *27*, 3689. [CrossRef]
11. Masoumifeshani, E.; Chojecki, M.; Rutkowska-Zbik, D.; Korona, T. Association Complexes of Calix[6]arenes with Amino Acids Explained by Energy-Partitioning Methods. *Molecules* **2022**, *27*, 7938. [CrossRef]
12. Wojtkowiak, K.; Jezierska, A. Exploring the Dynamical Nature of Intermolecular Hydrogen Bonds in Benzamide, Quinoline and Benzoic Acid Derivatives. *Molecules* **2022**, *27*, 8847. [CrossRef] [PubMed]
13. Podjed, N.; Modec, B. Hydrogen Bonding and Polymorphism of Amino Alcohol Salts with Quinaldinate: Structural Study. *Molecules* **2022**, *27*, 996. [CrossRef] [PubMed]

Disclaimer/Publisher's Note: The statements, opinions and data contained in all publications are solely those of the individual author(s) and contributor(s) and not of MDPI and/or the editor(s). MDPI and/or the editor(s) disclaim responsibility for any injury to people or property resulting from any ideas, methods, instructions or products referred to in the content.

Article

High-Density "Windowpane" Coordination Patterns of Water Clusters and Their NBO/NRT Characterization

Frank Weinhold

Theoretical Chemistry Institute and Department of Chemistry, University of Wisconsin-Madison, Madison, WI 53706, USA; weinhold@chem.wisc.edu

Abstract: Cluster mixture models for liquid water at higher pressures suggest the need for water clusters of higher coordination and density than those commonly based on tetrahedral H-bonding motifs. We show here how proton-ordered water clusters of increased coordination and density can assemble from a starting cyclic tetramer or twisted bicyclic (Möbius-like) heptamer to form extended *Aufbau* sequences of stable two-, three-, and four-coordinate "windowpane" motifs. Such windowpane clusters exhibit sharply reduced (~90°) bond angles that differ appreciably from the tetrahedral angles of idealized crystalline ice I_h. Computed free energy and natural resonance theory (NRT) bond orders provide quantitative descriptors for the relative stabilities of clusters and strengths of individual coordinative linkages. The unity and consistency of NRT description is demonstrated to extend from familiar supra-integer bonds of the molecular regime to the near-zero bond orders of the weakest linkages in the present H-bond clusters. Our results serve to confirm that H-bonding exemplifies resonance–covalent (fractional) bonding in the sub-integer range and to further discount the dichotomous conceptions of "electrostatics" for intermolecular bonding vs. "covalency" for intramolecular bonding that still pervade much of freshman-level pedagogy and force-field methodology.

Keywords: supramolecular chemistry; hydrogen bonding; water clusters; natural bond orbitals; natural resonance theory; natural bond orders; Grotthuss proton ordering; water wires; glassy water; quantum cluster equilibrium

1. Introduction

The earliest applications of ab initio natural bond orbital (NBO) analysis [1–4] consistently revealed a "donor–acceptor" (resonance–covalency-type "charge transfer") picture of hydrogen bonding that was sharply at odds with then-prevalent "electrostatic" conceptions of intermolecular interactions [5,6]. Although the IUPAC *Gold Book* definition of H-bonding was subsequently revised to acknowledge the importance of covalency in H-bonding [7], superficial "dipole–dipole" rationalizations of H bonding continue to survive in many freshman-level expositions [8]. Arguments against the charge-transfer picture or in support of classical-type long-range, multipole, or "electrostatically driven" conceptions of H-bonding continue to appear [9,10] (vs. replies in [11–13]) in the research literature, and similar simplifying approximations persist in the empirical force fields of popular molecular dynamics (MD) simulation methods [14] that are commonly adopted to describe H-bonding in condensed phases.

The daunting task of describing macroscopic phases of liquid water or other H-bonded fluids may seem to demand the drastic long-range approximations of intermolecular ("noncovalent") interactions as compared to the exchange-type ("covalent") interactions of the short-range molecular regime. However, a more practical and accurate approach to describing intermolecular H-bonding is achieved by adopting *supramolecular clusters* [15] $\{C_n\}$ as the conceptual "building blocks" of the macroscopic liquid-phase description, based on the known *continuity* of high-density liquid and low-density gaseous phases around the fluid critical point [16]. More specifically, quantum cluster equilibrium (QCE) theory [17–19] provides a practical numerical implementation of such "cluster mixture" [20–25] modeling of macroscopic phase properties, based on accurate values of electronic and vibrational properties of H-bonded {C_n} clusters that can be obtained at any chosen ab initio or density functional theory (DFT) level. The key input for QCE-based thermodynamic modeling of an aqueous phase is the data set of supramolecular clusters whose self-consistent (T,P)-dependent equilibrium populations are determined from the computed partition functions for each cluster by the standard methods of quantum statistical thermodynamics [26].

Among the many H-bonded fluids of practical interest, water itself presents the most studied yet still most perplexing phase behavior of the terrestrial regime [27]. Even the microscopic structure and properties of "ordinary" liquid water under near-ambient conditions remain matters of controversy [28]. Further mysteries surround the phase behavior of water at higher temperatures and pressures, where both theory [29–32] and experiments [33–36] have suggested the existence of an alternative high-density phase of liquid water that could lead to a liquid–liquid critical point and an exotic new domain of thermodynamic behavior near 220 K and 1–2 kbar.

The primary goal of present work was to computationally search for a new class of water clusters {W_n} based on the quadrilateral ("windowpane") coordination motif of the cyclic tetramer (Figure 1) that might contribute to equilibrium QCE populations in the neighborhood of the proposed high-density phase. In each case, we restricted attention to clusters that maintain maximal Grotthuss-type proton ordering for the powerful effects of cooperative stabilization [37,38], as exemplified by the clockwise ordering of in-ring OH bonds in the view of Figure 1. The near −90° coordination angles of the windowpane class correspond to reduced next-neighbor distances and increased mass/volume ratios compared to the characteristic tetrahedral angles and chair–hexagon coordination motifs of ice-I-like clusters. The search for cooperatively stabilized windowpane clusters is organized in *Aufbau* fashion toward increasing numbers of fully four-coordinate sites that more adequately sample the intermolecular interactions expected to dominate in the phase behavior of the low-temperature and high-pressure regime. The resulting windowpane clusters can serve as computational input for subsequent QCE studies to examine their possible role in the equilibrium cluster distributions of the water-phase diagram.

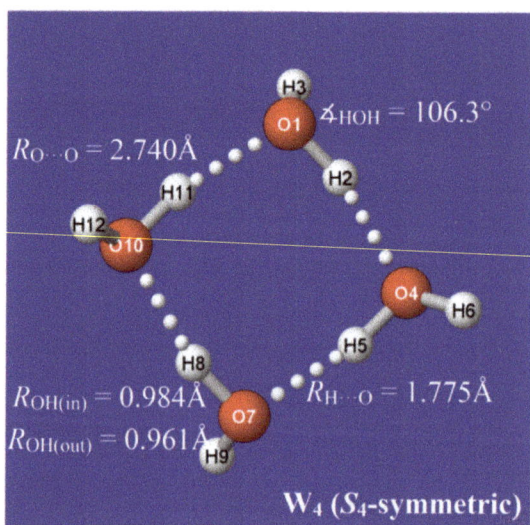

Figure 1. Equilibrium structural properties of cyclic $(H_2O)_4$ "windowpane" cluster (B3LYP/6-311++G** level).

A secondary goal of this study was to characterize each computed cluster in deeper conceptual terms that can clarify distinctive features of the underlying H-bond interactions. Such characterization should include aspects of overall cluster stability, strengths of individual coordinative linkages, shifts in atomic charge distribution, and other orbital-level features of free vs. coordinated water molecules. For these purposes, we employed NBO analysis [39,40] to obtain localized descriptors of molecular and intermolecular bonding features. Of particular interest are natural resonance theory (NRT) bond orders [41], which are expected to exhibit useful correlations with bond lengths [42,43], bond energies [44,45], bond stretching frequencies [46–48], NMR ^{1}J and ^{1h}J spin-coupling constants [49], and other experimentally measurable properties.

Although the present study of novel water clusters was primarily directed toward equilibrium thermodynamic properties, it is important to note that such studies can also yield information on the kinetics and mechanisms of water cluster reactions. This is particularly true when, as in the present case, each cluster of the class is created in a sequential *Aufbau* manner from a previous member, e.g., by successive dimer additions of the form

$$W_k + W_2 \rightleftharpoons W_{k+2} \qquad (1)$$

where $W_k = (H_2O)_k$ is a k-mer of a chosen coordination pattern. Analogous to elementary $A + B \rightleftharpoons C$ chemical reactions, one can compute the transition state $(W_k \cdots W_2)^\ddagger$ and other features of the intrinsic reaction coordinate [50] (IRC) for each such cluster reaction. Similarly, for other cluster species satisfying the simultaneous QCE equilibrium conditions,

$$W_j + W_k \rightleftharpoons W_{j'} + W_{k'} \; (j + k = j' + k') \qquad (2)$$

standard quantum chemical methods can be employed to determine transition-state features and associated absolute rate constants along the associated IRC [51]. However, such deeper mechanistic aspects of cluster formation were not addressed in the present work.

2. Computational Methods

For direct comparisons with many previous chemical applications in the NBO/NRT literature [52,53], we employed the familiar B3LYP/6-311++G** level of hybrid density functional theory for all geometry optimizations and energy evaluations of the present work. As shown elsewhere [54,55], realistic treatment of thermodynamic properties requires balanced treatment of energetic (primarily electronic) and entropic (primarily vibrational) contributions to free energy. All species were fully optimized and checked for vibrational stability with standard options of the *Gaussian-16* program [56]. NBO/NRT analyses were completed with the *NBO7* program [57,58] in interactive *G16/NBO7* configuration. Structural and orbital graphics were obtained with the *NBOPro7@Jmol* utility program [59]. For NRT analyses of larger clusters, keyword selections for enlarged dynamic memory and the number of resonance structures were required to obtain fully converged bond orders. Ready-to-run input files containing optimized cartesian coordinates and keyword input for each cluster are included in the Supporting Information (SI). As shown particularly in ref. [54], many DFT variants and additional "corrections" (for dispersion, counterpoise, etc.) give qualitatively similar results for individual cluster structures and relative energies, even if some choices prove "best" for a particular thermodynamic comparison. The provided SI files allow re-optimization of cluster structures for alternative method/basis levels of choice.

3. Sequential Aufbau of 2-, 3-, 4-Coordinate Windowpane Water Clusters

The properties of each water cluster W_k of an envisioned class are dictated by its specific H-bond coordination pattern. As primary descriptors of this pattern, we expect that each water molecule may generally be involved in two-, three-, or four-coordinate H-bonding to other molecules of the cluster (with singly coordinated "dangling" molecules excluded in leading clusters of the equilibrium thermodynamic distribution). For labeling purposes, the coordination pattern of each cluster may be usefully described by the number of quadruply (*q*), triply (*t*), or doubly (*d*) coordinated sites, appended as pre-superscripts (viz., $^{q,t,d}W_n$) to the cluster symbol. In this notation, the cyclic water tetramer of Figure 1 is labeled $^{0,0,4}W_4$, with each monomer doubly coordinated in chain-like linkages to the substrate.

The structural logic for sequential *Aufbau* construction of windowpane clusters is straightforward. Starting from an existing cluster of this class, such as the cyclic water tetramer of Figure 1, one can choose any edge-type coordination (such as that between O(1) and O(10) in Figure 1) as a "base" for a new windowpane by attaching a water dimer in parallel fashion with two new H-bonds, as shown in the left panel of Figure 2. For maximum stabilization in forming this new H-bond attachment (e.g., from emanating H(12) at O(10)), the Grotthuss-type proton ordering should be continued around the edges of the newly formed windowpane that joins to O(1). The net result of this particular attachment is that sites O(10) and O(1) become tri-coordinate ($t \to t + 2$), while other sites remain di-coordinate, leading to an overall $^{0,0,4}W_4 \to {}^{0,2,4}W_6$ change in labeling. Some of these clusters, such as $^{0,0,4}W_4$ itself or the cubane-like $^{0,8,0}W_8$ described below, are featured in many previous cluster investigations, but the emphasis here is on hierarchical families of clusters that can be associated with a well-defined mechanistic *Aufbau* sequence of dimer additions, particularly leading to higher four-coordinate (*q*-type) motifs.

By alternating the sign of folding angles between panes, such additions can be continued indefinitely in "ladder-like" procession, as shown in successive panels of Figure 2. Each panel of Figure 2 includes (in parentheses) the per-monomer energy and standard-state Gibbs free energy change with respect to free water molecules, which serve to exhibit the important cooperative (nonadditive) effects of Grotthuss-ordered coordination patterns. The first four panels ($^{0,0,4}W_4$, $^{0,2,4}W_6$, $^{0,4,4}W_8$, $^{0,6,4}W_{10}$) show the addition of successive rungs to the ladder pattern, up to the four-pane member. The ensuing $^{1,4,4}W_9$ (row 3, left) is the alternative "2 × 2" four-pane cluster, which adopts a buckled saddle-shape deformation from planarity with a central four-coordinate monomer. From the starting two-pane ladder ($^{0,2,4}W_6$) at the upper right, one can also attempt to add another rung that curls backward (*E*-like) rather than forward (*Z*-like), but this optimizes to the cubane-like $^{0,8,0}W_8$ cluster (row 3, right). The cubane motif becomes an evident building block for extensions to two-cube ($^{4,8,0}W_{12}$), three-cube ($^{8,8,0}W_{16}$), or longer rod-like clusters, as illustrated in the final row of the figure.

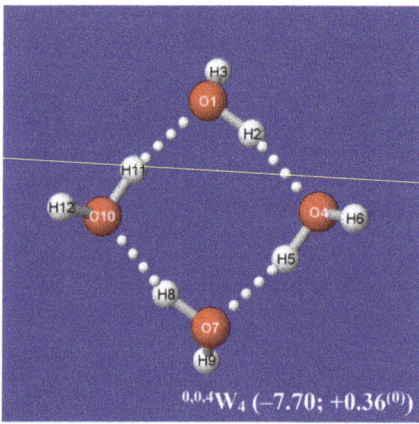 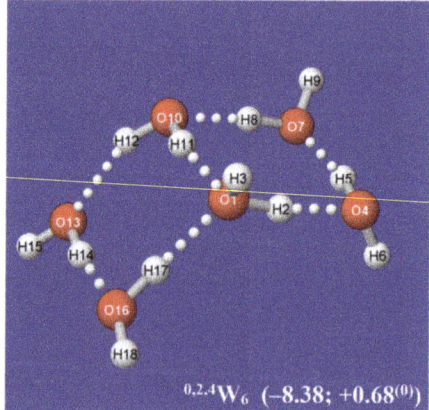

$^{0,0,4}W_4$ (−7.70; +0.36$^{(0)}$) $\qquad$ $^{0,2,4}W_6$ (−8.38; +0.68$^{(0)}$)

Figure 2. *Cont.*

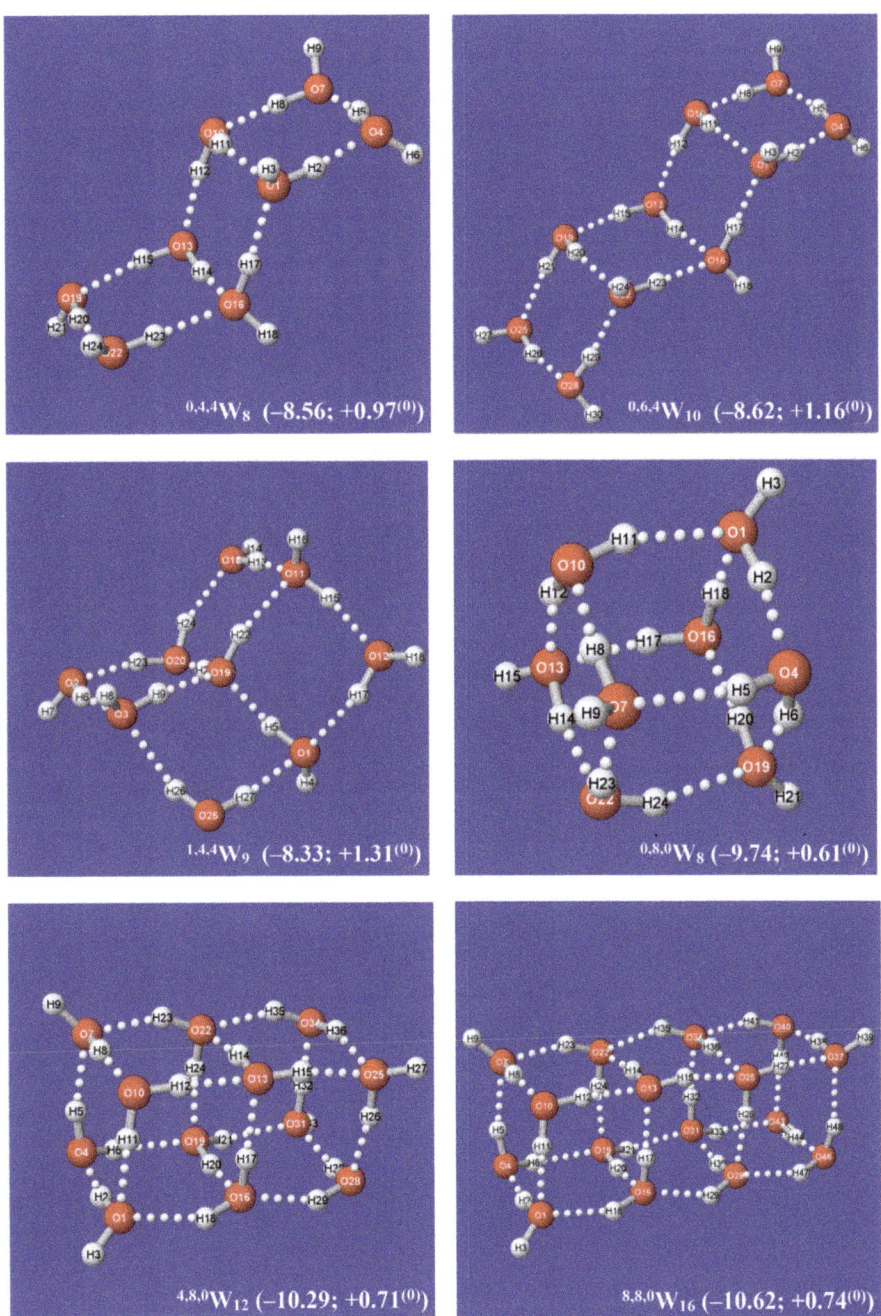

Figure 2. Calculated *Aufbau* sequence of windowpane clusters $^{q,t,d}W_k$ from starting cyclic tetramer $^{0,0,4}W_4$ (**upper left**), showing parenthesized per-monomer changes (kcal/mol) in energy (ΔE) and Gibbs free energy ($\Delta G^{(0)}$) from free water molecules in each panel.

An alternative *Aufbau* starting point is provided by the twisted two-pane ($^{1,0,6}W_7$) cluster shown in the upper-left panel of Figure 3. This cluster features "Möbius-like" coordination with a continuous Grotthuss-ordered chain passing twice through the unique four-coordinate central monomer to form a closed loop. Remaining panels of Figure 3 show selected clusters that are obtained by successive Grotthuss-ordered dimer additions to $^{1,0,6}W_7$, aimed at increasing q numbers of saturated four-coordinate sites. The resulting structures all incorporate the higher density coordination angles of the windowpane motif, but they exhibit irregular overall shapes that appear suitable as possible contributions to bulk liquid or amorphous solid phases. As seen in Figures 2 and 3, the $^{8,8,0}W_{16}$ cluster (Figure 2, lower right) achieves the largest number of three- and four-coordinate sites ($q = t = 8$) and the deepest per-monomer energy (-10.62 kcal/mol) in the depicted sequences. However, whether some or all of these clusters contribute significantly to known roots of the QCE equations, or whether (like the buckyball-type clathrate clusters previously studied [60]) they can serve as leading contributors to entirely new roots (phases) of the QCE phase diagram remains to be investigated.

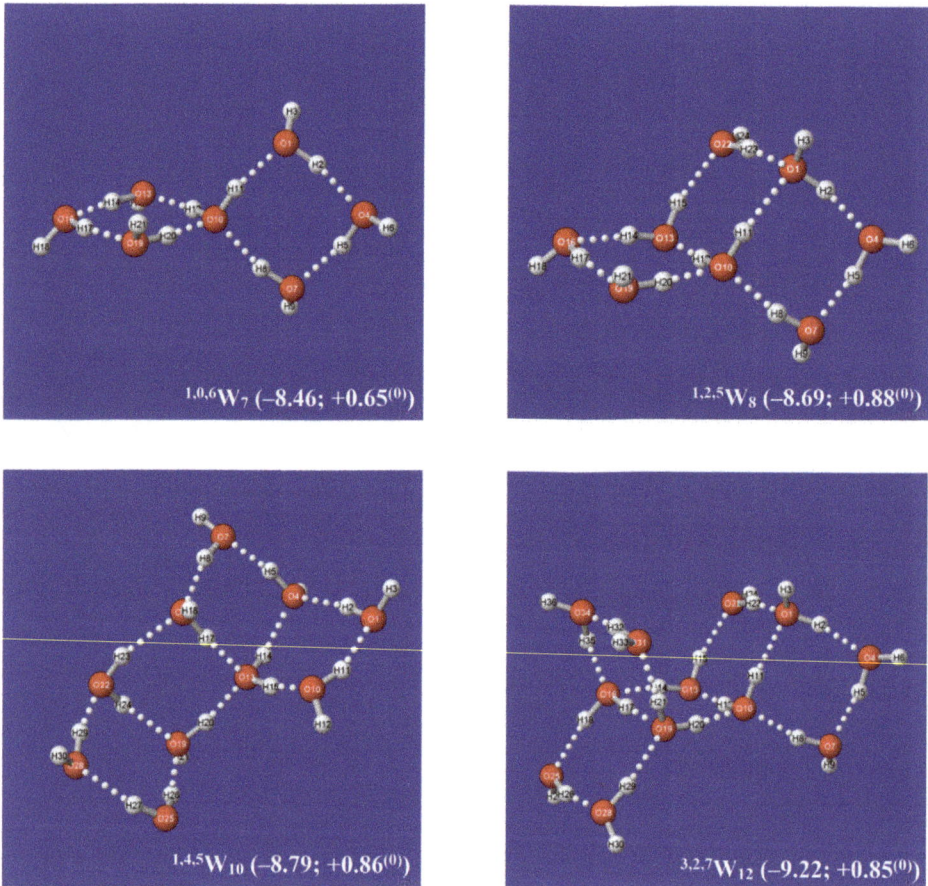

Figure 3. *Cont.*

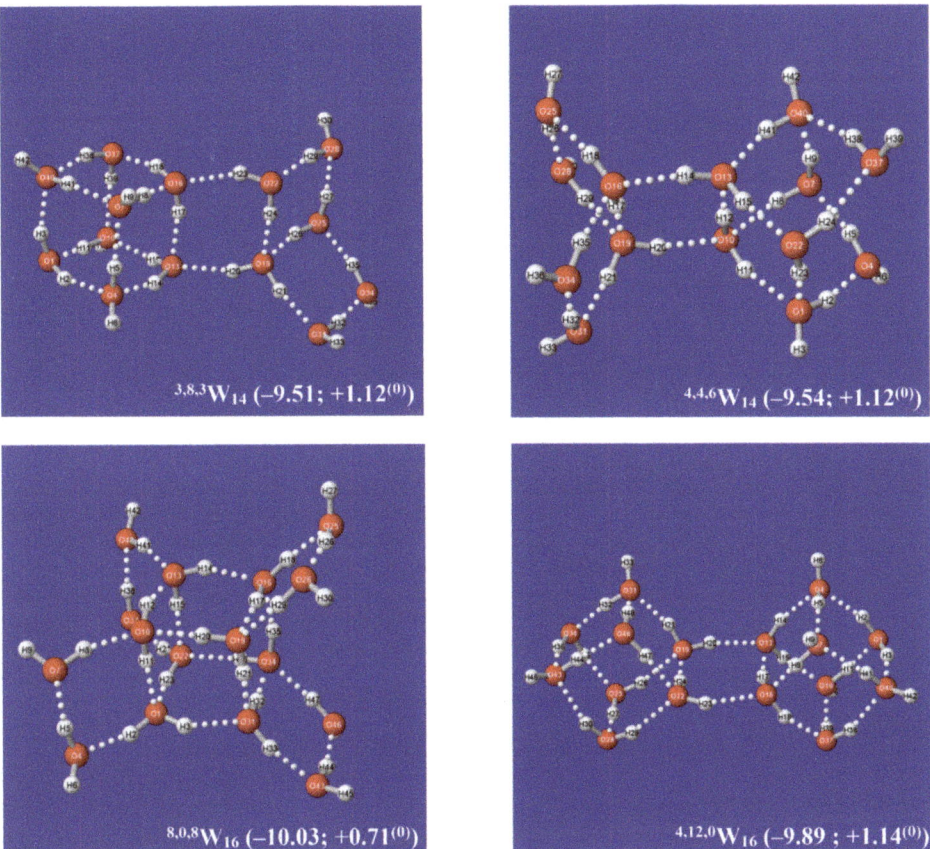

Figure 3. Similar to Figure 2, for successive $^{q,t,d}W_k$ windowpane clusters built from the Möbius-like $^{1,0,6}W_7$ cluster (**upper left**).

It is evident that each *Aufbau* cluster shown in Figures 2 and 3 may have alternative isomeric rearrangements of the proton network without altering the $q/t/d$ descriptors of O⋯(H)⋯O coordination linkages. Such alternative $^{q,t,d}W_n^{(alt)}$ isomers may have higher point group symmetry, different proton orderings (e.g., Grotthuss cycles around individual panes rather than overall periphery), and higher or lower energy than the *Aufbau*-derived clusters described above. Figure 4 displays two such alternative high-symmetry forms of the $^{0,2k,4}W_{2k+4}^{(sym)}$ sequence ($k = 1, 2$), with respective C_s ($k = 1$), C_i ($k = 2$) symmetry. The C_s-symmetric $^{0,2,4}W_6^{(Cs)}$ structure (Figure 4, left) is slightly higher in energy than $^{0,2,4}W_6$ of Figure 2, but C_i-symmetric $^{0,4,4}W_8^{(Ci)}$ (Figure 4, right) is slightly lower in energy than its low-symmetry counterpart in Figure 2. The inherent chirality of the coordination pattern about *each* O atom of higher-coordinated water clusters of Figures 2 and 3 indicates that reduced symmetry (net chirality) is a high-probability feature of equilibrium water cluster distributions in any phase involving their participation.

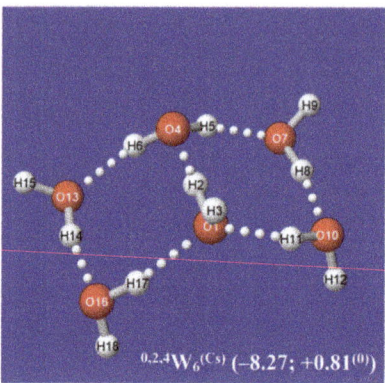
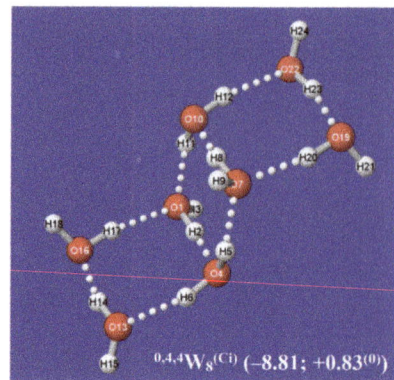

Figure 4. Alternative higher-symmetry $^{0,2k,4}W_{2k+4}^{(sym)}$ clusters (k = 1,2), one (C_s) of higher energy, the other (C_i) of lower energy than the corresponding low-symmetry structure of Figure 2.

Note that although H-bonds are considered weak noncovalent attractions, the cumulative energy release from larger cluster formation (viz., $\Delta E \approx 170$ kcal/mol for the $^{8,8,0}W_{16}$ cluster) can readily exceed that necessary to dissociate a strong covalent bond, as in the ion pair clusters involved in self-dissociation (pH) of liquid water [54,55]. The per-monomer free energies of formation shown in Figures 2 and 3 remain slightly positive under standard-state conditions, but the windowpane clusters are expected to gain increased stability relative to the ice-like clusters of the near-ambient regime as pressure increases. Full thermochemical and vibrational spectroscopic values for each cluster are included with the optimized coordinates in SI.

4. Natural Atomic Charge and Bond Order Characterizations

Among the many descriptors provided by NBO analysis, the natural atomic charges $\{Q_A\}$ and interatomic bond orders $\{b_{AB}\}$ are most intimately associated with traditional empirical concepts of chemical bonding theory. Long-held perceptions of *dichotomy* between intra- vs. intermolecular forces (viz., "covalency" for chemical bond formation (b_{AB} = 1, 2, 3,...) vs. "electrostatics" for H-bond formation ($b_{H\cdots O} \approx 0.1$–0.2)) have long impeded true progress in the supramolecular domain. Demonstrations of how quantal Q_A, b_{AB} descriptors extend seamlessly across the supposed divide can therefore serve to refute the obsolete dipole–dipole conceptions of H-bonding (and other so-called "noncovalent" interactions) that still pervade freshman-level pedagogy and classical force-field methodology. In the present section, we wish to test the usefulness of NBO/NRT-based Q_A, b_{AB} descriptors when applied to the large data base of windowpane water clusters as described above.

4.1. General Features of Donor–Acceptor Interactions in Water Clusters

In every H-bond of every water cluster, NBO analysis reveals the characteristic $n_O \rightarrow \sigma^*_{OH}$ donor–acceptor ("charge transfer") interaction that transfers a slight electronic charge (Q_{CT}) from the oxygen lone pair (n_O) of the Lewis base (LB) site into the valence antibond (σ^*_{OH}) of the proximal Lewis acid (LA) site. Figure 5 depicts the n_O-σ^*_{OH} interaction for one of the H-bonds of W_{4c}, showing the strongly overlapping forms of pre-orthogonal PNBOs deep inside van der Waals contact. The insets show details of the interaction that are routinely provided in NBO output, including (in kcal/mol; upper right) the second-order perturbative estimate of n_O-σ^*_{OH} donor–acceptor attraction ($\Delta E_{CT}^{(2)}$), the corresponding steric opposition of n_O-σ_{OH} donor–donor repulsion (ΔE_{steric}), and the net binding energy (ΔE_{net}). The known high transferability of NBOs [61] then assures that the individual n_O, σ^*_{OH} orbitals are quite similar to those in water monomer and dimer as well as other windowpane clusters. However, one can also recognize the slight misalignments of ring strain that lower PNBO overlaps throughout the windowpane series and lead to the nuances in charge distribution, structure, and bond strength discussed below.

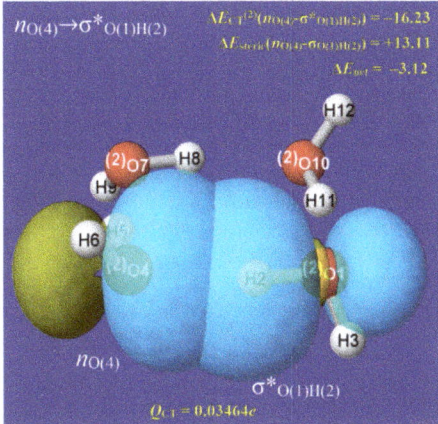

Figure 5. Pre-orthogonal (PNBO) depiction of $n_O \rightarrow \sigma^*_{OH}$ orbital interaction in one H-bond of W_{4c}, with energetic (kcal/mol) and charge transfer (e) details as insets (see text).

Alternatively, the effects of $n_{O(4)} \rightarrow \sigma^*_{O(1)H(2)}$ interaction can be quantified by *deleting* this single specific matrix element from the DFT calculation (with standard $DEL keylist options [62]) and recalculating the energy and reoptimized geometry as though it were absent in nature. As shown in Figure 6, this single deletion "breaks" the $O(4) \cdots H(2)-O(1)$ hydrogen bond (and initial S_4 symmetry) to give an open-chain structure with $R_{O(1)\cdots O(4)}$ separation increased by ~0.5 Å to near-van der Waals contact distance. The monomers at each chain terminus also reorient to near coplanarity (contrary to the ~120° dihedral twisting of the two remaining monomers), thereby allowing partial re-gain of $n^{(\sigma)}_{O(4)} \rightarrow \sigma^*_{O(1)H(2)}$ attraction with the weaker *in*-plane $n^{(\sigma)}_{O(4)}$ lone pair of O(4). By such $DEL deletion searches, one verifies that the specific $n_{O(4)} \rightarrow \sigma^*_{O(1)H(2)}$ interaction is the unique "smoking gun" that is both *necessary and sufficient* for characteristic H-bonding between O(1) and O(4) monomers.

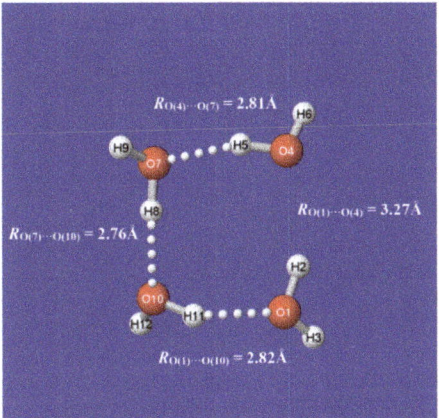

Figure 6. $DEL (partially)-reoptimized structure of original W_{4c} cluster (Figure 1), showing effects of deleting the single $n_{O(4)} \rightarrow \sigma^*_{O(1)H(2)}$ interaction of Figure 5 (at the point where the maximum number of optimization steps was completed).

All such NBO-based energetic and $DEL deletion descriptors can be obtained for other windowpane clusters of Figures 2 and 3. In the following, we focus instead on subtleties of the charge distributions and H-bond strengths that relate to the interesting cooperative effects of the highly ordered proton patterns ("water wires") formed by the H-bond networks.

4.2. Natural Atomic Charge Distributions

In principle, the simple water dimer (W_2) might be seen as the fundamental conceptual building block for studies of electronic charge distribution and stability in clusters of higher complexity. However, Figure 7 exhibits the detailed comparisons of H atom (italic) and O atom (plain text) natural charges in W_2 vs. cubane-like $^{0,8,0}W_8$ to show the surprising *contrasts* between these species. In the two panels of Figure 7, the O(1) and O(16) monomers of the cubane cluster (right) are, respectively, the direct analogs of O(4) and O(2) monomers in the dimer (left), yet the net charges on the monomers of the dimer are directly *opposite* those in the cluster. Similar contrasts between charge distributions of the supposed "building block" dimer and those of higher coordination complexes are found throughout the clusters of Figures 2 and 3.

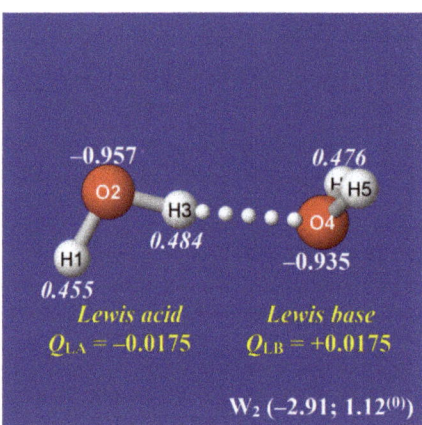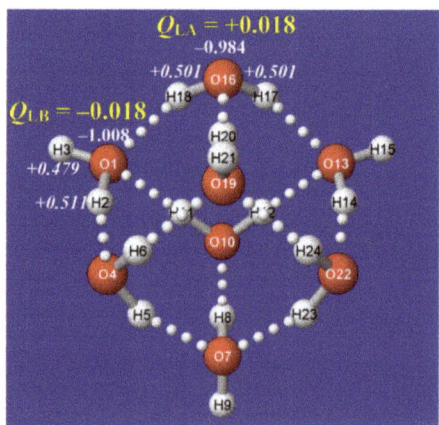

Figure 7. Natural atomic charges for H (italics; white) and O (plain text; white) atoms of water dimer (**left**) and cubane-like $^{0,8,0}W_8$ cluster (**right**), with corresponding net charges (yellow) of formal Lewis acid (*e*-acceptor) and Lewis base (*e*-donor) water molecules in each species, showing the *reversal* of apparent charge flow in the two cases. (Parenthesized per-monomer energy and free energy for W_2 also allow direct stability comparisons with clusters of Figures 2–4).

How can the conflicting charge patterns of Figure 7 be rationalized? At the termini of each H-bond are two water monomers that can be identified as the LB (formal *e*-pair "donor") and LA (formal σ*$_{OH}$ "acceptor" vacancy). In the simple water dimer, the $n_O \rightarrow$ σ*$_{OH}$ donor–acceptor interaction necessarily results in net charge transfer (ca. 0.017*e*) from LB to LA (Figure 7, left), resulting in the LB$^{\delta+}$···LA$^{\delta-}$ charge pattern. However, in more complex water clusters, the surroundings of any chosen H-bond may be seen as a network of "water wires" that allow charge to redistribute as necessary to optimize overall cluster stability. Specifically, the multiple network connections allow electronic charge to be redistributed to achieve near *neutrality* at *q*- or *d*-coordinated sites, whose equal numbers of donor and acceptor interactions can be tuned to avoid capacitive build-up. However, at *t*-coordinated sites, which necessarily have an imbalance of donor (t_d, LB) vs. acceptor (t_a, LA) connections, it becomes advantageous to confer excess *anionic* charge on t_d sites (increasing LB strength) and *cationic* charge on t_a sites (increasing LA strength), thus leading to the commonly observed LB$^{\delta-}$···LA$^{\delta+}$ charge pattern.

To illustrate these propensities of cluster charge distribution, Figure 8 displays selected Q_O (plain text) and Q_H (italic) atomic charges of the $^{8,0,8}W_{16}$ cluster for two q-type sites (centered at O(1), O(13)) and one d-type site (at O(46)), showing the significantly reduced net monomer charges compared to those of the water dimer. More complete listings of monomer charge values and coordination type at each O atom for all clusters of this study are included in SI, as illustrated for the $^{8,0,8}W_{16}$ cluster in Table 1. The subtle variations in molecular charge indicate the extreme "feedback" sensitivity to every detail of the surrounding H-bond network, showing that overall network topology has taken precedence over characteristics of the water dimer (single H-bond) "building block" of which the network is composed.

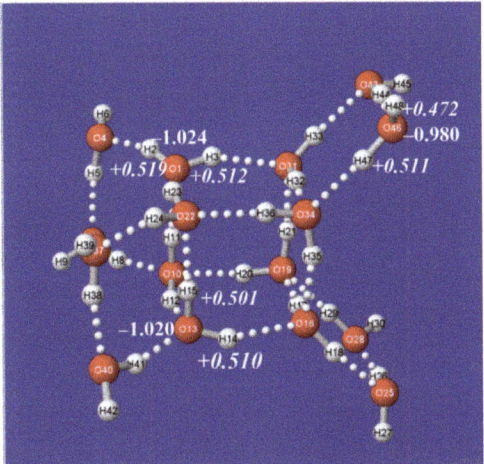

Figure 8. Similar to Figure 7, for representative quadruply (q-type) coordinated O(1)H(2)H(3) and O(13)H(14)H(15) molecules of the cubane-like core, and doubly (d-type) coordinated O(46)H(47)H(48) molecule on a bridged wing of the $^{8,0,8}W_{16}$ cluster.

Table 1. Total natural charge Q_i and $q/t/d$ coordination type for each water monomer (centered on O(i)) of the $^{4,4,6}W_{16}$ cluster. (Similar tables are found in SI for each $^{q,t,d}W_n$ cluster of the present work).

Cluster	O_i	Q_i	$q/t/d$
$^{4,4,6}W_{14}$	1	−0.00662	t_d
	4	−0.00324	d
	7	+0.00963	t_a
	10	−0.00636	q
	13	−0.00107	q
	16	−0.00414	q
	19	+0.00264	q
	22	+0.01427	t_a
	25	+0.00079	d
	28	−0.00357	d
	31	+0.00465	d
	34	−0.00250	d
	37	+0.00087	d
	40	−0.00536	t_d

4.3. Natural Bond Order Correlations

The distended shapes of windowpane clusters provide clear evidence of the severe effects of "ring strain" in altering the network O—H⋯O bonds from the idealized geometries of isolated H-bonds in binary complexes. Nevertheless, one expects that network H-bonds should continue to exhibit the robust correlations with NBO/NRT measures of bond order and charge transfer that were previously demonstrated for free binary H-bonded species [63]. We now turn to examining the supramolecular extension of such correlations for the classical bond order–bond length (BOBL) relationships that have long been fruitfully employed in the integer (single-, double-, triple-, etc., bond) range of covalent bonding in molecules [42,43].

A simple example of such BOBL correlations is illustrated in Figure 9 for the $^{1,4,4}W_9$ windowpane cluster of Figure 2. For each O⋯H—O linkage, the total $b_{O \cdots O}$ bond order is obtained as the sum of $b_{O \cdots H}$ (major) and "long-bond" [64] $b_{O \cdot O}$ (minor) contributions,

$$b_{O \cdots O} = b_{O \cdots H} + b_{O \cdot O} \quad (3)$$

with sub-integer values ranging from 0.02 to 0.18 in this simple cluster. As shown in the right panel, the BOBL correlation is of excellent quality (Pearson correlation coefficient $\chi \approx -0.97$), and the least-squares regression line (shown in the inset) allows close prediction of $R_{O \cdots O}$ distances to near the 0.01Å level(!), despite the fact that NBO/NRT descriptors receive *no* input from real-space molecular geometry or spatial distribution of electron density. Thus, the resonance–covalency concepts underlying NRT bond order evaluations appear to extend seamlessly into this *sub*-integer range of weak H-bonding in clusters, practically as well as the familiar *supra*-integer range of strong covalent bonding and resonance in molecules.

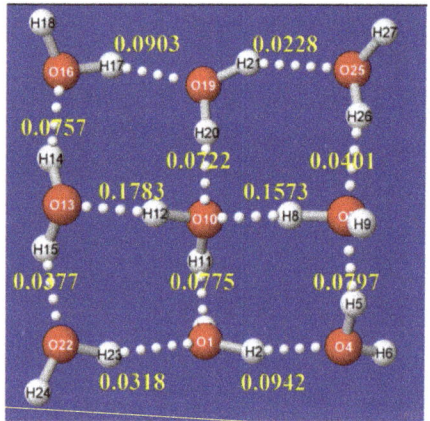 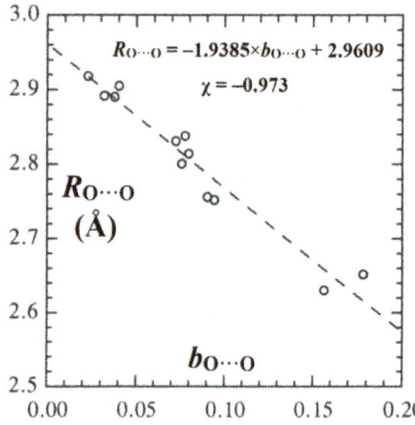

Figure 9. Calculated NRT bond orders $b_{O \cdots O}$ of the $^{1,4,4}W_9$ windowpane cluster (**left panel**), showing their excellent BOBL correlation (Pearson $\chi = -0.973$) with optimized $R_{O \cdots O}$ bond lengths (**right panel**).

More complex three-dimensional structures of windowpane clusters obstruct clear visual representation of all relevant $b_{O \cdots O}$ bond orders and tend to show additional effects of ring strain. Comprehensive listings of $b_{O \cdots O}$ bond orders and $R_{O \cdots O}$ distances (Å) for all H-bonds in all clusters (keyed to the atom numberings of Figures 2 and 3) are presented as tables in SI, as exemplified for the $^{4,4,6}W_{14}$ cluster in Table 2. In this case, the $b_{O \cdots O}$-$R_{O \cdots O}$ correlation is found to be weaker, but still of reasonably high quality ($\chi \approx -0.91$), reflecting the heterogeneities of higher-coordination motifs.

Table 2. NRT bond orders b_{ij} and bond lengths R_{ij} (Å) for all $O(i)\cdots O(j)$ H-bonds of the $^{4,4,6}W_{14}$ cluster (with atom numberings as shown in Figure 3). (See SI for similar tables for all clusters of the present work).

$^{4,4,6}W_{14}$	i	j	b_{ij}	R_{ij}
	1	4	0.1223	2.6972
	1	10	0.0638	2.8562
	1	22	0.0495	2.9173
	4	7	0.1301	2.7127
	7	10	0.0655	2.8861
	7	40	0.0490	2.9332
	10	13	0.1148	2.7789
	13	16	0.0923	2.7808
	13	40	0.1148	2.6940
	16	19	0.0937	2.8005
	16	25	0.0901	2.7772
	16	34	0.0923	2.7725
	19	28	0.0997	2.7415
	19	31	0.0984	2.7440
	22	37	0.0666	2.8167
	25	28	0.0839	2.7599
	31	34	0.0900	2.7594
	37	40	0.0616	2.8367

It is also of interest to examine the global BOBL correlations for *all* windowpane clusters of the present work, covering ca. 250 individual $b_{O\cdots O}$-$R_{O\cdots O}$ H-bonded pairs in a broad variety of coupled coordination motifs. Figure 10 displays the BOBL scatter plot, least-squares regression line, and Pearson correlation coefficient for this entire data set of hydrogen bonds, showing the strong correlation ($\chi = -0.90$) that persists in spite of increasingly heterogeneous cluster topologies.

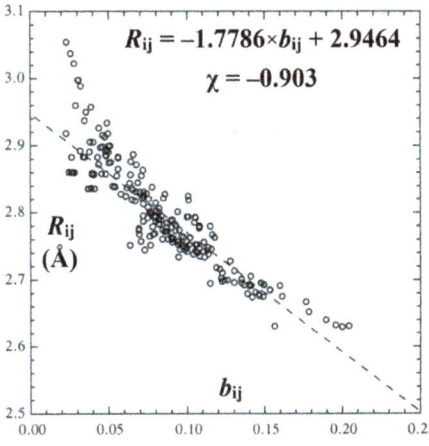

Figure 10. Scatter plot, least-squares regression line, and Pearson correlation coefficient (χ) for b_{ij}-R_{ij} BOBL correlations of all (~250) H-bonds in the clusters of Figures 2 and 3.

The degraded accuracy of the linear least-squares regression fit in Figure 10 (compared, e.g., to that in Figure 9) can be primarily attributed to the upward deviations from linearity that are evident near $b_{ij} \to 0$. However, it is important to recognize that these deviations are *required* on physical grounds, because intermonomer separation should asymptotically *diverge* ($R_{ij} \to \infty$) as bond order vanishes ($b_{ij} \to 0$). Indeed, only the higher-order connectivity of the H-bond network prevents such asymptotic dissociation when any single H-bond is severed, so the proper appearance of such nonlinearity in the $b_{ij} \to 0$ limit serves to further confirm the resonance–covalent nature of H-bonding even in this range of interaction strengths near the limit of chemical interest.

5. Conclusions

In the present work, we have employed standard density functional methods to computationally characterize a broad variety of unusual "windowpane" clusters that may play a role in the high-density fluid phase(s) of water. Despite their diverse topological forms and unusual angular features, we have demonstrated that these clusters are fully compliant with water's known facility in forming doubly (*d*-type), triply (*t*-type), and quadruply (*q*-type) coordinative linkages to other water molecules, leading to multiply connected ("water-wired") networks of increasing energetic stability when proper Grotthuss-type proton ordering is maintained. The *Aufbau* construction approach also suggests the mechanistic sequence by which such Grotthuss-ordered clusters can readily form from successive aggregation with water dimers.

We have also employed natural bond orbital (NBO) and natural resonance theory (NRT) analysis tools to demonstrate the consistency and accuracy with which H-bonding in these clusters conforms to the general conceptual picture of *resonance–covalency* ("charge transfer") as the authentic origin of intermolecular O−H···O attractions. The charge flows and adaptive bond order and structural shifts in these clusters are shown to obey familiar bond order–bond length (BOBL) correlations with high accuracy ($|\chi| > 0.9$). Moreover, the BOBL correlations also exhibit the expected *deviations* from linearity in the asymptotic limit of vanishing bond order where $R_{O\cdots O}$ distance becomes divergent. Although connections can be shown between NBO and Bader-type descriptors [65], we believe that the NRT bond orders of the present work provide broader predictive utility and more nuanced inclusion of resonance effects than the topological descriptors as employed in previous studies of water clustering (e.g., [66]).

The reader is reminded that "correlation is not causation." Nevertheless, the *continuity* of robust BOBL correlations that stretch across the broad extremes of supramolecular (sub-integer) vs. molecular (multi-integer) bond orders strongly implies their *shared* origin in unified "covalency" concepts, contrary to the dichotomous viewpoint that still dominates freshman-level teaching of chemical principles and many facets of force-field methodology.

Supplementary Materials: The following supporting information can be downloaded at: https://www.mdpi.com/article/10.3390/molecules27134218/s1. The Supporting Information (SI) file contains optimized geometrical coordinates, NBO/NRT keyword input, and other computational details in ready-to-run Gaussian input files for all equilibrium water clusters described in the paper. The file also contains tables of computed natural atomic charges and natural bond orders for all water clusters of the work.

Funding: Support for computational facilities was provided in part by the National Science Foundation Grant CHE-0840494.

Institutional Review Board Statement: Not applicable.

Informed Consent Statement: Not applicable.

Data Availability Statement: Not applicable.

Acknowledgments: Thanks are due to Eric Glendening (Indiana State University) for assistance in wrangling consistent NRT bond orders for the many challenging H-bonding interactions in this study.

Conflicts of Interest: The author declares no conflict of interest.

Sample Availability: Samples of the clusters are not available from the author.

References

1. Reed, A.E.; Weinhold, F. Natural bond orbital analysis of near-Hartree-Fock water dimer. *J. Chem. Phys.* **1983**, *78*, 4066–4073. [CrossRef]
2. Reed, A.E.; Weinhold, F.; Curtiss, L.A.; Pochatko, D.J. Natural bond orbital analysis of molecular interactions: Theoretical studies of binary complexes of HF, H_2O, NH_3, N_2, O_2, F_2, CO, and CO_2 with HF, H_2O, and NH_3. *J. Chem. Phys.* **1986**, *84*, 5687–5705. [CrossRef]
3. Curtiss, L.A.; Melendres, C.A.; Reed, A.E.; Weinhold, F. Theoretical studies of O_2^-:$(H_2O)_n$ clusters. *J. Comput. Chem.* **1986**, *7*, 294–305. [CrossRef]
4. Reed, A.E.; Curtiss, L.A.; Weinhold, F. Intermolecular interactions from a natural bond orbital, donor-acceptor viewpoint. *Chem. Rev.* **1988**, *88*, 899–926. [CrossRef]
5. Fowler, P.W.; Buckingham, A.D. The long-range model of intermolecular forces. *Mol. Phys.* **1983**, *50*, 1349–1361. [CrossRef]
6. Buckingham, A.D.; Fowler, P.W.; Stone, A.J. Electrostatic predictions of shapes and properties of van der Waals molecules. *Int. Rev. Phys. Chem.* **1986**, *5*, 107–114. [CrossRef]
7. Arunan, E.; Desiraju, G.R.; Klein, R.A.; Sadlej, J.; Scheiner, S.; Alkorta, I.; Clary, D.C.; Crabtree, R.H.; Dannenberg, J.J.; Hobza, P.; et al. Defining the hydrogen bond: An account (IUPAC Technical Report). *Pure Appl. Chem.* **2011**, *83*, 1619–1636. [CrossRef]
8. Weinhold, F.; Klein, R.A. What is a hydrogen bond? Resonance covalency in the supramolecular domain. *Chem. Educ. Res. Pract.* **2014**, *15*, 276–285. [CrossRef]
9. Stone, A.J. Natural bond orbitals and the nature of the hydrogen bond. *J. Phys. Chem. A* **2017**, *121*, 1531–1534. [CrossRef]
10. Murray, J.S.; Lane, P.; Clark, T.; Riley, K.E.; Politzer, P. Sigma-holes, pi-holes and electrostatically-driven interactions. *J. Mol. Model.* **2012**, *18*, 541–548. [CrossRef]
11. Weinhold, F.; Glendening, E.D. Comment on "Natural bond orbitals and the nature of the hydrogen bond". *J. Phys. Chem. A* **2018**, *122*, 724–732. [CrossRef]
12. Weinhold, F. Anti-electrostatic pi-hole bonding: How covalency conquers Coulombics. *Molecules* **2022**, *27*, 377. [CrossRef]
13. Herbert, J.M. Neat, simple, and wrong: Debunking electrostatic fallacies regarding noncovalent interactions. *J. Phys. Chem. A* **2021**, *125*, 7125–7137. [CrossRef]
14. Brooks, C.L., III; Case, D.A.; Plimpton, S.; Roux, B.; Van Der Spoel, D.; Tajkhorshid, E. Classical molecular dynamics. *J. Chem. Phys.* **2021**, *154*, 100401. [CrossRef]
15. Ludwig, R. Water: From clusters to the bulk. *Angew. Chem. Int. Ed.* **2011**, *40*, 1808–1827. [CrossRef]
16. Andrews, T. On the continuity of the gaseous and liquid states of matter. *Philos. Trans. R. Soc. Lond.* **1869**, *159*, 575–590. [CrossRef]
17. Weinhold, F. Quantum cluster equilibrium theory of liquids: General theory and computer implementation. *J. Chem. Phys.* **1998**, *109*, 367–372. [CrossRef]
18. Weinhold, F. Quantum cluster equilibrium theory of liquids: Illustrative application to water. *J. Chem. Phys.* **1998**, *109*, 373–384. [CrossRef]
19. Kirchner, B.; Weinhold, F.; Friedrich, J.; Perlt, E.; Lehmann, S.B.C. Quantum Cluster Equilibrium. In *Many-Electron Approaches in Physics, Chemistry and Mathematics*; Bach, V., Site, L.D., Eds.; Springer Mathematical Physics Studies: New York, NY, USA, 2014; pp. 77–96.
20. Mayer, J.E.; Mayer, M.G. *Statistical Mechanics*; Wiley: New York, NY, USA, 1940.
21. Frank, H.S.; Wen, W. Ion-solvent interaction. Structural aspects of ion-solvent interaction in aqueous solutions: A suggested picture of water structure. *Discuss. Faraday Soc.* **1957**, *24*, 133–140. [CrossRef]
22. Nemethy, G.; Scheraga, H.A. Structure of water and hydrophobic bonding in proteins. I. A model for the thermodynamic properties of liquid water. *J. Chem. Phys.* **1962**, *36*, 3382–3400. [CrossRef]
23. Walrafen, G.E. Raman and infrared spectral investigations of water structure. In *Water: A Comprehensive Treatise*; Franks, F., Ed.; Plenum: New York, NY, USA, 1972; Volume 1, pp. 151–214.
24. Symons, M.C.R. Water structure and reactivity. *Acc. Chem. Res.* **1981**, *14*, 179–187. [CrossRef]
25. Benson, S.W.; Siebert, E.D. A simple two-structure model for liquid water. *J. Am. Chem. Soc.* **1992**, *114*, 4269–4276. [CrossRef]
26. McQuarrie, D.A. *Statistical Mechanics*; Harper & Row: New York, NY, USA, 1976.
27. Pettersson, L.G.M.; Henchman, R.H.; Nilsson, A. Water: The most anomalous liquid. *Chem. Rev.* **2016**, *116*, 7459–7462. [CrossRef]
28. Wernet, P.; Nordlund, D.; Bergmann, U.; Cavalleri, M.; Odelius, M.; Ogasawara, H.; Näslund, L.Å.; Hirsch, T.K.; Ojamäe, L.; Glatzel, P.; et al. The structure of the first coordination shell in liquid water. *Science* **2004**, *304*, 995–999. [CrossRef]
29. Speedy, R.; Angell, C. Isothermal compressibility of supercooled water and evidence for a thermodynamic singularity at $-45\,°C$. *J. Chem. Phys.* **1976**, *65*, 851–858. [CrossRef]
30. Mishima, O.; Stanley, H.E. The relationship between liquid, supercooled and glassy water. *Nature* **1998**, *396*, 329–335. [CrossRef]
31. Gallo, P.; Amann-Winkel, K.; Angell, C.A.; Anisimov, M.A.; Caupin, F.; Chakravarty, C.; Lascaris, E.; Loerting, T.; Panagiotopoulos, A.Z.; Russo, J.; et al. Water: A tale of two liquids. *Chem. Rev.* **2016**, *116*, 7463–7500. [CrossRef]

32. Hestand, N.J.; Skinner, J.L. Perspective: Crossing the Widom line in no man's land: Experiments, simulations, and the location of the liquid-liquid critical point in supercooled water. *J. Chem. Phys.* **2018**, *149*, 140901. [CrossRef] [PubMed]
33. Nilsson, A.; Pettersson, L.G.M. Perspective on the structure of liquid water. *Chem. Phys.* **2011**, *389*, 1–34. [CrossRef]
34. Manka, A.; Pathak, H.; Tanimura, S.; Wölk, J.; Strey, R.; Wyslouzil, B.E. Freezing water in no-man's land. *Phys. Chem. Chem. Phys.* **2012**, *14*, 4505–4516. [CrossRef] [PubMed]
35. Xu, X.; Petrik, N.G.; Smith, R.S.; Kay, B.D.; Kimmel, G.A. Growth rate of crystalline ice and the diffusivity of supercooled water from 126 to 262 K. *Proc. Natl. Acad. Sci. USA* **2016**, *113*, 14921–14925. [CrossRef]
36. Kim, K.H.; Späh, A.; Pathak, H.; Perakis, F.; Mariedahl, D.; Amann-Winkel, K.; Sellberg, J.A.; Lee, J.H.; Kim, S.; Park, J.; et al. Maxima in the thermodynamic response and correlation functions of deeply supercooled water. *Science* **2017**, *358*, 1589–1593. [CrossRef]
37. Weinhold, F. Nature of H-bonding in clusters, liquids and enzymes: An ab initio, natural bond orbital perspective. *J. Mol. Struct. (THEOCHEM)* **1997**, *398–399*, 181–197. [CrossRef]
38. Weinhold, F. Resonance character of hydrogen-bonding interactions in water and other H-bonded species. In *Peptide Solvation and H-Bonds: Advances in Protein Chemistry*; Baldwin, R.L., Baker, D., Eds.; Elsevier: San Diego, CA, USA, 2006; Volume 72, pp. 121–155.
39. Weinhold, F.; Landis, C.R.; Glendening, E.D. What is NBO analysis and how is it useful? *Int. Rev. Phys. Chem.* **2016**, *35*, 399–440. [CrossRef]
40. Glendening, E.D.; Landis, C.R.; Weinhold, F. Natural Bond Orbital Theory: Discovering Chemistry with NBO7. In *Complementary Bonding Analysis*; Grabowsky, S., Ed.; Walter de Gruyter GmbH & Co KG: Amsterdam, The Netherlands, 2021; pp. 129–156.
41. Glendening, E.D.; Landis, C.R.; Weinhold, F. Resonance theory reboot. *J. Am. Chem. Soc.* **2019**, *141*, 4156–4166. [CrossRef]
42. Coulson, C.A.; Lennard-Jones, J.E. The electronic structure of some polyenes and aromatic molecules. VII. Bonds of fractional order by the molecular orbital method. *Proc. R. Soc. Lond.* **1939**, *A169*, 413–428.
43. Mishra, P.C.; Rai, D.K. Bond order-bond length relationship in all-valence-electron molecular orbital theory. *Mol. Phys.* **1972**, *23*, 631–634. [CrossRef]
44. Johnston, H.S.; Parr, C. Activation energies from bond energies. I. Hydrogen transfer reactions. *J. Am. Chem. Soc.* **1963**, *85*, 2544–2551. [CrossRef]
45. Johnstone, R.A.W.; Loureiro, R.M.S.; Lurdes, M.; Cristiano, S.; Labat, G. Bond energy/bond order relationships for NO linkages and a quantitative measure of ionicity: The role of nitro groups in hydrogen bonding. *Arkivoc* **2010**, *2010*, 142–169. [CrossRef]
46. Badger, R.M. A relation between internuclear distances and bond force constants. *J. Chem. Phys.* **1934**, *2*, 128–131. [CrossRef]
47. Boyer, M.A.; Marsalek, O.; Heindel, J.P.; Markland, T.E.; McCoy, A.B.; Xantheas, S.S. Beyond Badger's rule: The origins and generality of the structure-spectra relationship of aqueous hydrogen bonds. *J. Phys. Chem. Lett.* **2019**, *10*, 918–924. [CrossRef]
48. Bürgi, H.; Dunitz, J.D. Fractional bonds: Relations among their lengths, strengths, and stretching frequencies. *J. Am. Chem. Soc.* **1987**, *109*, 2924. [CrossRef]
49. Gründemann, S.; Limbach, H.-H.; Buntkosky, G.; Sabo-Etienne, S.; Chaudret, B. Distance and scalar HH-coupling correlations in transition metal dihydrides and dihydrogen complexes. *J. Phys. Chem. A* **1999**, *103*, 4752–4754. [CrossRef]
50. Fukui, K. The path of chemical reactions—The IRC approach. *Acc. Chem. Res.* **1981**, *14*, 363–368. [CrossRef]
51. Weinhold, F. Kinetics and mechanism of water cluster equilibria. *J. Phys. Chem. B* **2014**, *118*, 7792–7798. [CrossRef]
52. Weinhold, F.; Landis, C.R. *Valency and Bonding: A Natural Bond Orbital Donor-Acceptor Perspective*; Cambridge University Press: Cambridge, UK, 2005.
53. Weinhold, F.; Landis, C.R. *Discovering Chemistry with Natural Bond Orbitals*; John Wiley: Hoboken, NJ, USA, 2012.
54. Perlt, E.; von Domaros, M.; Kirchner, B.; Ludwig, R.; Weinhold, F. Predicting the ionic product of water. *Sci. Rep.* **2017**, *7*, 10244. [CrossRef]
55. Kirchner, B.; Ingenmey, J.; von Domaros, M.; Perlt, E. The ionic product of water in the eye of the quantum cluster equilibrium. *Molecules* **2022**, *27*, 1286. [CrossRef]
56. Frisch, M.E.; Trucks, G.W.; Schlegel, H.B.; Scuseria, G.E.; Robb, M.A.; Cheeseman, J.R.; Scalmani, G.; Barone, V.P.; Petersson, G.A.; Nakatsuji, H.J.; et al. *Gaussian 16*; Gaussian Inc.: Wallingford, CT, USA, 2016.
57. Glendening, E.D.; Badenhoop, J.K.; Reed, A.E.; Carpenter, J.E.; Bohmann, J.A.; Morales, C.M.; Karafiloglou, P.; Landis, C.R.; Weinhold, F. *NBO 7.0*; Theoretical Chemistry Institute, University of Wisconsin: Madison, WI, USA, 2018; Available online: https://nbo7.chem.wisc.edu/ (accessed on 2 May 2022).
58. Glendening, E.D.; Landis, C.R.; Weinhold, F. NBO 7.0: New vistas in localized and delocalized chemical bonding theory. *J. Comput. Chem.* **2019**, *40*, 2234–2241. [CrossRef]
59. Weinhold, F.; Phillips, D.; Glendening, E.D.; Foo, Z.Y.; Hanson, R.M. *NBOPro7@Jmol*; Theoretical Chemistry Institute, University of Wisconsin: Madison, WI, USA, 2018.
60. Ludwig, R.; Weinhold, F. Quantum cluster equilibrium theory of liquids: Freezing of QCE/3-21G water to tetrakaidecahedral "Bucky-ice". *J. Chem. Phys.* **1999**, *110*, 508–515. [CrossRef]
61. Carpenter, J.E.; Weinhold, F. Transferability of natural bond orbitals. *J. Am. Chem. Soc.* **1988**, *110*, 368–372. [CrossRef]
62. Weinhold, F.; Glendening, E.D. NBO 7.0 Program Manual, Sec. B-5. Available online: https://nbo7.chem.wisc.edu/nboman.pdf (accessed on 2 May 2022).

63. Weinhold, F.; Klein, R.A. What is a hydrogen bond? Mutually consistent theoretical and experimental criteria for characterizing H-bonding interactions. *Mol. Phys.* **2012**, *110*, 565–579. [CrossRef]
64. Landis, C.R.; Weinhold, F. 3c/4e $\hat{\sigma}$-type long-bonding: A novel transitional motif toward the metallic delocalization limit. *Inorg. Chem.* **2013**, *52*, 5154–5166. [CrossRef]
65. Weinhold, F. Natural bond critical point analysis: Quantitative relationships between NBO-based and QTAIM-based topological descriptors of chemical bonding. *J. Comput. Chem.* **2012**, *33*, 2440–2449. [CrossRef]
66. Castor-Villegas, V.M.; Guevara-Vela, J.M.; Vallejo Narváez, W.E.; Pendása, A.M.; Rocha-Rinza, T.; Fernández-Alarcón, A. On the strength of hydrogen bonding within water clusters on the coordination limit. *J. Comput. Chem.* **2020**, *41*, 2266–2277. [CrossRef]

Article

A QCT View of the Interplay between Hydrogen Bonds and Aromaticity in Small CHON Derivatives

Miguel Gallegos [1], Daniel Barrena-Espés [1], José Manuel Guevara-Vela [2], Tomás Rocha-Rinza [2] and Ángel Martín Pendás [1,*]

[1] Department of Analytical and Physical Chemistry, University of Oviedo, 33006 Oviedo, Spain
[2] Institute of Chemistry, National Autonomous University of Mexico, Circuito Exterior, Ciudad Universitaria, Delegación Coyoacán, Mexico City C.P. 04510, Mexico
* Correspondence: ampendas@uniovi.es

Abstract: The somewhat elusive concept of aromaticity plays an undeniable role in the chemical narrative, often being considered the principal cause of the unusual properties and stability exhibited by certain π skeletons. More recently, the concept of aromaticity has also been utilised to explain the modulation of the strength of non-covalent interactions (NCIs), such as hydrogen bonding (HB), paving the way towards the in silico prediction and design of tailor-made interacting systems. In this work, we try to shed light on this area by exploiting real space techniques, such as the Quantum Theory of Atoms in Molecules (QTAIM), the Interacting Quantum Atoms (IQA) approaches along with the electron delocalisation indicators Aromatic Fluctuation (FLU) and Multicenter (MCI) indices. The QTAIM and IQA methods have been proven capable of providing an unbiased and rigorous picture of NCIs in a wide variety of scenarios, whereas the FLU and MCI descriptors have been successfully exploited in the study of diverse aromatic and antiaromatic systems. We used a collection of simple archetypal examples of aromatic, non-aromatic and antiaromatic moieties within organic molecules to examine the changes in π delocalisation and aromaticity induced by the Aromaticity and Antiaromaticity Modulated Hydrogen Bonds (AMHB). We observed fundamental differences in the behaviour of systems containing the HB acceptor within and outside the ring, e.g., a destabilisation of the rings in the former as opposed to a stabilisation of the latter upon the formation of the corresponding molecular clusters. The results of this work provide a physically sound basis to rationalise the strengthening and weakening of AMHBs with respect to suitable non-cyclic non-aromatic references. We also found significant differences in the chemical bonding scenarios of aromatic and antiaromatic systems in the formation of AMHB. Altogether, our investigation provide novel, valuable insights about the complex mutual influence between hydrogen bonds and π systems.

Keywords: hydrogen bond; interacting quantum atoms; aromaticity; QTAIM

1. Introduction

The hydrogen bond (HB) is one of the the most important non-covalent interactions (NCI) in nature. Since its first appearance in the chemistry parlance, back in the second decade of the twentieth century [1], HB interactions have been recognised as key factors determining the properties and structure of a wide variety of molecules and materials. Indeed, the role of HBs is known to affect countless systems, from simple molecular liquids and solids, such as water or hydrogen fluoride, to complex and intricate biomolecules. Furthermore, in recent years, a renewed interest for HB interactions has arisen within the scientific community owing to the importance of these contacts in emerging technologies such as (i) CO_2 capture [2–5], (ii) rechargeable aqueous zinc [6,7] and aprotic Li-O_2 batteries [8], (iii) photovoltaic cells [9,10], (iv) asymmetric catalysis [11], or (v) hydrogen production [12], among others.

As it often happens in the context of inter-molecular bonding scenarios, the complex interplay between different kinds of interactions drives the global properties of supramolecular systems. Therefore, the combination of HB with other similar or drastically different NCIs is of particular importance. We can consider, for instance, the prototypical example of water clusters. The existence of single HB donors and acceptors in H_2O clusters has been associated with the mutual strengthening (cooperativity) of HBs (Figure 1a), whereas the occurrence of double HB donors and acceptors has been related with the reciprocal weakening (anticooperativity) of HBs (Figure 1b) [13–17]. Additionally, there are other instances of non-additive effects of hydrogen bonding reported in the literature, e.g., charge assisted HBs [18,19] and ion-dipole contacts [20].

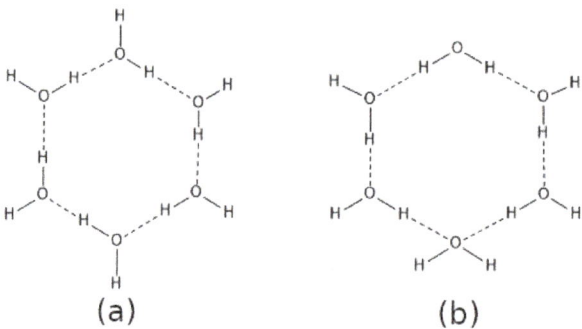

Figure 1. (a) Homodromic and (b) antidromic cycles within the structure of $(H_2O)_6$. These two motifs are, respectively, related to cooperative and anticooperative hydrogen bonding effects.

As a general result, the above mentioned cooperative and anticooperative effects are the result of subtle electron fluctuations that accompany the formation of non-covalently bonded systems [21]. Some of these electron redistributions take place through σ bonds and, it is thus common to refer to them as σ-cooperative or σ-anticooperative HB effects. However, such charge transfers might also occur throughout π systems particularly those found in conjugated moieties [14,22–27]. Well-known examples of the interplay between H-bonds and conjugated π systems are Resonance-Assisted Hydrogen Bonds (RAHB) as originally proposed by Gilli et al. [28,29]. RAHBs are understood usually as the result of π-cooperative effects, which considerably strengthen HBs coupled with π bonds. On the other hand, conjugated systems and hydrogen bonds can also reveal anticooperative effects as those found, for instance, in the bicyclic fused rings of malondialdehyde [23,30] or in Resonance-Inhibited Hydrogen Bonds (RIHB) [25–27].

Another particularly relevant interplay between H-bonds and π systems can be found in the case of the more recently proposed Aromaticity and Antiaromaticity Modulated Hydrogen Bonds (AMHB) [31,32]. The concept of AMHB was first introduced to rationalise the apparent strengthening or weakening of HB interactions modulated by changes in aromaticity and antiaromaticity in the involved systems. Although clearly intuitive and useful, the ideas of aromaticity and antiaromaticity are built upon elusive and ill-defined chemical concepts, which hinder a quantitative and rigorous analysis. Fortunately, state-of-the-art wave function analysis methods have proved very useful in the study of electron delocalisation, which is a critical aspect in the study of aromaticity and antiaromaticity. In particular, and in the context of Quantum Chemical Topology (QCT), the Quantum Theory of Atoms in Molecules (QTAIM) [33] and the Interacting Quantum Atoms (IQA) [34] methods have been successfully exploited to investigate the mutual influence of HBs and π systems [22–25].

In this work, we make usage of the QTAIM and IQA approaches as well as electronic delocalisation indices developed within the conceptual framework of QCT to provide a detailed real-space-based picture of AMHB. For this purpose, we compared the energetics and studied the chemical bonding scenario using QCT in the formation of different AMHB molecular clusters shown in Figure 2. We emphasize the effects of the formation of different molecular clusters on pairwise inter-atomic interactions. For the sake of convenience, and considering the large computational cost of some QCT analyses, derivatives of the simple, but representative, azete and pyridinde molecules will be used as model systems in this work. It should be noticed that these molecules have already been successfully employed in the literature [31,32] as minimal models to study hydrogen bond driven dimerisation phenomena. The manuscript is organised as follows. First, we provide a brief background of the QTAIM and IQA approaches. Then, we discuss the electronic and energetic changes accompanying the dimerisation of a collection of organic scaffolds. Later, we consider the interplay between the above mentioned changes and the aromatic character of the monomers. Lastly, we examine somewhat atypical systems to finally gather the main conclusions of this work.

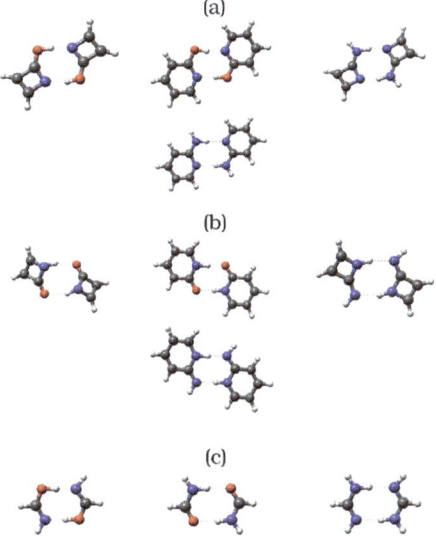

Figure 2. Systems examined throughout this investigation. (**a**) Dimers with the hydrogen bond acceptor contained within the ring (ACR): azet-2-ol (AZH), 2- hydroxypyridine (2HP), azet-2-amine (AZA) and 2-aminopyridine (2AP). (**b**) Dimers with the hydrogen bond donor contained within the ring (DCR): azet-2(1H)-one (AZH), pyridin-2(1H)-one (2HP), azet-2-amine (AZA) and pyridin-2(1H) imine (2AP). (**c**) Representation of the ACR and DCR dimers of formamide (NCO) and formamidine (NCN), used as reference. In NCN, the ACR and DCR tautomers are indistinguishable.

2. Theoretical Framework
2.1. Real Space Wavefunction Analyses

The QTAIM theory, as originally formulated by Bader [33], is a method of wave function analysis based on the topology of the electron density $\rho(r)$, in which the real space is fragmented in a collection of attraction basins (Ω) induced by the topology of $\rho(r)$. In QTAIM, traditional chemical ideas, such as the concept of chemical groups or fragments, atomic charges or bond orders, emerge naturally without the need of any reference. Moreover, the QTAIM partition can be performed starting either from theoretical (electronic structure calculations) or experimental (high-resolution X-ray diffraction data [35]) determinations of the electron density of the system. This combination of robustness and practicality has made QTAIM to be widely employed to shed light into a large variety of phenomena including catalysis [36–38], electrical conductivity [39–41] and aromaticity [42–44], to name a few.

Based on a 3D partition as that defined by QTAIM, the IQA methodology [34] divides a fully interacting non-separable quantum mechanical system into chemically meaningful interacting entities. The total electronic energies in IQA can be written as a sum of one-body (intra-atomic) and two-body (inter-atomic) terms [34,45], as:

$$E = \sum_A E_{\text{self}}^A + \sum_{A>B} E_{\text{int}}^{AB}, \quad (1)$$

where E_{self}^A is the energy of atom A, which includes the electron–nucleus attraction, the inter-electronic repulsion, and the kinetic energy within atom A. Additionally, E_{int}^{AB} is the total interaction energy between atoms A and B. This term encompasses all the available interaction terms between the nucleus and electrons within atoms A and B. The constituting terms of the total inter-atomic interaction between two atoms, E_{int}^{AB}, can be regrouped to express the latter as a sum of purely covalent (i.e., exchange-correlation, V_{xc}^{AB}) and ionic (i.e., classical, V_{cl}^{AB}) components:

$$E_{\text{int}}^{AB} = V_{\text{cl}}^{AB} + V_{\text{xc}}^{AB}. \quad (2)$$

Indeed, the IQA energy decomposition provides a particularly convenient way to study and characterise the chemical nature of the interaction among atoms in an electronic system.

2.2. Aromaticity

Aromaticity is a multi-factorial concept which is thought to modify and even to determine the structural, energetic, electronic, and magnetic properties of some molecules. Due to lack of an inherent Dirac observable defining it, aromaticity is usually described in terms of its effects on conjugated systems, such as enhanced thermodynamic stability or structural rigidity. Although the idea of aromaticity was conceived solely upon the interpretation of experimental results, the birth of quantum and computational chemistry motivated the development of multiple tools and techniques aiming at its quantitative analysis. One of the most common approaches to study and measure aromaticity is the nucleus-independent chemical shifts (NICS) [46,47] method, as originally proposed in 1996 by Schleyer and coworkers [46]. The NICS approach has been used for several decades to study numerous π skeletons in a variety of fields. Nevertheless, some results obtained through the NICS descriptor have turned to be highly questionable [48–50], even contradicting, in some cases, other aromaticity measures based on reactivity [51]. Among many other criteria exploited to quantify aromaticity we can find (i) structural indices, which evaluate bond equalisation [52], (ii) energy decomposition analyses which require a reference molecule [53], and (iii) electronic descriptors which evaluate the amount of electron delocalisation among the atoms forming a cyclic structure. The last-mentioned set includes a number of methods that have been developed relying on the partition of the electronic density offered by QTAIM, such as the Para Delocalisation Index (PDI) [54], the Aromatic Fluctuation Index (FLU) [55], or the Multicenter Index (MCI) [56]. The PDI and FLU approaches provide an estimate of the aromaticity of a system in terms of the electron delocalisation within the cyclic skeleton, whereas the MCI method arises from a generalised population analysis leading to a many-atoms bond index. In the present work, we have made use of some electronic delocalisation-based descriptors, such as the MCI or the FLU indices, to quantify the changes in aromaticity and antiaromaticity of each molecule upon the formation of the corresponding dimer. We have chosen the FLU and MCI indicators to account for the changes in aromaticity and antiaromaticity upon interaction of the monomers under consideration given their proven accuracy and reliability [57]. Furthermore, these methods are fully compatible with the rest of the QCT analyses performed in this report. Unfortunately, the PDI method is not applicable to some of the examined systems herein, because it can only be used for six-membered rings. Finally, we also used the IQA partitioning to study in detail the energetic changes accompanying the generation of the molecular clusters shown in Figure 2, with a particular emphasis on their role in HB formation.

3. Computational Details

The structures of the hydrogen-bonded dimers in Figure 2 were optimised in the gas phase and the resultant approximate wavefunctions and electron densities were afterwards dumped for further analysis. All geometry optimisations were performed using the ORCA quantum chemistry package version 5.0.3 [58] using the PBE0 hybrid functional [59] along with the Def2-TZVP basis set [60] and the atom-pairwise dispersion correction with the Becke–Johnson damping scheme [61,62]. For the sake of computational efficiency, the Resolution of Identity (RI) approximation was used for the Coulomb integrals with the default COSX grid for HF exchange, as implemented in ORCA [58]. On the other hand, the auxiliary Def2 basis set was used for the RI-J approximation. The combination of such an exchange-correlation functional and basis set has proven [32] suitable for the characterisation of the systems under study. Moreover, DNLPO-CCSD(T)/def2-QZVPP single point calculations were performed on the optimised geometries of the monomers and dimers in order to test the accuracy of our DFT results. An extrapolation to the complete basis set limit was performed through the def2-SVP/def2-TZVP scheme as implemented in ORCA [58] so as to ameliorate the Basis Set Superposition Error (BSSE). The nature of the stationary points (corresponding to local minima of the potential energy surface) was characterised through the computation of the corresponding harmonic frequencies. QTAIM and IQA calculations were performed using the AIMALL [63] and PROMOLDEN codes [64]. The exchange-correlation energy was partitioned as indicated in reference [65]. Finally, all aromaticity indices discussed along this manuscript were computed using the ESI-3D code [66]. We have denoted the dimers with (i) the hydrogen bond Acceptor Contained within the Ring and (ii) the hydrogen bond Donor Contained within the Ring, as ACR and DCR, respectively. For the ACR azet-2-1H-one (AZH) dimer, we performed a geometry-constrained optimisation to ensure the attainment of the right tautomer of the constituting monomers. On the other hand we observed for the DCR-AZH dimer a structure considerably deviated from planarity as opposed to the rest of the systems. Additional information such as optimised structures, electronic energies, and a more complete survey of IQA and QTAIM (e.g., IQA energy of different groups as well as other QTAIM descriptors) can be found in the electronic supporting information.

4. Results and Discussion

4.1. General Energetic Changes Induced by HB Formation

We consider first the differences in dimerisation energies of the different systems shown in Figure 2. Table 1 reports the values of $\Delta\Delta E$,

$$\Delta\Delta E(Y_2) = \Delta E(Y_2) - \Delta E(R_2), \tag{3}$$

in which $\Delta E(Y_2)$ is the energy change associated to the process

$$Y + Y \rightleftharpoons Y \cdots Y, \quad \Delta E(Y_2), \tag{4}$$

and R is the corresponding reference system used for ACR and DCR, namely the dimers of formamide (NCO) and formamidine (NCN) in the corresponding ACR and DCR form. A negative/positive value of $\Delta\Delta E(X_2)$ indicates a stronger/weaker interaction in $X\cdots X$ with respect to the reference complex $R\cdots R$.

The straightforward comparison of the DFT and CC values for $\Delta\Delta E$ reveals that both levels of theory are in good agreement concerning the sign and magnitude of $\Delta\Delta E$. These observations indicate that our DFT results offer a reliable picture of the energetics of the binding phenomena under study. As the footnote of Table 1 reports, all the values for ΔE are negative, pointing, as expected, to stabilising dimerisation contacts in all the investigated dimers. Furthermore, the easiness of the complexation seems to be driven, as expected, by the hydrogen bond formation as reflected by the correlation of the binding energies with the ρ at the bond critical point of the HB contacts (see SI Figure S3). We also note that the N–C=N bonding pattern leads to lower binding energies, due to the larger acidity of H atoms bonded to oxygen. The AMHB [32] interpretation of the sign of $\Delta\Delta E$ in Table 1 states that the ACR dimers AZH and AZA display either an increase in aromaticity or a decrease in antiaromaticity, respectively, as a consequence of the formation of the investigated H-bonds. Ditto for the DCR clusters 2HP and 2AP. On the other hand, the ACR complexes 2HP and AZA along with the DCR systems AZH and AZA exhibit the opposite behaviour. We consider now QCT analyses to further dissect these energetic trends.

Table 1. Values of $\Delta\Delta E$, as defined in Equation (3), computed in the DFT and CC approximations described in the main text. NCO and NCN denote formamide and formamidine, respectively, the reference systems shown in Figure 2c. All values are reported in kcal/mol.

System	$\Delta\Delta E^{DFT}$	$\Delta\Delta E^{CC}$	System	$\Delta\Delta E^{DFT}$	$\Delta\Delta E^{CC}$
2HP (ACR)	5.06	3.01	2HP (DCR)	−7.11	−7.03
NCO (ACR)	0.00	0.00	NCO (DCR)	0.00	0.00
AZH (ACR)	−2.82	−3.34	AZH (DCR)	6.40	5.38
2AP (ACR)	3.22	2.58	2AP (DCR)	−8.16	−7.50
NCN (ACR)	0.00	0.00	NCN (DCR)	0.00	0.00
AZA (ACR)	−10.05	−9.58	AZA (DCR)	2.13	0.89

The corresponding values for $\Delta E / \text{kcal} \cdot \text{mol}^{-1}$ for the references in Formula (4) are NCO (ACR): $\Delta E^{DFT} = -35.45$, $\Delta E^{CC} = -30.10$; NCN (ACR): $\Delta E^{DFT} = -16.75$, $\Delta E^{CC} = -13.64$; NCO (DCR): $\Delta E^{DFT} = -15.93$, $\Delta E^{CC} = -13.59$; NCN (DCR): $\Delta E^{DFT} = -16.75$, $\Delta E^{CC} = -13.64$.

4.2. Quantum Chemical Topology Analyses

In order to further deepen into the origin of the observed trends in the evolution of the binding energies reported in Table 1, we examined the non-covalent interactions established between both monomers using QCT techniques. For the sake of convenience, the nomenclature shown in Figure 3 will be used to refer to the atoms involved in the intermolecular bonding pattern of these dimers. We first consider the electron redistribution of electron charge due to the formation of the investigated H-bonds. We point out that the formation of an HB is associated with a reorganisation of the electronic density of the moieties involved in this interaction. There is, indeed, a transfer of electron charge from the HB acceptor to the HB donor, with the proton acting as a bridge. For small HB dimers such as $(H_2O)_2$ or $(HF)_2$, two of the simplest HBs, such charge displacement makes the HB acceptor a better proton donor. Ditto for the HB donor becoming a better proton acceptor. Notwithstanding, the present work deals with dimers where each molecule acts simultaneously as an HB donor and an HB acceptor and hence, there is no effective charge transfer between the monomers. However, the presence of an HB induces a rearrangement of the electron density that interacts with the π clouds in each molecule of the studied systems.

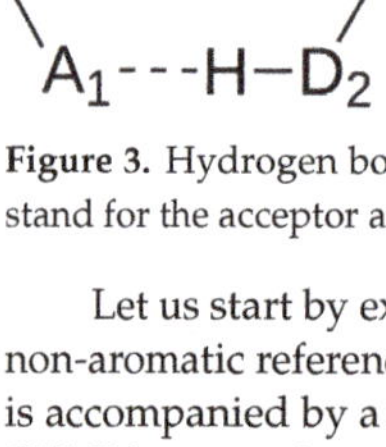

$$\begin{array}{c} D_1\text{-H---}A_2 \\ \langle \quad \rangle \\ A_1\text{---H-}D_2 \end{array}$$

Figure 3. Hydrogen bond connectivity pattern involved in the formation of the dimers. A and D stand for the acceptor and donor HB moieties, respectively.

Let us start by examining the major changes undergone upon the dimerisation of the non-aromatic reference systems: formamide and formamidine. The complexation process is accompanied by a significant electron redistribution, as reflected by the change in the QTAIM atomic charges, collected in Table 2. The formation of the non-covalent interactions leads, in both cases, to a noticeable electron enrichment of the D and A atoms (between

0.03 and 0.09 electrons) at the cost of decreasing the electron population of the H atom by ≈0.04–0.09 electrons. On the other hand, the central C atom undergoes a noticeable change in its average electron number of −0.09–0.01 a.u. depending on the nature of the acceptor moiety.

Table 2. Change in the QTAIM electron populations of the atoms involved in the HB contacts upon the formation of the dimers for (i) the hydrogen bond Acceptor Contained within the Ring (ACR) and (ii) the hydrogen bond Donor Contained within the Ring (DCR) cases. The labelling of the atoms is shown in Figure 3. All values are reported relative to the monomers which were used as reference. Atomic units are used throughout.

	ACR			
System	$\Delta Q(D)$	$\Delta Q(H)$	$\Delta Q(A)$	$\Delta Q(C)$
2HP	−0.087	+0.049	−0.067	+0.069
NCO	−0.085	+0.041	−0.072	+0.085
AZH	−0.089	+0.065	−0.077	+0.045
2AP	−0.040	+0.070	−0.021	+0.019
NCN	−0.051	+0.088	−0.024	+0.008
AZA	−0.061	+0.113	−0.063	−0.002
	DCR			
System	$\Delta Q(D)$	$\Delta Q(H)$	$\Delta Q(A)$	$\Delta Q(C)$
2HP	−0.070	+0.105	−0.024	−0.040
NCO	−0.052	+0.085	−0.025	−0.007
AZH	−0.037	+0.064	−0.024	−0.011
2AP	−0.063	+0.101	−0.031	−0.012
NCN	−0.051	+0.088	−0.024	+0.008
AZA	−0.030	+0.074	−0.034	−0.007

These observations, and with the particular exception of the bridging C atom, are very similar for both NCO and NCN bonding patterns and evidence a conspicuous rise in the polarisation of the system due to the formation of the corresponding dimers. Such an increase in the local polarisation of the terminal atoms enhances the electrostatic interaction in the HB contacts, as reflected by the large classical components of the A···H interaction as reported in Figure S2.

We also considered the change in the number of electrons shared among bonded atoms, as measured by the delocalisation index (DI) (see SI Figure S5 for more details), as gathered in Table 3. The D–H bond order decreases significantly (≈0.16–0.25) upon dimerisation, thus weakening the covalent component to the D–H interaction, as evidenced by the prominent destabilisation of ≈30–40 kcal/mol found for V_{xc}^{D-H}. A similar, yet more subtle, weakening of the covalent component can also be observed for the C–A bond. On the other hand, the DI(D–C) is increased by 0.07–0.10 electron pairs, going from a single D–C bond to a slightly higher bond order (≈1.1 in the general case). These results point out that hydrogen bonding reinforces the D–C double bond character at the expense of decreasing that of the C–A interaction. We observed a similar effect in our analysis of RAHB in which the DI corresponding to double bonds decrease while that of single bonds have the opposite behaviour after the formation of the RAHB [22].

This last observation is fulfilled for all the systems and suggests that the formation of the dimers may trigger two opposed effects. Because the A–C bond is contained within the ring in ACR dimers, and ΔDI(C–A) < 0 as indicated in Table 3, we would expect that the formation of the H-bond would decrease the number of π electrons in the associated ACR cyclic structures as represented in Figure 4. On the contrary, the D–C bond is included in the cyclic structures of DCR dimers, and ΔDI(D–C) > 0 (Table 3), then the number of π electrons must increase in the DCR dimers due to the formation of the H-bond. Accordingly, Table S8 indicates that the group energy of the ACR/DCR rings increase/decrease upon the formation of the corresponding dimers. These changes in electron delocalisation affect the aromaticity and antiaromaticity of the investigated systems as discussed below.

Table 3. Change in the electron delocalisation index of the atoms involved in the HB contacts (Figure 3) upon the formation of the dimers with (i) the hydrogen bond acceptor contained within the ring (ACR) and (ii) the hydrogen bond donor contained within the ring (DCR). These changes are computed with respect to the values of the monomers, which were used as references. Atomic units are used throughout.

	ACR			
System	ΔDI(D–H)	ΔDI(D–C)	ΔDI(C–A)	ΔDI(H–A)
2HP	−0.207	+0.056	−0.064	+0.152
NCO	−0.246	+0.100	−0.171	+0.197
AZH	−0.216	+0.079	−0.136	+0.134
2AP	−0.151	+0.051	−0.034	+0.103
NCN	−0.181	+0.073	−0.076	+0.117
AZA	−0.262	+0.131	−0.167	+0.165
	DCR			
System	ΔDI(D–H)	ΔDI(D–C)	ΔDI(C–A)	ΔDI(H–A)
2HP	−0.209	+0.065	−0.091	+0.132
NCO	−0.162	+0.069	−0.070	+0.099
AZH	−0.103	+0.034	−0.043	+0.066
2AP	−0.234	+0.067	−0.106	+0.163
NCN	−0.181	+0.073	−0.076	+0.117
AZA	−0.152	+0.043	−0.053	+0.099

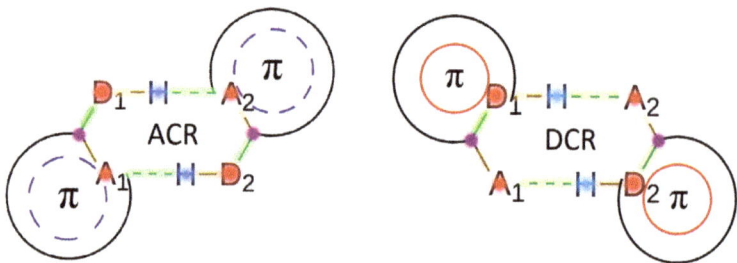

Figure 4. Representation of the major electronic changes induced by the dimerisation process in systems with (i) the hydrogen bond acceptor contained within the ring (ACR) and (ii) the hydrogen bond donor contained within the ring (DCR) displayed in the left and right parts of the Figure, respectively. Red and blue colors indicate QTAIM atoms with electronic charge accumulation and depletion, respectively, due to the formation of the molecular cluster. Ditto for green and orange bonds, employed to highlight an increase or decrease in the DI values. Purple is used to show not clearly established scenarios in this regard.

4.3. Perturbation of the Aromaticity of the π Skeleton

We consider now the interplay between aromaticity and antiaromaticity with the inter-molecular HB contacts of Figure 2. Table 4 gathers the change in the aromaticity indices of the intra-molecular π skeleton upon dimerisation, as measured by the MCI and FLU indices (further details about these indices can be found in Section 1 of the SI).

Table 4. Change in the MCI and FLU aromaticity indices, along with the change in aromatic/antiaromatic character (Γ), induced by the formation of the dimers in Figure 2. If ΔΓ > 0, there is either (i) an increase of aromaticity or (ii) a reduction of antiaromaticity; vice versa when ΔΓ < 0.

System	ΔMCI (au)	ΔFLU (au)	ΔΓ	System	ΔMCI (au)	ΔFLU (au)	ΔΓ
2HP (ACR)	−0.008	+0.001	−	2HP (DCR)	+0.007	−0.010	+
AZH (ACR)	+0.001	−0.017	+	AZH (DCR)	−0.001	−0.005	−
2AP (ACR)	−0.004	+0.001	−	2AP (DCR)	+0.007	−0.010	+
AZA (ACR)	+0.004	−0.021	+	AZA (DCR)	−0.001	−0.007	−

Before discussing in detail the changes in the aromatic character of the spectator groups, it may be enlightening to provide a grasp of the FLU and MCI aromaticity indices. The former measures the electron sharing between neighbouring atoms in a ring as well as its similarities between the constituents of the cyclic structure. Thus, a FLU value of zero corresponds to an "ideal" aromatic system, while positive values evidence a deviation from aromaticity. On the other hand, the MCI index measures the collective electron delocalisation along a collection of M centres. As opposed to the FLU, large MCI values suggest a high aromatic character, whereas any other situation usually results in vanishing MCI indexes. Although these metrics were specifically designed to measure aromaticity, they have been successfully used to study antiaromaticity as well [67]. We used the Hückel rule to assign the aromatic or antiaromatic character of the examined monomers as shown in Figure 5. The aromaticity metrics of the monomers (see Table S1) are in agreement with the aromaticity or antiaromaticity label as determined by the Hückel rule. Indeed, the DCR form of AZH and AZA is more aromatic than their ACR counterparts. Likewise, the ACR tautomer of 2HP and 2AP is more aromatic than the corresponding DCR structures.

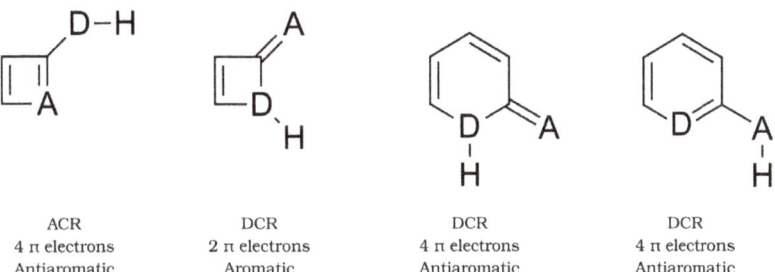

ACR	DCR	DCR	DCR
4 π electrons	2 π electrons	4 π electrons	4 π electrons
Antiaromatic	Aromatic	Antiaromatic	Antiaromatic

Figure 5. Aromatic and antiaromatic character of the monomers shown in Figure 2 determined with Hückel rule.

We discuss now how the aromaticity and antiaromaticity of the corresponding monomers change due to the formation of the HB interactions. We mention that apart from the DCR-AZH scaffold, all dimers adopt a nearly fully planar disposition which is key for the delocalisation of π electrons and it is optimal for the formation of the AMHB. Except for a slight discrepancy within the results of (i) the FLU on one hand and (ii) MCI and NICS [32] on the other for the change in the aromatic character between the DCR systems AZH and AZA, there is a good agreement between the computed sign of $\Delta\Delta E$ and the changes of aromaticity and antiaromaticity in the examined systems, as shown in Figure 6.

ACR

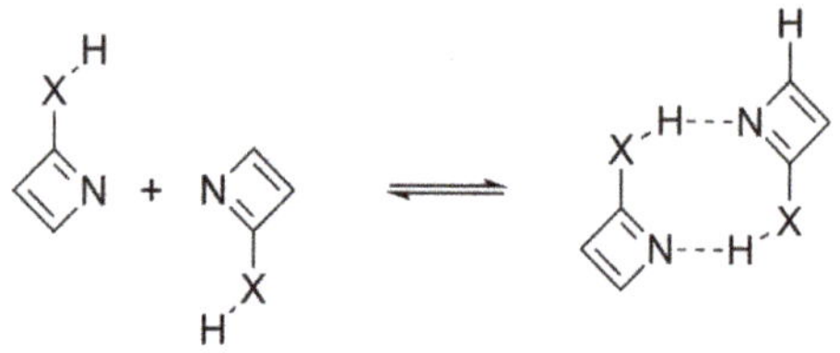

X= O (AZH): $\Delta\Delta E^{cc}$ = −3.34 kcal/mol ($\Delta E > 0$)
X= N-H (AZA): $\Delta\Delta E^{cc}$ = −9.58 kcal/mol ($\Delta E > 0$)
(Reduction in antiaromaticity)

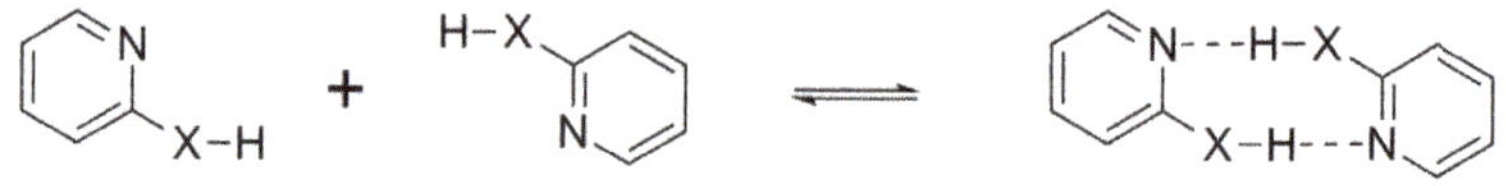

X= O (2HP): $\Delta\Delta E^{cc}$ = 3.01 kcal/mol ($\Delta E < 0$)
X= N-H (2AP): $\Delta\Delta E^{cc}$ = 2.58 kcal/mol ($\Delta E < 0$)
(Reduction in aromaticity)

DCR

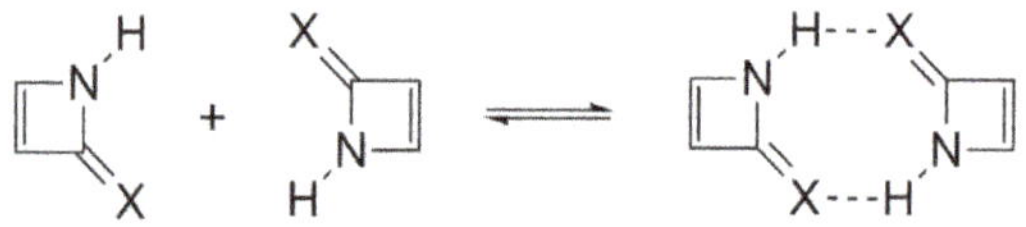

X= O (AZH): $\Delta\Delta E^{cc}$ = 5.38 kcal/mol ($\Delta E < 0$)
X= N-H (AZA): $\Delta\Delta E^{cc}$ = 0.89 kcal/mol ($\Delta E < 0$)
(Reduction in aromaticity)

X= O (2HP): $\Delta\Delta E^{cc}$ = −7.03 kcal/mol ($\Delta E > 0$)
X= N-H (2AP): $\Delta\Delta E^{cc}$ = −7.50 kcal/mol ($\Delta E > 0$)
(Reduction in antiaromaticity)

Figure 6. Values of $\Delta\Delta E$ and changes in aromaticity and antiaromaticity ($\Delta\Gamma$) for the AMHB investigated herein.

In short, we observe that the condition $\Delta\Delta E < 0$, i.e., a more favourable formation of the HB with respect to the reference is related with a reduction in the antiaromatic character of the monomers. Correspondingly, when $\Delta\Delta E > 0$, i.e., a less favourable HB with respect to the reference is accompanied by a reduction of the aromaticity of the monomers. These observations based on QCT are consistent with those of Wu and coworkers [32]. We note some flexibility in the interpretation of these results. Namely, the decrease in aromaticity in the HB formation of the AZA and AZH dimers in DCR configuration was interpreted in Reference [32] as an intensification of antiaromaticity (Figure 7). Similarly, the decrease in antiaromaticity of the DCR systems 2HP and 2AP after the generation of the corresponding dimers was interpreted as an HB reinforced by an increase in aromaticity [32]. This observation suggests that aromaticity and antiaromaticity can be put in a similar scale using electron delocalisation tools within the conceptual framework of QCT.

We consider now a further QCT description of the aromatic and antiaromatic moieties considered herein (Figure 5). We observed important differences concerning the atoms directly involved in the H-bond depending on the aromatic or antiaromatic character of the interacting monomers. For the sake of clarity, we will generally refer to the changes of QTAIM and IQA properties upon dimerisation, with respect to the NCN or NCO reference systems.

As expected, the relative change in the atomic charges and delocalisation indices reported in Tables 2 and 3 have a notable impact on the covalent and ionic components to the total IQA interaction energies in the atoms directly involved in the H-bond. Figure 8 collects the change in the classical and exchange-correlation components of E_{int} for the atoms entailed directly in the inter-molecular contact, reported relative to the NCO or NCN reference systems. The aromatic scaffolds (see Figure 5) seem to consistently stabilise the V_{xc}^{C-A} and V_{xc}^{D-H} contributions over the rest of the interactions. The classical term of other interactions shows, on the other hand, trends that are more complicated to interpret. The interplay between these two contribution results in a net stabilisation of the C–A component at the cost of partially disrupting the D–C bond, when measured with respect to their references.

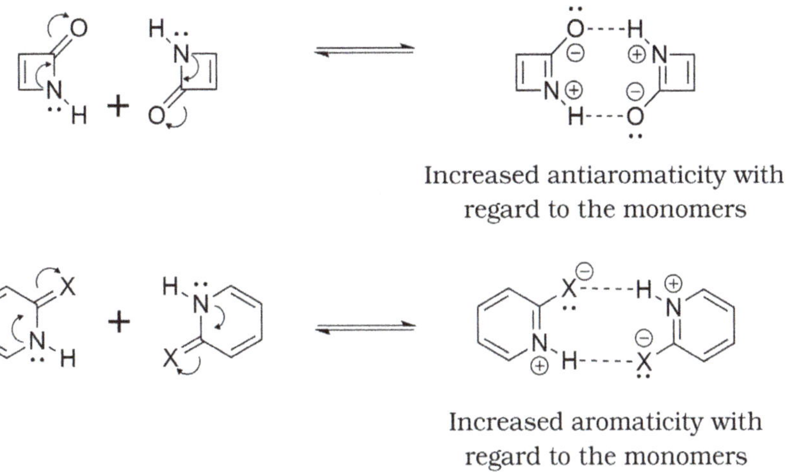

Increased antiaromaticity with regard to the monomers

Increased aromaticity with regard to the monomers

Figure 7. Alternative interpretation of the AMHB in the DCR dimers compared to that offered in Figure 6 as increases in aromaticity and antiaromaticity in the (i) 2HP and 2AP on one hand and (ii) the AZA and AZH on the other, respectively.

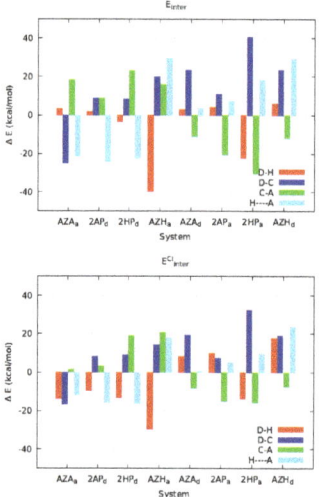
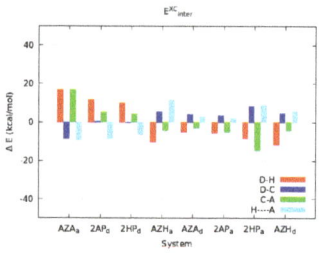

Figure 8. Relative changes in the IQA interaction energies upon dimerisation, all of the values are reported relative to their NCO or NCN reference systems. The tautomeric forms ACR and DCR are indicated by the a and d subscripts, respectively.

Antiaromatic systems have a different behaviour concerning the weakened and strengthened interactions in the inter-molecular region due to the formation of the examined H-bonds. These monomers (Figure 5) generally strengthen both the classical and exchange-correlation components of the H⋯A contact further than the reference systems. This fortifying of the HB interaction is accompanied by a noticeable destabilisation of the covalent component of the C–A bond, which, as previously discussed, is more strengthened in the aromatic compounds than it is in the reference compounds.

4.4. The Peculiar Case of the AZH (DCR) Dimer

As previously mentioned, all of the dimers with the exception of AZH (DCR) exhibit a planar or quasi-planar structure. However, the lowest local energy minimum found for the last-mentioned compound adopts a distorted conformation. Based on the aforementioned observations, one might conjecture at first glance that such a geometrical distortion could be understood as a way to alleviate the reduction of aromaticity induced by the formation of the dimers. To explore this idea, two additional conformational isomers were studied, as represented in Figure 9. Both the bent–trans bent–cis structures are bona fide local minima with a non-planar geometry. We performed a constraint optimisation in order to obtain the corresponding planar AZH (DCR) isomer. As shown in Table 5, the bent isomers are almost degenerate in terms of energy, being the bent–cis structure slightly more stable by ≈0.5 kcal/mol. On the other hand, the planar structure is ≈7 kcal/mol less stable than the latter.

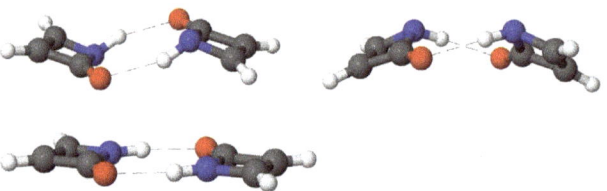

Figure 9. Different isomers of the AZH (DCR) dimers: bent–trans, bent–cis, planar.

The aromaticity metrics reported in Table 5 indicate that the restriction of the 4-membered rings in AZH to remain in a plane would lead to a further reduction of the aromaticity of the dimer. Moreover, the distortion of the low energy (bent) conformations has a dramatic impact on the QTAIM descriptors. Indeed, the changes in the atomic charges and the delocalisation indices are drastically increased, as reflected by the trends in ΔQ and ΔDI values, gathered in Table 6. As it can be seen from the changes in the delocalisation indices in the same chart, the planar isomers lead to a very prominent decrease of the D–H and C–A bond orders while promoting the delocalisation of electrons involved in the D–C and H–A interactions.

Table 5. Energies and aromaticity indices for the different isomers of the AZH (DCR) dimers. All values are reported relative to the most stable isomer (bent–cis).

Isomer	ΔE (kcal/mol)	Δ FLU(a.u.)	Δ MCI(a.u.)
bent–trans	+0.48	+0.0006	−0.0001
bent–cis	0.00	0.0000	0.0000
planar	+7.38	+0.0094	−0.0038

Further information can be obtained through the analysis of the IQA interaction energies, as gathered at the bottom of Table 6. The trends in the E_{int} energies reveal that distorting the more stable bent geometry stabilises all the pairwise interactions involved between the terminal atoms participating in the binding. This observation is particularly prominent for the D–C and H–A bonds. The interplay between the exchange-correlation and electrostatic contributions also leads to a moderate stabilisation of the D–H and C–A bonds despite the already mentioned decrease in the DI index. Thus, and in agreement with the aforementioned trends, forcing the planarity of the system further boosts the HB contacts as well as the π cloud of electrons through the promotion of the "in-ring" resonant structure. Such an effect is consistent with the enhancement of the anti-aromatic character of the system (top of Figure 7) along with the decrease of the net binding affinities, despite the more favourable hydrogen bonding established between the monomers.

Table 6. Change in the QTAIM charges and electron delocalisation indices, in atomic units, upon the formation of the different AZH (DCR) dimers. Some selected changes of IQA interaction energies (in kcal/mol) are reported as well.

	QTAIM charges			
System	$\Delta Q(D)$	$\Delta Q(H)$	$\Delta Q(A)$	$\Delta Q(C)$
bent–trans	−0.039	+0.063	−0.024	−0.007
bent–cis	−0.037	+0.064	−0.024	−0.011
planar	−0.185	+0.127	−0.030	+0.065
	Delocalisation indices			
System	$\Delta DI(D-H)$	$\Delta DI(D-C)$	$\Delta DI(C-A)$	$\Delta DI(H-A)$
bent–trans	−0.099	+0.035	−0.043	+0.064
bent–cis	−0.103	+0.034	−0.043	+0.066
planar	−0.169	+0.070	−0.068	+0.074
	IQA energy partition			
System	ΔE_{int}^{D-H}	ΔE_{int}^{D-C}	ΔE_{int}^{C-A}	E_{int}^{H-A}
bent–trans	−12.80	−21.21	+11.99	−87.41
bent–cis	−12.87	−18.43	+13.47	−88.66
planar	−40.98	−121.17	−1.07	−106.99

5. Conclusions

We presented an analysis of aromaticity and antiaromaticity modulated hydrogen bonds using quantum chemical topology tools, namely the QTAIM, the IQA energy partition as well as the electronic delocalisation indicators FLU and MCI. For this purpose, we considered rings containing either the H-bond acceptor (ACR) or the H-bond donor (DCR). Our results show how the formation of the investigated H-bonds can trigger subtle electronic rearrangements with a quite significant impact in the stability and properties of the involved interacting systems. We described large changes in QTAIM charges and electron delocalisation indices along with their accompanying classical and exchange-correlation components of the IQA interaction energies related with the formation of these HB clusters. We also found fundamental differences within the ACR and DCR systems, for example, the weakening and strengthening of double bonds within the cyclic structures of ACR and DCR, a condition which leads to the destabilisation and stabilisation of the rings in these systems. Additionally, we related the enhancement and impairment of the examined H-bonds with respect to non-aromatic (i.e., non-cyclic) structures with changes in the aromatic and antiaromatic character of the system. We observe that reductions in aromaticity can be interpreted as increases in antiaromaticity and vice versa. Therefore, our results indicate that aromaticity and antiaromaticity can be considered on a common scale using QCT tools. Our results also point that the deviation from planarity of specific AMHB clusters could be related with a trend of the system to ameliorate a reduction in aromaticity. Overall, we expect the results of our investigation to provide novel useful insights about the intricate interplay among H-bond and π systems.

Supplementary Materials: The following supporting information can be downloaded at: https://www.mdpi.com/article/10.3390/molecules27186039/s1, Table S1. Aromaticity metrics, Table S2. Electronic energies, Tables S3 and S4. Binding energies, Tables S5 and S6. IQA and QCT descriptors, Table S7. IQA group energies, Table S8. IQA ring energies, Table S9. Atomic charges, Tables S10–S22. Optimised geometries, Table S23. Electronic energies for the isomers of the AZH (DCR) dimer, Table S24. Atomic charges for the isomers of the AZH (DCR) dimer, Tables S25–S28. IQA and QCT descriptors for the isomers of the AZH (DCR) dimer, Figures S1 and S2. Change in the IQA interaction energies upon dimerisation, Figure S3. DFT dimerisation energy as a function of the electron density at the BCP of the intermolecular HB contacts.

Author Contributions: Conceptualization and methodology, J.M.G.-V., M.G. and T.R.-R.; software, Á.M.P.; validation, formal analysis; investigation, resources, data curation, D.B.-E., M.G. and J.M.G.-V.; writing and visualization, M.G., J.M.G.-V. and T.R.-R.; supervision, project administration, J.M.G.-V., T.R.-R. and Á.M.P.; funding acquisition, T.R.-R. and Á.M.P. All authors have read and agreed to the published version of the manuscript.

Funding: We gratefully acknowledge financial support the Spanish MICINN (grant PGC2018-095953-B-I00). M. Gallegos specially acknowledges the Spanish MICIU/MIU for the FPU19/02903 grant. We are also grateful to DGTIC/UNAM (project LANCAD-UNAM-DGTIC-250) for computer time.

Data Availability Statement: Structures of the studied systems are reported in the Electronic Supplementary Information.

Conflicts of Interest: The authors declare no conflict of interest.

Sample Availability: Samples of compounds are not available from the authors.

References

1. Moore, T.S.; Winmill, T.F. CLXXVII—The state of amines in aqueous solution. *J. Chem. Soc. Trans.* **1912**, *101*, 1635–1676. [CrossRef]
2. Luo, X.Y.; Fan, X.; Shi, G.L.; Li, H.R.; Wang, C.M. Decreasing the Viscosity in CO_2 Capture by Amino-Functionalized Ionic Liquids through the Formation of Intramolecular Hydrogen Bond. *J. Phys. Chem. B* **2016**, *120*, 2807–2813. [CrossRef] [PubMed]
3. Malhotra, D.; Cantu, D.C.; Koech, P.K.; Heldebrant, D.J.; Karkamkar, A.; Zheng, F.; Bearden, M.D.; Rousseau, R.; Glezakou, V.A. Directed Hydrogen Bond Placement: Low Viscosity Amine Solvents for CO_2 Capture. *ACS Sustain. Chem. Eng.* **2019**, *7*, 7535–7542. [CrossRef]
4. Altamash, T.; Amhamed, A.; Aparicio, S.; Atilhan, M. Effect of Hydrogen Bond Donors and Acceptors on CO2 Absorption by Deep Eutectic Solvents. *Processes* **2020**, *8*, 1533. [CrossRef]

5. Saeed, U.; Khan, A.L.; Gilani, M.A.; Bilad, M.R.; Khan, A.U. Supported liquid membranes comprising of choline chloride based deep eutectic solvents for CO2 capture: Influence of organic acids as hydrogen bond donor. *J. Mol. Liq.* **2021**, *335*, 116155. [CrossRef]
6. Cao, L.; Li, D.; Soto, F.A.; Ponce, V.; Zhang, B.; Ma, L.; Deng, T.; Seminario, J.M.; Hu, E.; Yang, X.Q.; et al. Highly Reversible Aqueous Zinc Batteries enabled by Zincophilic–Zincophobic Interfacial Layers and Interrupted Hydrogen-Bond Electrolytes. *Angew. Chem. Int. Ed.* **2021**, *60*, 18845–18851. [CrossRef]
7. Sun, T.; Zheng, S.; Nian, Q.; Tao, Z. Hydrogen Bond Shielding Effect for High-Performance Aqueous Zinc Ion Batteries. *Small* **2022**, *18*, 2107115. [CrossRef]
8. Xiong, Q.; Li, C.; Li, Z.; Liang, Y.; Li, J.; Yan, J.; Huang, G.; Zhang, X. Hydrogen-Bond-Assisted Solution Discharge in Aprotic Li–O$_2$ Batteries. *Adv. Mater.* **2022**, *34*, 2110416. [CrossRef]
9. Lu, X.; Cao, L.; Du, X.; Lin, H.; Zheng, C.; Chen, Z.; Sun, B.; Tao, S. Hydrogen-Bond-Induced High Performance Semitransparent Ternary Organic Solar Cells with 14% Efficiency and Enhanced Stability. *Adv. Opt. Mater.* **2021**, *9*, 2100064. [CrossRef]
10. Li, X.; Zhou, L.; Lu, X.; Cao, L.; Du, X.; Lin, H.; Zheng, C.; Tao, S. Hydrogen bond induced high-performance quaternary organic solar cells with efficiency up to 17.48% and superior thermal stability. *Mater. Chem. Front.* **2021**, *5*, 3850–3858. [CrossRef]
11. Doyle, A.G.; Jacobsen, E.N. Small-Molecule H-Bond Donors in Asymmetric Catalysis. *Chem. Rev.* **2007**, *107*, 5713–5743. [CrossRef] [PubMed]
12. Guzmán, J.; Urriolabeitia, A.; Polo, V.; Fernández-Buenestado, M.; Iglesias, M.; Fernández-Alvarez, F.J. Dehydrogenation of formic acid using iridium-NSi species as catalyst precursors. *Dalton Trans.* **2022**, *51*, 4386–4393. [CrossRef] [PubMed]
13. Liu, K.; Cruzan, J.D.; Saykally, R.J. Water Clusters. *Science* **1996**, *271*, 929–933. [CrossRef]
14. Steiner, T. The Hydrogen Bond in the Solid State. *Angew. Chem. Int. Ed.* **2002**, *41*, 48–76. [CrossRef]
15. Guevara-Vela, J.M.; Chávez-Calvillo, R.; García-Revilla, M.; Hernández-Trujillo, J.; Christiansen, O.; Francisco, E.; Martín Pendás, Á.; Rocha-Rinza, T. Hydrogen-Bond Cooperative Effects in Small Cyclic Water Clusters as Revealed by the Interacting Quantum Atoms Approach. *Chem. Eur. J.* **2013**, *19*, 14304–14315. [CrossRef]
16. Guevara-Vela, J.M.; Romero-Montalvo, E.; Gómez, V.A.M.; Chávez-Calvillo, R.; García-Revilla, M.; Francisco, E.; Martín Pendás, Á.; Rocha-Rinza, T. Hydrogen bond cooperativity and anticooperativity within the water hexamer. *Phys. Chem. Chem. Phys.* **2016**, *18*, 19557–19566. [CrossRef]
17. Castor-Villegas, V.M.; Guevara-Vela, J.M.; Narváez, W.E.V.; Martín Pendás, Á.; Rocha-Rinza, T.; Fernández-Alarcón, A. On the strength of hydrogen bonding within water clusters on the coordination limit. *J. Comput. Chem.* **2020**, *41*, 2266–2277. [CrossRef]
18. Dalrymple, S.A.; Shimizu, G.K.H. Crystal Engineering of a Permanently Porous Network Sustained Exclusively by Charge-Assisted Hydrogen Bonds. *J. Am. Chem. Soc.* **2007**, *129*, 12114–12116. [CrossRef]
19. Wang, H.; Gurau, G.; Shamshina, J.; Cojocaru, O.A.; Janikowski, J.; MacFarlane, D.R.; Davis, J.H.; Rogers, R.D. Simultaneous membrane transport of two active pharmaceutical ingredients by charge assisted hydrogen bond complex formation. *Chem. Sci.* **2014**, *5*, 3449. [CrossRef]
20. Pedzisa, L.; Hay, B.P. Aliphatic C-H Anion Hydrogen Bonds: Weak Contacts or Strong Interactions? *J. Org. Chem.* **2009**, *74*, 2554–2560. [CrossRef]
21. Scheiner, S. *Hydrogen Bonding*; Oxford University Press: New York, NY, USA, 1997.
22. Guevara-Vela, J.M.; Romero-Montalvo, E.; Costales, A.; Martín Pendás, Á.; Rocha-Rinza, T. The nature of resonance-assisted hydrogen bonds: A quantum chemical topology perspective. *Phys. Chem. Chem. Phys.* **2016**, *18*, 26383–26390. [CrossRef] [PubMed]
23. Romero-Montalvo, E.; Guevara-Vela, J.M.; Costales, A.; Martín Pendás, Á.; Rocha-Rinza, T. Cooperative and anticooperative effects in resonance assisted hydrogen bonds in merged structures of malondialdehyde. *Phys. Chem. Chem. Phys.* **2017**, *19*, 97–107. [CrossRef] [PubMed]
24. Guevara-Vela, J.M.; Gallegos, M.; Valentín-Rodríguez, M.A.; Costales, A.; Rocha-Rinza, T.; Martín Pendás, Á. On the Relationship between Hydrogen Bond Strength and the Formation Energy in Resonance-Assisted Hydrogen Bonds. *Molecules* **2021**, *26*, 4196. [CrossRef] [PubMed]
25. Guevara-Vela, J.M.; Romero-Montalvo, E.; del Río-Lima, A.; Martín Pendás, Á.; Hernández-Rodríguez, M.; Rocha-Rinza, T. Hydrogen-Bond Weakening through π Systems: Resonance-Impaired Hydrogen Bonds (RIHB). *Chem. Eur. J.* **2017**, *23*, 16605–16611. [CrossRef]
26. Nowroozi, A.; Sarhadinia, S.; Masumian, E.; Nakhaei, E. A comprehensive theoretical study of tautomeric and conformeric preferences, intramolecular hydrogen bonding, and π-electron delocalisation in β-selenoaminoacrolein with its thio and oxo analogs. *Struc. Chem.* **2014**, *25*, 1359–1368. [CrossRef]
27. Masumian, E.; Nowroozi, A. Computational investigation on the intramolecular resonance-inhibited hydrogen bonding: A new type of interaction versus the RAHB model. *Theor. Chem. Acc.* **2015**, *134*. [CrossRef]
28. Gilli, G.; Bellucci, F.; Ferretti, V.; Bertolasi, V. Evidence for resonance-assisted hydrogen bonding from crystal-structure correlations on the enol form of the .beta.-diketone fragment. *J. Am. Chem. Soc.* **1989**, *111*, 1023–1028. [CrossRef]
29. Bertolasi, V.; Gilli, P.; Ferretti, V.; Gilli, G. Evidence for resonance-assisted hydrogen bonding. 2. Intercorrelation between crystal structure and spectroscopic parameters in eight intramolecularly hydrogen bonded 1, 3-diaryl-1, 3-propanedione enols. *J. Am. Chem. Soc.* **1991**, *113*, 4917–4925. [CrossRef]

30. Bertolasi, V.; Pretto, L.; Gilli, G.; Gilli, P. π-Bond cooperativity and anticooperativity effects in resonance-assisted hydrogen bonds (RAHBs). *Acta Crystallogr. B Struct. Sci.* **2006**, *62*, 850–863. [CrossRef]
31. Wu, J.I.; Jackson, J.E.; von Ragué Schleyer, P. Reciprocal Hydrogen Bonding–Aromaticity Relationships. *J. Am. Chem. Soc.* **2014**, *136*, 13526–13529. [CrossRef]
32. Kakeshpour, T.; Wu, J.I.; Jackson, J.E. AMHB: (Anti)aromaticity-Modulated Hydrogen Bonding. *J. Am. Chem. Soc.* **2016**, *138*, 3427–3432. [CrossRef] [PubMed]
33. Bader, R.F.W. *Atoms in Molecules. A Quantum Theory*; Oxford University Press: Oxford, UK, 1995.
34. Blanco, M.A.; Martín Pendás, A.; Francisco, E. Interacting Quantum Atoms: A Correlated Energy Decomposition Scheme Based on the Quantum Theory of Atoms in Molecules. *J. Chem. Theory Comput.* **2005**, *1*, 1096–1109. [CrossRef] [PubMed]
35. Farrugia, L.J.; Evans, C.; Lentz, D.; Roemer, M. The QTAIM Approach to Chemical Bonding Between Transition Metals and Carbocyclic Rings: A Combined Experimental and Theoretical Study of (η5-C5H5)Mn(CO)3, (η6-C6H6)Cr(CO)3, and (E)-(η5-C5H4)CF=CF(η5-C5H4)(η5-C5H5)2Fe. *J. Am. Chem. Soc.* **2008**, *131*, 1251–1268. [CrossRef] [PubMed]
36. Teixeira, F.; Mosquera, R.; Melo, A.; Freire, C.; Cordeiro, M.N.D.S. Driving Forces in the Sharpless Epoxidation Reaction: A Coupled AIMD/QTAIM Study. *Inorg. Chem.* **2017**, *56*, 2124–2134. [CrossRef] [PubMed]
37. Hooper, T.N.; Garçon, M.; White, A.J.P.; Crimmin, M.R. Room temperature catalytic carbon–hydrogen bond alumination of unactivated arenes: Mechanism and selectivity. *Chem. Sci.* **2018**, *9*, 5435–5440. [CrossRef] [PubMed]
38. Escofet, I.; Armengol-Relats, H.; Bruss, H.; Besora, M.; Echavarren, A.M. On the Structure of Intermediates in Enyne Gold(I)-Catalyzed Cyclizations: Formation of trans-Fused Bicyclo[5.1.0]octanes as a Case Study. *Chem. Eur. J.* **2020**, *26*, 15738–15745. [CrossRef]
39. Martín Pendás, Á.; Guevara-Vela, J.M.; Crespo, D.M.; Costales, A.; Francisco, E. An unexpected bridge between chemical bonding indicators and electrical conductivity through the localization tensor. *Phys. Chem. Chem. Phys.* **2017**, *19*, 1790–1797. [CrossRef]
40. Astakhov, A.A.; Tsirelson, V.G. Spatially resolved characterisation of electron localization and delocalisation in molecules: Extending the Kohn-Resta approach. *Int. J. Quantum Chem.* **2018**, *118*, e25600. [CrossRef]
41. Gil-Guerrero, S.; Ramos-Berdullas, N.; Martín Pendás, Á.; Francisco, E.; Mandado, M. Anti-ohmic single molecule electron transport: Is it feasible? *Nanoscale Adv.* **2019**, *1*, 1901–1913. [CrossRef]
42. Matito, E.; Solà, M.; Salvador, P.; Duran, M. Electron sharing indexes at the correlated level. Application to aromaticity calculations. *Faraday Discuss.* **2007**, *135*, 325–345. [CrossRef]
43. Casademont-Reig, I.; Guerrero-Avilés, R.; Ramos-Cordoba, E.; Torrent-Sucarrat, M.; Matito, E. How Aromatic Are Molecular Nanorings? The Case of a Six-Porphyrin Nanoring. *Angew. Chem. Int. Ed.* **2021**, *60*, 24080–24088. [CrossRef] [PubMed]
44. del Rosario Merino-García, M.; Soriano-Agueda, L.A.; de Dios Guzmán-Hernández, J.; Martínez-Otero, D.; Rivera, B.L.; Cortés-Guzmán, F.; Barquera-Lozada, J.E.; Jancik, V. Benzene and Borazine, so Different, yet so Similar: Insight from Experimental Charge Density Analysis. *Inorg. Chem.* **2022**, *61*, 6785–6798. [CrossRef] [PubMed]
45. Francisco, E.; Martín Pendás, A.; Blanco, M.A. A Molecular Energy Decomposition Scheme for Atoms in Molecules. *J. Chem. Theory Comput.* **2005**, *2*, 90–102. [CrossRef]
46. Schleyer, P.v.R.; Maerker, C.; Dransfeld, A.; Jiao, H.; van Eikema Hommes, N.J.R. Nucleus-Independent Chemical Shifts: A Simple and Efficient Aromaticity Probe. *J. Am. Chem. Soc.* **1996**, *118*, 6317–6318. [CrossRef] [PubMed]
47. Chen, Z.; Wannere, C.S.; Corminboeuf, C.; Puchta, R.; Schleyer, P.V.R. Nucleus-Independent Chemical Shifts (NICS) as an Aromaticity Criterion. *Chem. Rev.* **2005**, *105*, 3842–3888. [CrossRef] [PubMed]
48. Poater, J.; Solà, M.; Viglione, R.G.; Zanasi, R. Local Aromaticity of the Six-Membered Rings in Pyracylene. A Difficult Case for the NICS Indicator of Aromaticity. *J. Org. Chem.* **2004**, *69*, 7537–7542. [CrossRef]
49. Aihara, J.-i. Incorrect NICS-Based Prediction on the Aromaticity of the Pentalene Dication. *Bull. Chem. Soc. Jpn* **2004**, *77*, 101–102. [CrossRef]
50. Faglioni, F.; Ligabue, A.; Pelloni, S.; Soncini, A.; Viglione, R.G.; Ferraro, M.B.; Zanasi, R.; Lazzeretti, P. Why Downfield Proton Chemical Shifts Are Not Reliable Aromaticity Indicators. *Org. Lett.* **2005**, *7*, 3457–3460. [CrossRef]
51. von Ragué Schleyer, P.; Manoharan, M.; Jiao, H.; Stahl, F. The Acenes: Is There a Relationship between Aromatic Stabilization and Reactivity? *Org. Lett.* **2001**, *3*, 3643–3646. [CrossRef]
52. Kruszewski, J.; Krygowski, T. Definition of aromaticity basing on the harmonic oscillator model. *Tetrahedron Lett.* **1972**, *13*, 3839–3842. [CrossRef]
53. Fernández, I.; Frenking, G. Aromaticity in Metallabenzenes. *Chem. Eur. J.* **2007**, *13*, 5873–5884. [CrossRef] [PubMed]
54. Poater, J.; Fradera, X.; Duran, M.; Solà, M. The delocalisation Index as an Electronic Aromaticity Criterion: Application to a Series of Planar Polycyclic Aromatic Hydrocarbons. *Chem. Eur. J.* **2003**, *9*, 400–406. [CrossRef] [PubMed]
55. Matito, E.; Duran, M.; Solà, M. The aromatic fluctuation index (FLU): A new aromaticity index based on electron delocalisation. *J. Chem. Phys.* **2005**, *122*, 014109. [CrossRef] [PubMed]
56. Bultinck, P.; Ponec, R.; Damme, S.V. Multicenter bond indices as a new measure of aromaticity in polycyclic aromatic hydrocarbons. *J. Phys. Org. Chem.* **2005**, *18*, 706–718. [CrossRef]
57. Feixas, F.; Matito, E.; Poater, J.; Solà, M. On the performance of some aromaticity indices: A critical assessment using a test set. *J. Comput. Chem.* **2008**, *29*, 1543–1554. [CrossRef]
58. Neese, F.; Wennmohs, F.; Becker, U.; Riplinger, C. The ORCA quantum chemistry program package. *J. Chem. Phys.* **2020**, *152*, 224108. [CrossRef]

59. Adamo, C.; Barone, V. Toward reliable density functional methods without adjustable parameters: The PBE0 model. *J. Chem. Phys.* **1999**, *110*, 6158–6170. [CrossRef]
60. Weigend, F.; Ahlrichs, R. Balanced basis sets of split valence, triple zeta valence and quadruple zeta valence quality for H to Rn: Design and assessment of accuracy. *Phys. Chem. Chem. Phys.* **2005**, *7*, 3297. [CrossRef]
61. Grimme, S.; Antony, J.; Ehrlich, S.; Krieg, H. A consistent and accurate ab initio parametrization of density functional dispersion correction (DFT-D) for the 94 elements H-Pu. *J. Chem. Phys.* **2010**, *132*, 154104. [CrossRef]
62. Grimme, S.; Ehrlich, S.; Goerigk, L. Effect of the damping function in dispersion corrected density functional theory. *J. Comput. Chem.* **2011**, *32*, 1456–1465. [CrossRef]
63. Keith, A.T. *TK Gristmill Software*; AIMALL (Version 19.02.13); AIMALL: Overland Park, KS, USA, 2019.
64. Martín Pendás, A.; Francisco, E. Promolden. A QTAIM/IQA code (Avaliable from the authors upon request).
65. Maxwell, P.; Martín Pendás, Á.; Popelier, P.L.A. Extension of the interacting quantum atoms (IQA) approach to B3LYP level density functional theory (DFT). *Phys. Chem. Chem. Phys.* **2016**, *18*, 20986–21000. [CrossRef] [PubMed]
66. Matito, E. ESI-3D: Electron Sharing Indexes Program for 3D Molecular Space Partitioning. 2020.
67. Feixas, F.; Matito, E.; Solà, M.; Poater, J. Patterns of π-electron delocalization in aromatic and antiaromatic organic compounds in the light of Hückel's 4n + 2 rule. *Phys. Chem. Chem. Phys.* **2010**, *12*, 7126–7137. [CrossRef] [PubMed]

Article

Hydrogen Bonds with Fluorine in Ligand–Protein Complexes-the PDB Analysis and Energy Calculations

Wojciech Pietruś, Rafał Kafel, Andrzej J. Bojarski and Rafał Kurczab *

Department of Medicinal Chemistry, Maj Institute of Pharmacology, Polish Academy of Sciences, Smetna 12, 31-343 Krakow, Poland; pietrus@if-pan.krakow.pl (W.P.); rafal.kafel@gmail.com (R.K.); bojarski@if-pan.krakow.pl (A.J.B.)
* Correspondence: kurczab@if-pan.krakow.pl; Tel.: +48-126-62-3301

Abstract: Fluorine is a common substituent in medicinal chemistry and is found in up to 50% of the most profitable drugs. In this study, a statistical analysis of the nature, geometry, and frequency of hydrogen bonds (HBs) formed between the aromatic and aliphatic C–F groups of small molecules and biological targets found in the Protein Data Bank (PDB) repository was presented. Interaction energies were calculated for those complexes using three different approaches. The obtained results indicated that the interaction energy of F-containing HBs is determined by the donor–acceptor distance and not by the angles. Moreover, no significant relationship between the energies of HBs with fluorine and the donor type was found, implying that fluorine is a weak HB acceptor for all types of HB donors. However, the statistical analysis of the PDB repository revealed that the most populated geometric parameters of HBs did not match the calculated energetic optima. In a nutshell, HBs containing fluorine are forced to form due to the stronger ligand–receptor neighboring interactions, which make fluorine the "donor's last resort".

Keywords: fluorine; PDB; hydrogen bonds; HBs

Citation: Pietruś, W.; Kafel, R.; Bojarski, A.J.; Kurczab, R. Hydrogen Bonds with Fluorine in Ligand–Protein Complexes-the PDB Analysis and Energy Calculations. *Molecules* 2022, 27, 1005. https://doi.org/10.3390/molecules27031005

Academic Editor: Miroslaw Jablonski

Received: 30 December 2021
Accepted: 29 January 2022
Published: 2 February 2022

Publisher's Note: MDPI stays neutral with regard to jurisdictional claims in published maps and institutional affiliations.

Copyright: © 2022 by the authors. Licensee MDPI, Basel, Switzerland. This article is an open access article distributed under the terms and conditions of the Creative Commons Attribution (CC BY) license (https://creativecommons.org/licenses/by/4.0/).

1. Introduction

Fluorine is the most electronegative element, and this property has a significant impact on the bioavailability, lipophilicity, metabolic stability, acidity/basicity, and toxicity [1]. Since the second half of the 20th century [2], researchers have been exploring the possibility of using fluorinated molecules in medicine [1,3]. The important position of fluorinated molecules in medicinal chemistry can be understood by the exceptionally large number of fluorine-containing drugs currently available on the pharmaceutical market (Figure 1). The share of fluorinated compounds rose from 2% in 1970 to 8% in 1980, 13% in 1990, and reached 18% at the beginning of the 21st century. Among them, six products were in the "top-12" list and employed as anticancer, anti-inflammatory, analgesic, or antidepressant agents in medicine [4]. About 20% of the drugs used in 2010 contained fluorine atom(s) or fluoroalkyl group(s) [3], whereas in the last decade (2011–2020) 114 out of the 410 drugs approved by the Food and Drug Administration (FDA) (data from the Center for Drug Evaluation and Research (CDER)) [5] contained fluorine (Figure 1). Currently, fluorinated pharmaceuticals account for over 50% of the most profitable drugs (blockbuster drugs), and are also recognized as the best among the drugs used in almost all therapeutic areas [6].

The biological activity of drugs is determined by intermolecular interactions. These interactions also play an important role in stabilizing the ligand–biomolecule system. Hydrogen bonds (HBs), in particular, are considered to significantly influence the action of drug molecules on their targets [7–9]. Interestingly, fluorine or substituents containing this element have been shown to tune the intermolecular interactions in ligand–protein complexes [1,10]. Although characterized by high electronegativity, fluorine is a weak acceptor of HBs and, unlike other halogens, it is not a halogen bond (XB) donor in aromatic systems [11,12]. However, the results of our previous study on small model systems (e.g., 2,6-difluro-4-halogenoanilines) indicated that fluorine can act as a competitive and attractive acceptor for HBs and XBs as well as form F⋯F interactions [13]. Additionally, it is considered that fluorine-containing HBs are not typical and do not behave like conventional HBs (e.g., O⋯H–O and N⋯H–N), as demonstrated by a more angular nature and preference for less electronegative donors [14].

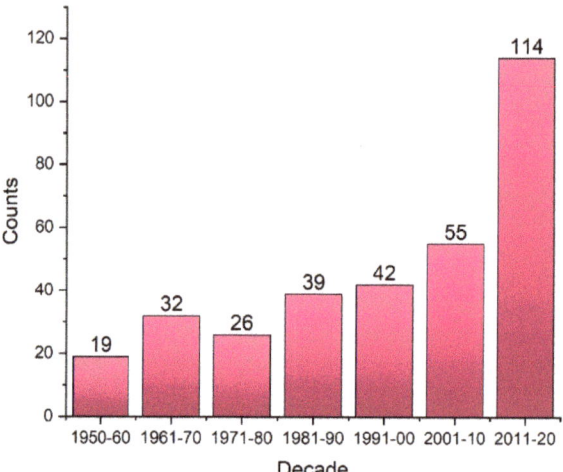

Figure 1. Number of marketed drugs containing fluorine per decade. Data were collected from the DrugCentral 2021 database (accessed 30 April 2021).

The biological activity of compounds can be tuned with the use of fluorine. However, there are no rules of thumb for predicting the preferred fluorine substitution sites in a molecule. Despite numerous studies on fluorine, the influence of this element on the pharmacodynamics properties of drugs remains unclear. A statistical analysis of the nature, geometry, and frequency of interactions occurring between fluorine in small molecules and the biological targets included in the Protein Data Bank (PDB) repository may allow understanding of the role of fluorine in ligand–receptor (L–R) complexes. Therefore, we carried out a wide statistical analysis and calculations to quantitatively and qualitatively explore the HBs (contacts) formed with fluorine in biological systems. The findings of this study may contribute to a thorough understanding of the effects of fluorine, to enable its rational use in drug design and for improving the efficiency of computational methods [15,16].

2. Results and Discussion

2.1. Choice of a Model System

We carefully chose the model systems used for database mining and quantum chemical calculations, with a focus on providing accurate representations of HBs with fluorine in the biological systems. Three types of HB donors were distinguished, namely hydroxyl, amine, and methyl group, and a pH of 7.4 ± 0.5 was considered to assess the protonation states of all entities. We extracted a ligand with interacting residue for performing high-level quantum chemical calculations in a reasonable time. This allowed us to determine the energy of isolated interaction with fluorine. For extracted amino acids from the main chain, the peptization reaction was reversed. Therefore, to maintain the proper structure of amino acids, missing atoms (hydroxyl to carboxylic group and hydrogen to nitrogen atom) were added and optimized with force field. Since 96% of the analyzed crystal structures were recorded with a resolution of <3 Å (Figure 2), we analyzed all the collected structures.

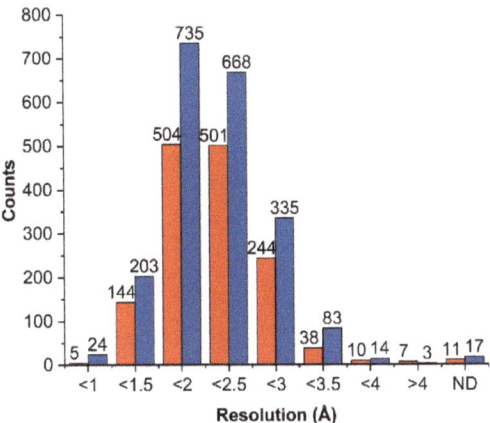

Figure 2. Statistical representation of the spectral resolution of the analyzed crystal structures deposited in the PDB database for L–R complexes with HB containing aliphatic fluorine (red bars) and aromatic fluorine (blue bars). "ND" (not determined) refers to the crystal structures obtained with the methods for which resolution was not specified (e.g., nuclear magnetic resonance).

2.2. General Statistics of HBs Containing Fluorine Atoms

It should be emphasized that the thresholds of HB geometric parameters considered in the statistical investigation based on the PDB data can significantly influence the results and conclusion. In this study, we assumed that the HB distance was <4Å and the HB angle was 90°–180°. Because we aimed to determine all contacts with fluorine atoms, these values can significantly exceed the standard geometric parameters of HBs; the distance can be below the sum of the van der Waals radii of interacting atoms, and the angle can differ by up to 120°. Complexes containing ligands with fluorine (from the LigandExpo repository) were extracted and analyzed. If the PDB entry contained more than one asymmetric unit (receptor oligomerization), the number of HBs with fluorine was multiplied by the number of occurrences of the same ligand. All measured HBs were used in further analysis, even if they came from the same PDB entry. The ligands were divided into two categories: molecules in which fluorine is bonded to an aliphatic carbon (F_{al}) and molecules containing fluorine bonded to an aromatic carbon (F_{ar}).

A total of 1787 (F_{al}) and 2324 (F_{ar}) unique PDB entries were found for fluorine-containing ligands. Based on the assigned boundaries and defined geometric thresholds, 165 interactions with a hydroxyl group, 612 with an amine group, and 3875 with a methyl group were identified for aliphatic fluorine (Figure 3A); and 121 interactions with a hydroxyl group, 606 with an amine group, and 6698 with H–C were identified for aromatic fluorine (Figure 3C). The number of F···H–O and F···H–N HBs was found to be larger for F_{al}, whereas the number of F···H–C HBs was two-fold higher for F_{ar}. For OH donors, three amino acids (SER, THR, TYR) were identified to be involved in HBs with F_{al}; however, for F_{ar}, it appeared that TYR participates less frequently in HBs, which may be attributed to the greater acidity of the OH group.

The more significant differences were observed for NH donors, in which the amino acids commonly involved in HBs with F_{al} were in the order ARG>ASN>GLN>GLY>LYS (Figure 3B) and in HBs with F_{ar} were in the order ARG>GLY>LYS>GLN>ASN (Figure 3D). Surprisingly, glycine is the second most common amino acid, forming HBs with F_{ar} because the others are polar amino acids with a free amino group in their side chain.

By contrast, no clear preferences of amino acids in the occurrence of HBs with F_{al} and F_{ar} were observed in the case of CH donors. However, the results highlighted that fluorine most frequently forms HBs with nonpolar amino acids (LEU, VAL, PHE, ILE, ALA), implying that it prefers hydrophobic areas of binding pockets (Figure 3B,D).

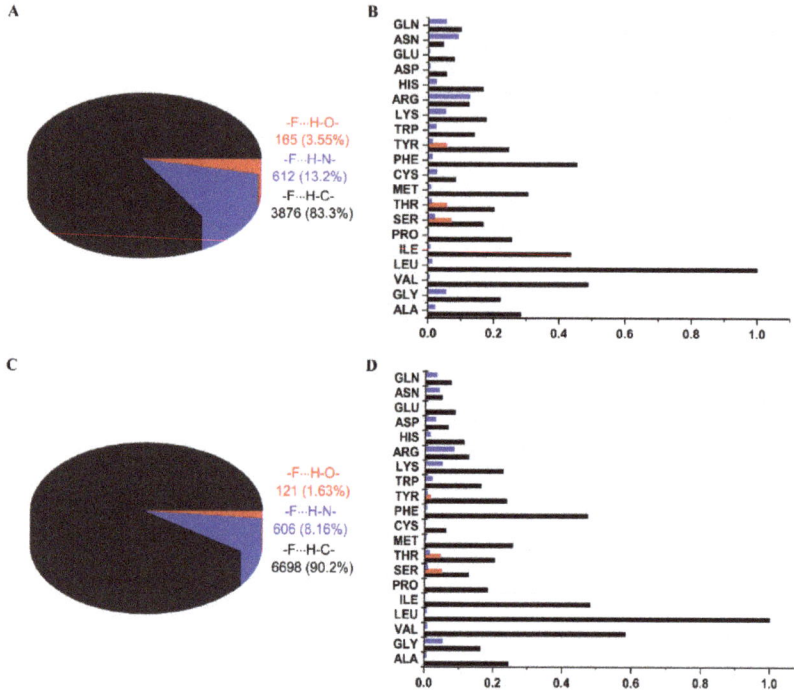

Figure 3. Number of hydrogen bonds found in the PDB repository to meet the boundaries for geometric parameters: (**A**) between F_{al} and OH, NH, and CH donors; (**B**) normalized division into interacting amino acids; (**C**) between F_{ar} and OH, NH, and CH donors; and (**D**) normalized division into interacting amino acids.

Based on this classification, we generated the density maps showing the geometric parameters of HBs (Figure 4). As only a small number of F···H–O HBs were identified, certain conclusions could not be drawn (Figure 4). The density maps of HB geometric parameters obtained for the NH and CH donors (as well as OH) revealed that fluorine prefers geometries with a distance of >3 Å and an angle of 100°–140° (Figure 4). However, it should be noted that more HBs were found for NH than OH donors, with a more linear geometry and short distances, but in many cases, those interactions are forced by the neighboring functional groups of a ligand interacting with amino acids.

In summary, fluorine-containing HBs reveal more angular geometric preferences than typical HBs (rather linear HBs O···H–O, N···H–N). Thus, in the next step, we explored the relationship between the geometry of F···H–X (X = O, N, C) bond and the energy contribution to the ligand–receptor complexes.

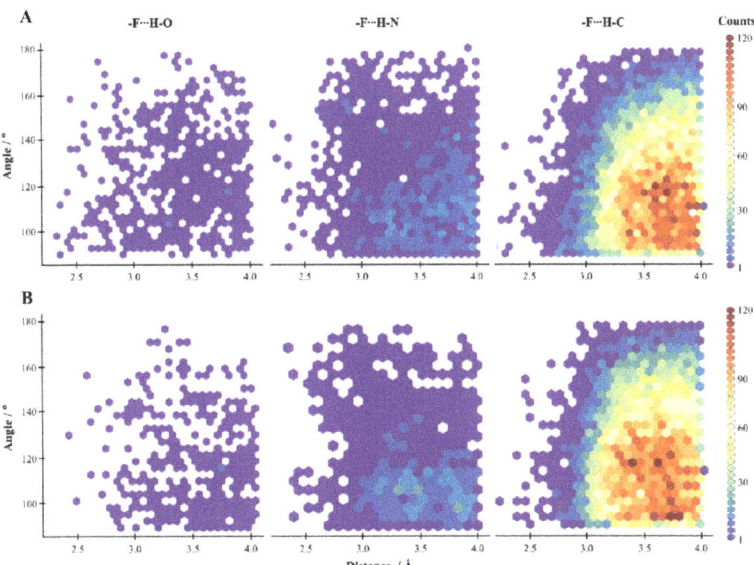

Figure 4. Density maps of geometric parameters of HBs with (**A**) F_{al} and (**B**) F_{ar}.

2.3. Energy of HBs with Fluorine

Ligand–receptor complexes are stabilized by various intermolecular forces, such as strong HBs (O···H–O, N···H–O, N···H–N), weak HBs (O···H–C, S···H–N), XBs, π-stacking, salt bridge, amide stacking, cation-π, and hydrophobic interactions and others [17]. Fluorine-containing HBs, especially those with hydroxyl and amine donors, are not common in biological systems (Figures 3 and 4). Therefore, it is important to determine the strength and geometric preferences of these HBs in biological systems. In this study, we attempted to evaluate the nature and energetic dependencies of HBs with fluorine in the theoretical background by performing quantum chemical calculations using small molecular systems extracted from ligand–biomolecule crystals (O–H, N–H, +N–H, and C–H were only considered to be HB donors). We determined the energy of HBs with fluorine found in crystal structures by applying three different methods as follows: (1) Diff—energy was calculated as the difference between the energy of the interacting molecules and the sum of the energies of isolated species calculated in Gaussian; (2) QTAIM—energy was calculated at BCP in AIMAll software; and (3) ETS—energy was calculated between two interacting molecules using the ETS-NOCV scheme implemented in ADF software.

At first, a simple statistical analysis was performed on the data obtained from the three approaches using the correlation coefficient and Pearson test in R (Supplementary Tables S1 and S2). The results of the analysis indicated the highest correlation between Diff and ETS methods ($p < 0.05$, correlation coefficient ~1) because they consider the energy of the entire system and approximately 70% of the calculated energy accounts for the same nature of interaction (attractive/repulsive). Additionally, the correlation decreased for stronger interactions (Supplementary Tables S1 and S2) due to the fact that the Diff method is intended for weak and medium HBs; strong HBs result in geometry distortion of the interacting molecules, which decreases the accuracy of the evaluation of the HB itself [18]. The energies of HBs calculated by QTAIM did not correlate with those determined by the remaining methods, since this method takes only isolated L–R interaction and neglects long-range interactions occurring between the atoms from separated fragments.

The distribution of calculated interaction energies for all selected complexes for a given type of HB with fluorine and method is illustrated in Figure 5. For HBs with F_{al}, the interaction energy calculated by QTAIM varied between 0 and −1.2 kcal/mol (the weakest HBs were found for $F_{al}\cdots$H–O, with an energy value of -0.64 kcal/mol. For $F_{al}\cdots$H–C HBs, the energy was (-0.69 kcal/mol). The strongest HBs (~-0.8 kcal/mol) were observed for $F_{al}\cdots$H–N (no significant difference was noted between $F_{al}\cdots$H–N HB and $F_{ar}\cdots$H–N$^+$ HB) (Figure 5A). The energy range determined by the Diff method was also between 0 and -1.2 kcal/mol, while the energy determined by the ETS method ranged from 0 to -8 kcal/mol (Figure 5A). Additionally, the Diff method indicated that the charge-assisted $F_{al}\cdots$H–N$^+$ was the weakest HB (-0.70 kcal/mol), while $F_{al}\cdots$H–N HB was stronger (-0.96 kcal/mol). Unlike Diff, the results of the ETS method showed that the charge-assisted $F_{al}\cdots$H–N$^+$ HB was the stronger (-5.11 kcal/mol), while $F_{al}\cdots$H–N HBs were weaker than $F_{al}\cdots$H–O HBs (-2.28 and -2.46 kcal/mol, respectively) (Figure 5A). For HBs with F_{ar}, a different trend was noted in the QTAIM method than for HBs with F_{al}, where the strongest HBs had OH as a donor (-0.94 kcal/mol). The $F_{ar}\cdots$H–N HBs were found to be slightly weaker with median energy values of ~-0.7 kcal/mol (for $^+$NH and NH donors), and the weakest was $F_{ar}\cdots$H–C HB (-0.62 kcal/mol) (Figure 5B). A comparison of Diff and ETS methods revealed a similar trend as in the case of F_{al}—the Diff method indicated that the $F_{ar}\cdots$H–N$^+$ HB was the weakest (-0.8 kcal/mol), while the ETS method indicated it as the strongest interaction (-5.3 kcal/mol) (Figure 5B). The QTAIM method showed that F$\cdots$H–O HBs with F_{ar} were stronger than those with F_{al}, while HBs with other donors were found at a similar energy level. However, it should be emphasized that both Diff and ETS methods revealed higher stabilization energy for HBs with F_{ar} compared to HBs with F_{al}.

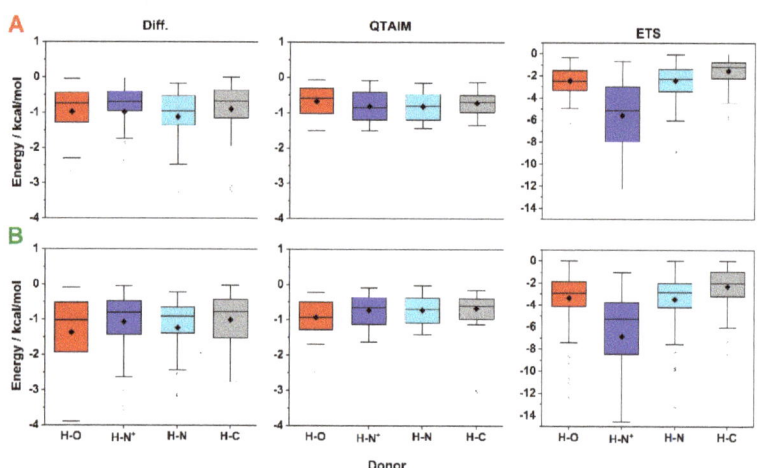

Figure 5. Box plots showing the distribution of the stabilization energy for HBs containing fluorine bonded to an (**A**) aliphatic or (**B**) aromatic carbon. A comparison is made for the individual donor groups (OH, $^+$NH, NH, and CH) as well as the calculation approaches used (Diff, QTAIM, and ETS).

The hydroxyl donor occurs in the side chains of three amino acid—tyrosine (TYR), threonine (THR), and serine (SER). The phenolic hydroxyl group (TYR) is significantly more acidic (pK_a of about 9.8 in polypeptides) than the aliphatic hydroxyl group (SER or THR, pK_a ~13.6) [19]. In addition, Graton et al. found in an analysis of the PDB repository that the distances and angle of HBs with a hydroxyl group decreased in the order THR > SER > TYR, which suggests that TYR forms the stronger HBs [20]. In the present study, the results obtained by the QTAIM method revealed that for F_{al} (Figure 6A), the energy of HBs does not depend on the F$\cdots$H–O angle, as the highest values were observed in the whole range of the analyzed angles. Instead, the energy of $F_{al}\cdots$H–O HBs closely correlated with the distance, as observed in the case of conventional hydrogen bonding. It should be mentioned that the QTAIM method showed higher energy for F$\cdots$H–O HBs with aromatic fluorine than for HBs with aliphatic fluorine (Figure 6). Most of the F···H–O HBs with distances shorter than 2.75 Å are repulsive (Figure 6), which shows that despite the stabilizing nature of the F···H–O HB itself, the interacting fragments have a positive energy contribution (repulsive character). This effect may be due to the interaction of the neighboring atoms, or high positive kinetic energy. All three methods showed that the highest stabilizing energies (red squares in Figure 6) were in the range of 2.85–3.45 Å (for both fluorine) and 150°–120° for F_{al} and 145°–120° for F_{ar}, suggesting that these are the optimal ranges of geometric parameters for F···H–O HBs. The analysis of the selected crystal structures did not show any significant differences between the F···H–O HBs of attractive and repulsive nature. The only differentiating factor identified was the F···O distance (Figure 6).

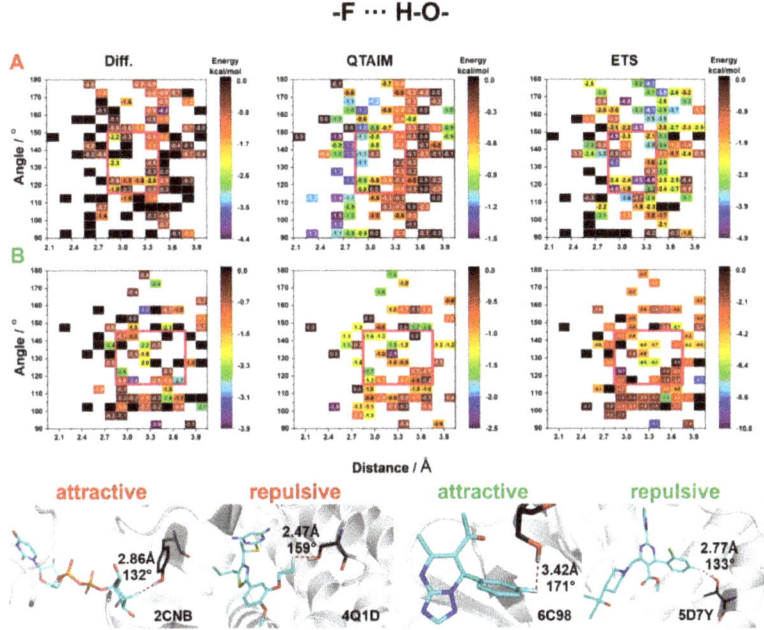

Figure 6. HB energy maps generated based on Diff, QTAIM, and ETS calculations of interaction energy between OH donor and fluorine attached to (**A**) an aliphatic fragment and (**B**) an aromatic ring for specific geometric parameters. The areas for which the highest stabilizing energy was observed in all three methods are marked with a red square.

Among amino acids, three (ARG, LYS, HIS) have additional amino groups. The side chain of arginine (ARG) is amphipathic because at physiological pH it contains a positively charged guanidine group (pK_a = 12.48). Another amphipathic amino acid is lysine (LYS), the side chain of which contains a positively charged primary amine group at the end of the long hydrophobic carbon tail (pK_a = 10.53). Histidine (HIS) contains an imidazole side chain. His pK_a is 6, above which one of the two protons is missing (in physiological pH, histidine has two tautomers). Since it is difficult to automatically protonate the appropriate nitrogen atom of histidine which forms an F···H–N$^+$ HB, we calculated the energy of F···H–N$^+$ HB only for LYS and ARG (Figure 7). The QTAIM calculations showed that the energy of F_{al}···H–N$^+$ and F_{al}···H–O HBs was similar. Additionally, F···H–N$^+$ was found to be strongly influenced by the distance between F···N and not by the angle (Figure 7) (as noticed for F···H–O HBs). A similar trend for both aliphatic and aromatic F was observed for F···H–N$^+$ HBs, but the highest interaction energy was mostly localized at higher values of the F_{ar}···H–N$^+$ angle (Figure 7). For aliphatic fluorine, the range of geometric parameters in which the three methods indicated the strongest $F_{al}\cdots H-N^+$ HBs was 140°–120° and 2.85–3.45 Å (as for $F\cdots H-O$ HB). In the case of F_{ar}, two ranges of geometric parameters were distinguished: (1) 170°–150° and 3.0–3.6 Å; and (2) < 120° and 2.4–3.6 Å. Due to its large volume, the guanidine group interacts not only with fluorine directly but also with neighboring atoms. Therefore, the energy calculated by Diff and ETS methods might be overestimated.

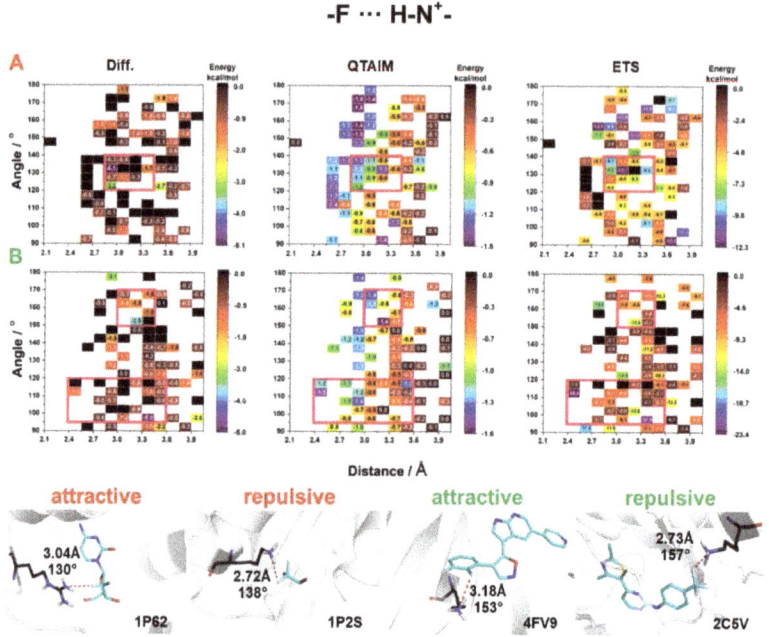

Figure 7. HB energy maps generated based on Diff, QTAIM, and ETS calculations of interaction energy between positively charged NH donor and fluorine attached to (A) an aliphatic fragment and (B) an aromatic ring for specific geometric parameters. The areas for which the highest stabilizing energy was observed in all three methods are marked with a red square.

Interestingly, the analysis of crystal structures with $F\cdots H-N^+$ HBs showed that for $F\cdots N^+$ distances of <2.8 Å, almost 70% of HBs exhibited a destabilizing character. Moreover, the interaction of positively charged nitrogen with the CF_3 group, with a partial positive charge on the carbon atom, is often repulsive (Figure 7).

The $F\cdots H-N$ HBs were found almost four times more frequently in PDB than $F\cdots H-O$ HBs (Figures 3 and 4). The energies of $F\cdots H-N$ HBs were calculated for all amino acids, except for arginine and lysine as these amino acids contain positively charged nitrogen atoms. Furthermore, whether the nitrogen atom was in the main chain or the side chains (ASN, GLN) was not considered in the analysis. The energies of $F\cdots H-N$ HBs determined by the QTAIM method showed no significant differences between F_{al} and F_{ar}. In addition, it must be noted that energy is inversely proportional to the HB distance and does not depend on the $F\cdots H-N$ angle (Figure 8). However, since the nitrogen atom was mostly present in the main chain, and was thus adjacent to different atoms, the energies calculated by Diff and ETS methods mostly had a destabilizing nature, which might be due to steric effects. On the other hand, for F_{al}, the areas where the energies were found to be high and exhibited a stabilizing character had a narrow range of 150°–125° and 2.4–3.45 Å, while for F_{ar} the areas were within the range of geometric parameters (165°–135° and 2.85–3.75 Å) (Figure 8). The analysis of selected crystal structures showed no significant differences between the systems with attractive energy and repulsive energy (Figure 8).

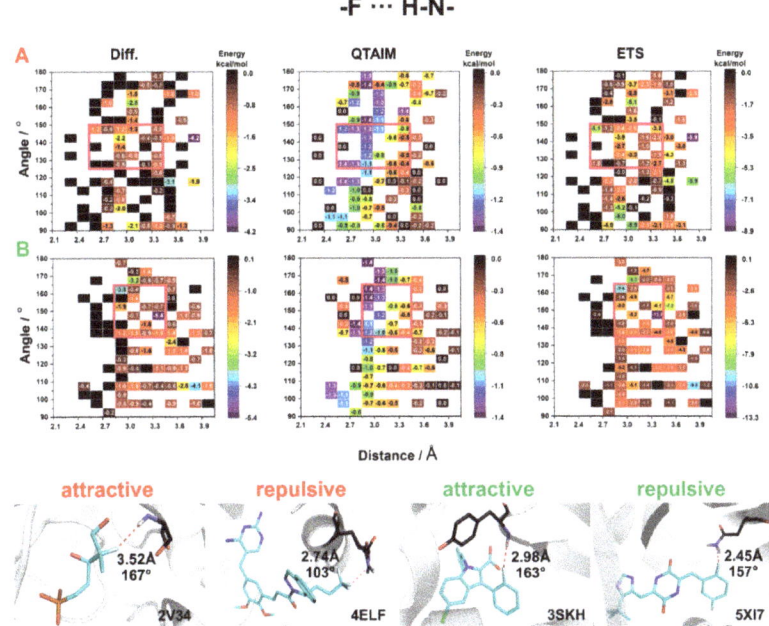

Figure 8. HB energy maps generated based on Diff, QTAIM, and ETS calculations of interaction energy between NH donor and fluorine attached to (**A**) an aliphatic fragment and (**B**) an aromatic ring for specific geometric parameters. The areas for which the highest stabilizing energy was observed in all three methods are marked with a red square.

The F···H–C HBs were found to be the most abundant in biological systems (Figures 3 and 4). The interaction energies calculated by the QTAIM method showed that F_{al}···H–C HBs are stronger than F_{ar}···H–C HBs. In addition, the interactions with an HB distance of <2.7 Å showed a destabilizing character (Figure 9). The results produced by Diff and ETS methods were quite divergent, and it is difficult to find any constant trend. Interestingly, the results obtained from all three methods indicate that HBs with F_{al} are stronger than those with F_{ar} (Figure 9). The energetically favorable HBs with F_{al} had an angle of 155°–145° and a distance of >2.7 Å, while HBs with F_{ar} had an angle of 160°–120° and a distance of >2.85 Å (red squares in Figure 9). The analysis of selected crystal structures showed that F···H–C HBs mostly exhibited a stabilizing character for distances longer than 3 Å (Figure 9).

Analysis of the ETS-NOCV decomposition results showed that for uncharged donors (OH, NH, CH) the contribution of Coulomb energy term has the greatest impact on the stabilization energy of HBs with fluorine, while for NH^+ the XC energy term has the largest contribution. The reason for the destabilizing nature of the shorter HBs with fluorine may be due to the high value of the kinetic energy contribution (Figure S2). To determine the significance of HBs with fluorine, the density maps of geometric parameters of HBs found in the PDB repository (Figures 3 and 4) were compared with the corresponding HB energy maps (Figures 6–9). The proposed areas (red squares) of favorable geometric parameters with the highest energy values did not match with the most occupied areas of geometric parameters. This suggests that HBs with fluorine do not play a significant role in the stabilization of the L–R system and are often formed under unfavorable geometric parameters.

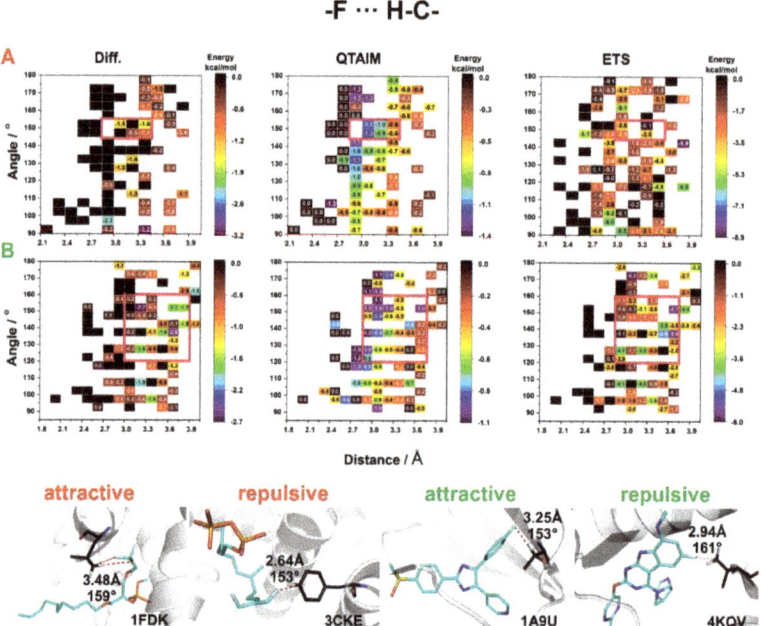

Figure 9. HB energy maps generated based on Diff, QTAIM, and ETS calculations of interaction energy between CH donor and fluorine attached to (**A**) an aliphatic fragment and (**B**) an aromatic ring for specific geometric parameters. The areas for which the highest stabilizing energy was observed in all three methods are marked with a red square.

3. Materials and Methods

3.1. PDB Analysis

We performed a statistical analysis of the structural data and investigated in detail the geometric parameters of the intermolecular HBs of fluorine in the structures deposited in the PDB repository. In the first step, all fluorine-containing ligands were identified in the LigandExpo database [21] and then divided into two groups: fluorine attached to an aliphatic carbon (F_{al}) and fluorine attached to an aromatic carbon (F_{ar}). In the next step, all crystal structures containing the abovementioned ligands were identified. The positions of hydrogen atoms were added, considering the stereochemical rules determining the most favorable position of hydrogens, using the Protein Preparation Wizard (Schrödinger Maestro Software) [22]. The in-house script was used to detect the interactions (contact) occurring between fluorine (as an acceptor) in the ligand and the neighboring HB donors (i.e., OH, NH, CH) in the receptor that met the following criteria: a distance of <4 Å and an angle of 90–180°.

Based on the obtained data, density maps showing the distribution of the geometric parameters of HBs were generated using R [23] environment as well as RColorBrewer [24], Hexbin [25], Rbokeh [26], and ggplot2 [27] libraries.

3.2. Calculation of Interaction Energy

To determine the energy of the studied intermolecular interactions with fluorine, all complexes obtained from the PDB were divided into subgroups based on the following: (i) angle of HB (ranged from 90° to 180° with a step of 10°), (ii) distance of HB (ranged from 2.5 to 4 Å with a step 0.1 Å), (iii) donor type of HB (OH, NH, $^+$NH, CH), and (iv) whether fluorine is bonded to an aromatic (F_{ar}) or aliphatic (F_{al}) carbon. Then, one representative PDB complex was randomly selected from each rectangle defined by unit distance and angle change. In the next step, all the selected systems were visually inspected to identify those in which the HB with fluorine was not the main stabilizing interaction and the number of adjacent supporting interactions was the smallest. These identified complexes were used to calculate the interaction energy of the fluorine-containing HB for the given geometric parameters. Figure S1 illustrates the distribution of complexes in a given HB distance–angle interval.

The appropriate ionization states at pH 7.4 ± 0.5 were determined using Epik v3.4 [28,29]. Using an in-house script, the structure of the ligand and the amino acid participating in the interaction was extracted, the missing atoms (included in the peptide bond) were added, and their positions were optimized (OPLS3 force field). The interaction energy of the identified complexes with fluorine was calculated using three commonly used approaches as follows.

The first method, named difference approach or Diff, works based on the assumption that the total interaction energy equals the energy required to separate two interacting molecules. Thus, the energy between the HB donor (X) and the acceptor (Y) is calculated as the difference between the total energy of the X···Y complex and the sum of the total energies of its frozen components [18,30]:

$$E_{int} = E(X \cdot Y) - [E(X) + E(Y)]$$

The energies of the separated molecules, as well as that of the complex, were calculated in Gaussian G16 software [31], using the Minnesota functional M06-2X [32,33] and Karlsruhe basis set def2-TZVP [34]. The polarizable continuum model (PCM) (solvent = water) was used for the calculation [35,36].

The second approach works based on Bader's quantum theory of "atoms in molecules" (QTAIM). In this approach, the topological analysis of electron density was carried out in AIMAll program [37]. The electron density calculated in Gaussian G16 at the M06-2X/def2-TZVP level and the PCM (solvent = water) were used in the analysis. The energy of the noncovalent bonds detected in the crystal structures was calculated using the Espinosa equation as follows:

$$E_{int} = \frac{1}{2}v(r)$$

where E_{int} is the energy of the interatomic interaction and $v(r)$ is the kinetic energy at the bond critical point (BCP). The above equation can be used for all types of HBs, van der Waals interactions, and weak interactions such as H···H and C–H···O [38].

The third approach works based on the energy decomposition analysis. Bonding was analyzed using the extended transition state (ETS) method [39], with the natural orbitals for the chemical valence (NOCV) scheme [40–42]. In this approach, the total energy of bonding between the interacting molecules (ΔE_{int}) is divided into different components as follows:

$$\Delta E_{int} = \Delta E_{dist} + \Delta E_{el} + \Delta E_{Pauli} + \Delta E_{orb} \tag{1}$$

where ΔE_{dist} is the energy required to promote the separated fragments from their equilibrium geometry to the structure they will take up in the complex, ΔE_{el} is the energy of the electrostatic interaction occurring between the two fragments in the supermolecule geometry, ΔE_{Pauli} is the energy of repulsion between the occupied orbitals of the two fragments, and ΔE_{orb} or the orbital interaction term refers to the energy of the stabilizing component due to the final orbital relaxation. All calculations were performed using the Amsterdam Density Functional (ADF) program [43–46], using the ETS-NOCV scheme. The Becke, Lee, Yang, and Parr exchange-correlation functional with the Grimme dispersion correction (B3LYP-D3) was used. A standard double-ζ STO basis containing one set of polarization functions was adopted for all the electrons (TZP). The total (electronic) bonding enthalpies ($\Delta E = \Delta E_{int}$) did not include the zero-point energy (ZPE) additions, finite temperature contributions or basis set superposition error corrections (BSSE).

4. Conclusions

Fluorine is a common substituent in medicinal chemistry and is found in the structure of several currently available blockbuster drugs. This element influences many pharmacokinetic and pharmacodynamic properties of drugs, but its role in stabilizing ligand–biomolecule systems still remains unclear. In this study, we performed a statistical and theoretical analysis of HBs with fluorine found in the PDB database, focusing on the different HB donors (hydroxyl, amine, and methyl groups). The energy range of distinct HBs (i.e., F$\cdots$H–O, F$\cdots$H–N$^+$, F$\cdots$H–N, F$\cdots$H–C) and optimal ranges of geometric parameters of HBs with fluorine were determined based on the selected PDB complexes.

The results of the analyses showed significant differences in the interaction of fluorine attached to an aliphatic carbon (F_{al}) and fluorine attached to an aromatic carbon (F_{ar}). The F$\cdots$H–O HBs with F_{ar} are more frequently formed with SER and THR, while those with F_{al} are formed by all amino acids with a polar hydroxyl group. Typically, F$\cdots$H–N HBs are formed with amino acids that have an amino group in their side chain (ARG, LYS, ASN, GLU). Hydrophobic amino acids most often form F$\cdots$H–C interactions, which suggests that fluorine prefers a hydrophobic environment in biological systems. It is worth noting that due to the three free electron pairs of fluorine, HBs are only influenced by the donor-acceptor distance and not by the angles. Although the three free electron pairs occupy the entire space around fluorine, F$\cdots$H–X HBs exhibit the characteristics of HBs, with exceeded standard angles. However, no significant differences were noted in the energies of HBs with fluorine depending on the donor type, which indicates that fluorine acts as a weak HB acceptor for all types of atoms. The optimal ranges of geometric parameters for HBs with fluorine were found to be 150°–120° and 2.9–3.6 Å. For F$\cdots$N$^+$ interactions, an HB distance shorter than 2.8 Å showed a destabilizing character in almost 70% of the cases.

It must be emphasized that all the analyzed crystal structures may not be crystallized at the lowest free energy form, and hence the observed interactions might not be in optimal geometries [47]. However, the results suggest that HBs with fluorine are forced to form, due to the stronger ligand–receptor neighboring interactions, which make fluorine the "donor's last resort" [48]. This is in line with Margareth Etter's rule that stronger HBs form first, and weaker donors and acceptors interact afterward [47]. All these findings suggest that fluorine does not form strong, stabilizing intermolecular interactions, and thus it seems that indirect influence of this element (electrostatic, inductive, and resonance effects) has a greater impact on the biological activity of compounds than his influences on the pharmacodynamics. The results of this study may contribute to a thorough understanding of hydrogen bonding with fluorine in biological systems which may serve to improve the tools currently available for the rational design of new fluorinated drugs.

Supplementary Materials: The following supporting information can be downloaded, Figure S1: Points of geometrical parameters of hydrogen bonds for which the energy value was calculated; Figure S2: Box plots showing the distribution of the different energy components: kinetic (red), electrostatic (cyan), Coulomb (blue), XC (olive), dispersion (magenta) energy for HB donors and fluorine attached to (A) an aliphatic fragment and (B) an aromatic ring. The energy values were calculated for specific geometric parameters using ETS-NOCV approach. Table S1: Correlation coefficient values calculated between results from every method; Table S2: Pearson test values calculated between results from every method.

Author Contributions: W.P.: Conceptualization, Methodology, Validation, Formal analysis, Investigation, Resources, Writing—original draft, Visualization, Project administration; R.K. (Rafał Kurczab): Conceptualization, Resources, Writing—original draft, Supervision, Project administration; R.K. (Rafał Kafel): Software, Visualization; A.J.B.: Investigation, Writing—original draft, Funding acquisition. All authors have read and agreed to the published version of the manuscript.

Funding: The authors acknowledge the financial support from the National Science Centre, Poland (grant no. 2019/35/N/NZ7/04312) and the statutory funding from the Maj Institute of Pharmacology, Polish Academy of Sciences, Poland.

Institutional Review Board Statement: Not applicable.

Informed Consent Statement: Not applicable.

Data Availability Statement: The data were obtained from the PDB repository and the DrugCentral 2021 database (accessed 30 April 2021).

Acknowledgments: Numerical simulations were performed by PLGrid Infrastructure (Prometheus, ACC Cyfronet, AGH). WP acknowledges the support of InterDokMed project no. POWR.03.02.00-00-I013/16. The authors thank Justyna Kalinowska-Tłuścik (Jagiellonian University), Mariusz Mitoraj (Jagiellonian University), and Filip Sagan (Jagiellonian University) for providing technical and substantive assistance in research.

Conflicts of Interest: The authors declare no conflict of interest.

Sample Availability: Samples of the compounds are not available from the authors.

References

1. Swallow, S. Fluorine in Medicinal Chemistry. In *Progress in Medicinal Chemistry*; Lawton, G., Witty, D.R., Eds.; Elsevier: Amsterdam, The Netherlands, 2015; Volume 54, pp. 65–133.
2. Fried, J.; Sabo, E.F. 9α-Fluoro derivatives of cortisone and hydrocortisone. *J. Am. Chem. Soc.* **1954**, *76*, 1455–1456. [CrossRef]
3. Zhou, Y.; Wang, J.; Gu, Z.; Wang, S.; Zhu, W.; Aceña, J.L.; Soloshonok, V.A.; Izawa, K.; Liu, H. Next Generation of Fluorine-Containing Pharmaceuticals, Compounds Currently in Phase II–III Clinical Trials of Major Pharmaceutical Companies: New Structural Trends and Therapeutic Areas. *Chem. Rev.* **2016**, *116*, 422–518. [CrossRef]
4. Tressaud, A. Fluorine, a key element for the 21st century. In *Fluorine*; Tressaud, A., Ed.; Elsevier: Amsterdam, The Netherlands, 2018; pp. 77–150.
5. Brown, D.G.; Wobst, H.J. A Decade of FDA-Approved Drugs (2010–2019): Trends and Future Directions. *J. Med. Chem.* **2021**, *64*, 2312–2338. [CrossRef] [PubMed]
6. Ursu, O.; Holmes, J.; Bologa, C.G.; Yang, J.J.; Mathias, S.L.; Stathias, V.; Nguyen, D.T.; Schürer, S.; Oprea, T. DrugCentral 2018: An update. *Nucleic Acids Res.* **2019**, *47*, D963–D970. [CrossRef] [PubMed]
7. Chen, D.; Oezguen, N.; Urvil, P.; Ferguson, C.; Dann, S.M.; Savidge, T.C. Regulation of protein-ligand binding affinity by hydrogen bond pairing. *Sci. Adv.* **2016**, *2*, e1501240. [CrossRef] [PubMed]
8. Bitencourt-Ferreira, G.; Veit-Acosta, M.; de Azevedo, W.F. Hydrogen bonds in protein-ligand complexes. *Methods Mol. Biol.* **2019**, *2053*, 93–107. [PubMed]
9. Sarkhel, S.; Desiraju, G.R. Hydrogen Bonds in Protein-Ligand Complexes: Strong and Weak Interactions in Molecular Recognition. *Proteins Struct. Funct. Genet.* **2004**, *54*, 247–259. [CrossRef] [PubMed]
10. Hogendorf, A.S.; Hogendorf, A.; Popiołek-Barczyk, K.; Ciechanowska, A.; Mika, J.; Satała, G.; Walczak, M.; Latacz, G.; Handzlik, J.; Kieć-Kononowicz, K.; et al. Fluorinated indole-imidazole conjugates: Selective orally bioavailable 5-HT7 receptor low-basicity agonists, potential neuropathic painkillers. *Eur. J. Med. Chem.* **2019**, *170*, 261–275. [CrossRef]
11. Howard, J.A.K.; Hoy, V.J.; O'Hagan, D.; Smith, G.T. How good is fluorine as a hydrogen bond acceptor? *Tetrahedron* **1996**, *52*, 12613–12622. [CrossRef]
12. Pietruś, W.; Kurczab, R.; Kafel, R.; Machalska, E.; Kalinowska-Tłuścik, J.; Hogendorf, A.; Żylewski, M.; Baranska, M.; Bojarski, A.J. How can fluorine directly and indirectly affect the hydrogen bonding in molecular systems?—A case study for monofluoroanilines. *Spectrochim. Acta Part A Mol. Biomol. Spectrosc.* **2021**, *252*, 119536. [CrossRef] [PubMed]
13. Pietruś, W.; Kurczab, R.; Kalinowska-Tłuścik, J.; Machalska, E.; Golonka, D.; Barańska, M.; Bojarski, A. Influence of fluorine substitution on nonbonding interactions in selected para-halogeno anilines. *ChemPhysChem* **2021**. [CrossRef]
14. Zhang, G.; He, W.; Chen, D. On difference of properties between organic fluorine hydrogen bond C-H F-C and conventional hydrogen bond. *Mol. Phys.* **2014**, *112*, 1736–1744. [CrossRef]
15. Meanwell, N.A. Fluorine and Fluorinated Motifs in the Design and Application of Bioisosteres for Drug Design. *J. Med. Chem.* **2018**, *61*, 5822–5880. [CrossRef]
16. Dunitz, J.D.; Taylor, R. Organic Fluorine Hardly Ever Accepts Hydrogen Bonds. *Chem. A Eur. J.* **1997**, *3*, 89–98. [CrossRef]
17. Ferreira De Freitas, R.; Schapira, M. A systematic analysis of atomic protein-ligand interactions in the PDB. *Medchemcomm* **2017**, *8*, 1970–1981. [CrossRef] [PubMed]
18. Afonin, A.V.; Vashchenko, A.V.; Sigalov, M.V. Estimating the energy of intramolecular hydrogen bonds from ^{1}H NMR and QTAIM calculations. *Org. Biomol. Chem.* **2016**, *14*, 11199–11211. [CrossRef] [PubMed]
19. Šolínová, V.; Kašička, V. Determination of acidity constants and ionic mobilities of polyprotic peptide hormones by CZE. *Electrophoresis* **2013**, *34*, 2655–2665. [CrossRef] [PubMed]
20. Graton, J.; Besseau, F.; Brossard, A.M.; Charpentier, E.; Deroche, A.; Le Questel, J.Y. Hydrogen-bond acidity of OH groups in various molecular environments (phenols, alcohols, steroid derivatives, and amino acids structures): Experimental measurements and density functional theory calculations. *J. Phys. Chem. A* **2013**, *117*, 13184–13193. [CrossRef] [PubMed]
21. Berman, H.; Henrick, K.; Nakamura, H. Announcing the worldwide Protein Data Bank. *Nat. Struct. Mol. Biol.* **2003**, *10*, 980. [CrossRef] [PubMed]

22. Madhavi Sastry, G.; Adzhigirey, M.; Day, T.; Annabhimoju, R.; Sherman, W. Protein, and ligand preparation: Parameters, protocols, and influence on virtual screening enrichments. *J. Comput. Aided. Mol. Des.* **2013**, *27*, 221–234. [CrossRef]
23. Team, R.C. Computational Many-Particle Physics. In *Lecture Notes in Physics*; Fehske, H., Schneider, R., Weiße, A., Eds.; Springer: Berlin/Heidelberg, Germany, 2008; Volume 739.
24. Neuwirth, E. *RColorBrewer: ColorBrewer Palettes*; R Package Version 1.1.2; R Package: Vienna, Austria, 2014. Available online: https://CRAN.R-project.org/package=RColorBrewer (accessed on 29 December 2021).
25. Lewin-Koh, N. *Hexagon Binning: An Overview*; R Core Team: Vienna, Austria, 2014.
26. Bokeh Development Team. Bokeh: Python Library for Interactive Visualization. 2014. Available online: http://www.bokeh.pydata.org (accessed on 29 December 2021).
27. Villanueva, R.A.M.; Chen, Z.J. ggplot2: Elegant Graphics for Data Analysis (2nd ed.). *Meas. Interdisc. Res. Perspect.* **2019**, *17*, 160–167. [CrossRef]
28. Greenwood, J.R.; Calkins, D.; Sullivan, A.P.; Shelley, J.C. Towards the comprehensive, rapid, and accurate prediction of the favorable tautomeric states of drug-like molecules in aqueous solution. *J. Comput. Aided. Mol. Des.* **2010**, *24*, 591–604. [CrossRef] [PubMed]
29. Shelley, J.C.; Cholleti, A.; Frye, L.L.; Greenwood, J.R.; Timlin, M.R.; Uchimaya, M. Epik: A software program for pKa prediction and protonation state generation for drug-like molecules. *J. Comput. Aided. Mol. Des.* **2007**, *21*, 681–691. [CrossRef]
30. Jabłoński, M. A Critical Overview of Current Theoretical Methods of Estimating the Energy of Intramolecular Interactions. *Molecules* **2020**, *25*, 5512. [CrossRef] [PubMed]
31. Frisch, M.J.; Trucks, G.W.; Schlegel, H.B.; Scuseria, G.E.; Robb, M.A.; Cheeseman, J.R.; Scalmani, G.; Barone, V.; Petersson, G.A.; Nakatsuji, H.; et al. *Gaussian 16 Revision C.01*; Gaussian Inc.: Wallingford, CT, USA, 2016.
32. Zhao, Y.; Truhlar, D.G. The M06 suite of density functionals for main group thermochemistry, thermochemical kinetics, noncovalent interactions, excited states, and transition elements: Two new functionals and systematic testing of four M06-class functionals and 12 other functions. *Theor. Chem. Acc.* **2008**, *120*, 215–241. [CrossRef]
33. Lin, Y.-S.; Li, G.-D.; Mao, S.-P.; Chai, J.-D. Long-Range Corrected Hybrid Density Functionals with Improved Dispersion Corrections. *J. Chem. Theory Comput.* **2013**, *9*, 263–272. [CrossRef]
34. Pritchard, B.P.; Altarawy, D.; Didier, B.; Gibson, T.D.; Windus, T.L. New Basis Set Exchange: An Open, Up-to-Date Resource for the Molecular Sciences Community. *J. Chem. Inf. Model.* **2019**, *59*, 4814–4820. [CrossRef]
35. Mennucci, B.; Cammi, R.; Tomasi, J. Analytical free energy second derivatives with respect to nuclear coordinates: Complete formulation for electrostatic continuum solvation models. *J. Chem. Phys.* **1999**, *110*, 6858–6870. [CrossRef]
36. Miertuš, S.; Scrocco, E.; Tomasi, J. Electrostatic interaction of a solute with a continuum. A direct utilizaion of AB initio molecular potentials for the prevision of solvent effects. *Chem. Phys.* **1981**, *55*, 117–129. [CrossRef]
37. Keith, T.A. *AIMAll*; TK Gristmill Software: Overland Park, KS, USA, 2015.
38. Espinosa, E.; Molins, E.; Lecomte, C. Hydrogen bond strengths revealed by topological analyses of experimentally observed electron densities. *Chem. Phys. Lett.* **1998**, *285*, 170–173. [CrossRef]
39. Ziegler, T.; Rauk, A. On the calculation of bonding energies by the Hartree Fock Slater method—I. The transition state method. *Theor. Chim. Acta* **1977**, *46*, 1–10. [CrossRef]
40. Michalak, A.; Mitoraj, M.; Ziegler, T. Bond orbitals from chemical valence theory. *J. Phys. Chem. A* **2008**, *112*, 1933–1939. [CrossRef] [PubMed]
41. Mitoraj, M.; Michalak, A. Applications of natural orbitals for chemical valence in a description of bonding in conjugated molecules. *J. Mol. Model.* **2008**, *14*, 681–687. [CrossRef] [PubMed]
42. Mitoraj, M.; Michalak, A. Natural orbitals for chemical valence as descriptors of chemical bonding in transition metal complexes. *J. Mol. Model.* **2007**, *13*, 347–355. [CrossRef]
43. Velde, G.; Bickelhaupt, F.M.; Baerends, E.J.; Fonseca Guerra, C.; van Gisbergen, S.J.A.; Snijders, J.G.; Ziegler, T. Chemistry with ADF. *J. Comput. Chem.* **2001**, *22*, 931–967. [CrossRef]
44. Baerends, E.J.; Ros, P. Self-consistent molecular Hartree-Fock-Slater calculations II. The effect of exchange scaling in some small molecules. *Chem. Phys.* **1973**, *2*, 52–59. [CrossRef]
45. Baerends, E.J.; Ellis, D.E.; Ros, P. Self-consistent molecular Hartree-Fock-Slater calculations I. The computational procedure. *Chem. Phys.* **1973**, *2*, 41–51. [CrossRef]
46. Velde, G.; Baerends, E.J. Numerical integration for polyatomic systems. *J. Comput. Phys.* **1992**, *99*, 84–98. [CrossRef]
47. Etter, M.C. Hydrogen bonds as design elements in organic chemistry. *J. Phys. Chem.* **1991**, *95*, 4601–4610. [CrossRef]
48. Taylor, R. The hydrogen bond between N-H or O-H and organic fluorine: Favourable yes, competitive no. *Acta Crystallogr. Sect. B Struct. Sci. Cryst. Eng. Mater.* **2017**, *73*, 474–488. [CrossRef]

Review

Isotope Effects on Chemical Shifts in the Study of Hydrogen Bonds in Small Molecules

Poul Erik Hansen

Department of Science and Environment, Roskilde University, Universitetsvej 1, DK-4000 Roskilde, Denmark; poulerik@ruc.dk

Abstract: This review is giving a short introduction to the techniques used to investigate isotope effects on NMR chemical shifts. The review is discussing how isotope effects on chemical shifts can be used to elucidate the importance of either intra- or intermolecular hydrogen bonding in ionic liquids, of ammonium ions in a confined space, how isotope effects can help define dimers, trimers, etc., how isotope effects can lead to structural parameters such as distances and give information about ion pairing. Tautomerism is by advantage investigated by isotope effects on chemical shifts both in symmetric and asymmetric systems. The relationship between hydrogen bond energies and two-bond deuterium isotope effects on chemical shifts is described. Finally, theoretical calculations to obtain isotope effects on chemical shifts are looked into.

Keywords: hydrogen bonding; isotope effects on ^{13}C chemical shifts; tautomerism; hydrogen bond energy; theoretical calculations

Citation: Hansen, P.E. Isotope Effects on Chemical Shifts in the Study of Hydrogen Bonds in Small Molecules. *Molecules* **2022**, *27*, 2405. https://doi.org/10.3390/molecules27082405

Academic Editor: Miroslaw Jablonski

Received: 11 March 2022
Accepted: 5 April 2022
Published: 8 April 2022

Publisher's Note: MDPI stays neutral with regard to jurisdictional claims in published maps and institutional affiliations.

Copyright: © 2022 by the author. Licensee MDPI, Basel, Switzerland. This article is an open access article distributed under the terms and conditions of the Creative Commons Attribution (CC BY) license (https://creativecommons.org/licenses/by/4.0/).

1. Introduction

Hydrogen bonding, both inter- and intramolecular, has profound effects on the properties of small molecules, reactivity, pK_a values, polarity, solubility and with that, e.g., penetration of membranes and biological effects in general. Most drugs and bioactive molecules are small molecules and often depend on binding to receptors partly via hydrogen bonds [1,2]. Related to intramolecular hydrogen bonding is tautomerism. One of the effective tools in the study of hydrogen bonding is isotope effects on NMR chemical shifts. Both primary and secondary isotope effects may be used. The most common isotope is ^2H, deuterium (D), but also ^{13}C, ^{15}N and ^{18}O as well as more rare isotopes are used. For observation of secondary isotope effects, the most common nuclei are ^1H, ^{13}C, ^{15}N, ^{19}F and ^{18}O. In case of primary isotope effects, the pair ^1H, ^2H is mostly used. The intrinsic isotope effects are defined as:

Primary: $^P\Delta H(D) = \delta(H) - \delta(D)$;
Secondary: $^n\Delta X(h) = \delta X(l) - \delta X(h)$.

l = light, h = heavy and X is the nucleus under investigation, n is the number of covalent bonds between the isotope and the investigated nucleus. For intramolecular hydrogen bonded cases, the isotope effects may be transmitted via the hydrogen bond. In the cases in which to opposite sign convention has been used, the sign is changed and the number is marked with an asterisk.

Deuterium isotope effects can be related to hydrogen bond strength and to hydrogen bond energies. Isotope effects are very useful in the equilibria of symmetric systems (lifting of degeneracy). They respond to nearby charges.

Most of the studies were performed in the liquid state, but isotope effects on chemical shifts can also in favorable cases be studied in the solid state.

The present review will primarily cover the last ten years. Other reviews that cover the subject are [3–6].

2. Techniques

2.1. Liquid State

As hydrogen bonding is the subject, the most relevant protons to exchange are those of OH, NH or SH groups. Deuteration can easily be achieved either by shaking with D_2O in, e.g., chloroform or methylene chloride and subsequent drying with water-free sodium sulphate or by dissolution in CH_3OD followed by evaporation. In case of exchange of CH protons this depends very much on the compound. An example is imidazolium acetate liquid ionic liquids which have been used to catalyze H-D exchange in 2-alkanones using CD_3OD as deuterium source. Long-range isotope effect on ^{13}C chemical shifts is measured [7]. For a different case see Section 3.1. Introduction of ^{18}O may be obtained by exchange, whereas ^{15}N usually requires a synthetic scheme. In the case of deuteration, a degree of deuteration of ~70% is desirable as some lines in the spectrum may become broad due to couplings to deuterium and a clear-cut difference in intensities between molecules with H and those with D is required in order to determine the signs of the isotope effects. As slow exchange is a prerequisite to measure isotope effects directly, low temperature may be necessary. For examples of very low temperature, see references in this paper and previous papers by the Limbach-Tolstoy group [6]. Cooling down may not only slow down the XD exchange but may also lead to observation of isotope effects of single rotamers as seen in, e.g., 2,6-dihydroxyacylaromatics (see Section 3.5). Exchange can be slowed down by use of a hydrogen bonding solvent such as DMSO [8]. In the case of ammonium ions, exchange may be slowed down by acidification.

For compounds only soluble in water, the deuterium isotope effects can be determined by recording spectra with varying ratios of H_2O/D_2O followed by extrapolations to 0 and 100% D_2O.

Isotope effects can in most cases be determined by simple 1D NMR spectra. However, for $^1\Delta N(D)$ isotope effects it is an advantage to use indirect techniques. D.F. Hansen et al. have developed a ^{13}C-detected ^{15}N double-quantum NMR experiment [9]. In cases with exchange, the HISQC technique is useful [10].

Assignment of isotope effects in molecules with more than one site of e.g., deuteration can be achieved by utilizing different degrees of incorporation or different rates of incorporation. Examples are in β-diketones and in enaminones. In these cases, the hydrogen at the central carbon exchange much more slowly than the XH protons. In other cases, either a spatial separation or a comparison with similar compounds with fewer sites of deuteration can be used (see Figure 1)

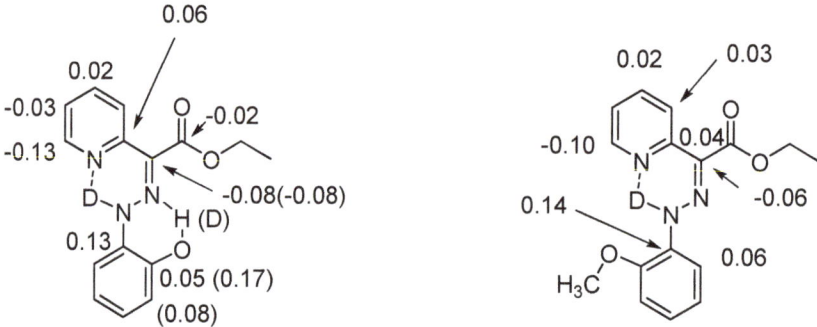

Figure 1. Deuterium isotope effects on ^{13}C chemical shifts. Values in brackets are due to the deuterium in brackets. Data from Ref. [11].

2.2. Solid State

Isotope effects on chemical shifts can in principle also be measured in solid-state magic angle spinning spectra. However, as the lines typically are broad only large isotope effects can be resolved. The isotope effects analyzed recently are deuterium isotope effects on ^{13}C

chemical shifts in a tautomeric case as isotope effects may be large in tautomeric systems. An example is the pyridoyl benzoyl β-diketones see Section 4. Additionally, deuterium isotope effects at ^{15}N chemical shifts could be observed in the triclinic phase of the complex between pentachlorophenol and 4-methylpyridine (Section 4). Another recent case is deuterium isotope effects on ^{15}N chemical shifts of ammonium ions (see Section 3.2). For early examples, see Refs. [5,6].

3. Static Cases
3.1. Ionic Liquids

Over the past two decades, ionic liquids have become very versatile and "green" solvents in which hydrogen bonding is also important. This has spurred research into the properties, but also led to new types of ionic liquids.

Moyna et al. [12] studied the deuterium isotope effects at ^{19}F chemical shifts of the counter ions PF_6^- and BF_4^- caused by deuterium of deuterated side-chains of 1-n-butyl-3-methylimidazolium (Figure 2).

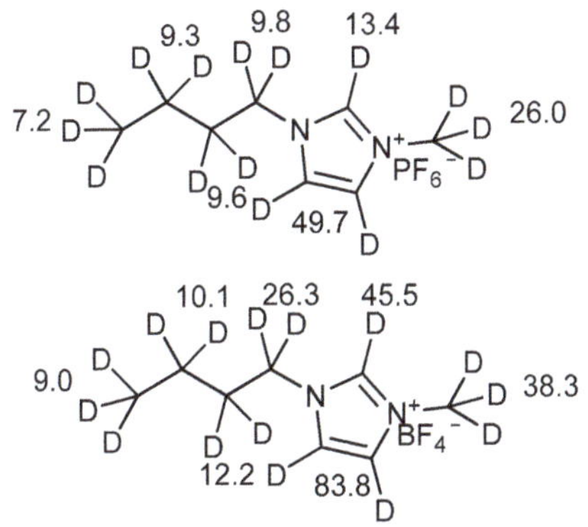

Figure 2. 1-n-butyl-3-methylimidazolium PF_6^- on top and the BF_4^- below. The numbers are the deuterium isotope effects from that particular set of deuterium seen at the fluorine signals of the counter ions. In the case of BF_4^-, the effects caused by D-2 and D-3 are actually also caused by D-5. Data for the aliphatic deuterations from Ref. [12] and those caused by deuteration at C-2, C-3 and C-5 from Ref. [13].

The effects caused by deuteration at the aliphatic chain are largest closest to the ring. The effects correlate with the change in polarization of the C-H bonds. The effects are largest for BF_4^- as the electron density is largest at the fluorines of this counter ion.

In a similar vein, deuterium isotope effects have been observed at the Cl^- chemical shifts when Cl^- is the counter ion [14] (Figure 3).

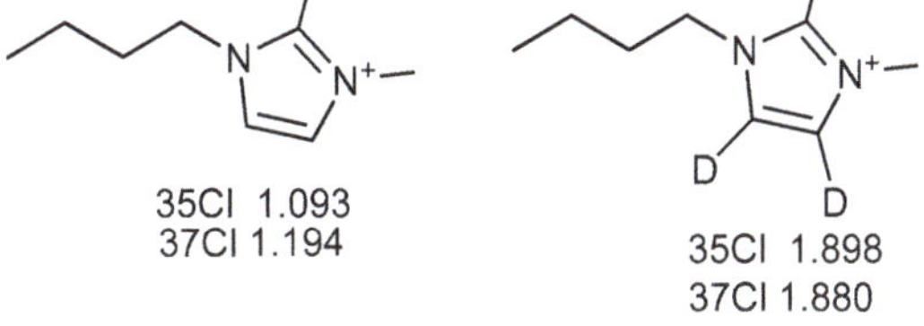

Figure 3. Numbers are deuterium isotope effects observed at the Cl resonance caused by the deuterium. Uncertainties ~0.1 ppm. Data from Ref. [14].

Combined o-hydroxy Schiff bases and ionic liquids based on amino acids (Figure 4) showed based on $^2\Delta$C-2(D) isotope effects that the COO^- group stabilizes the NH-form [15].

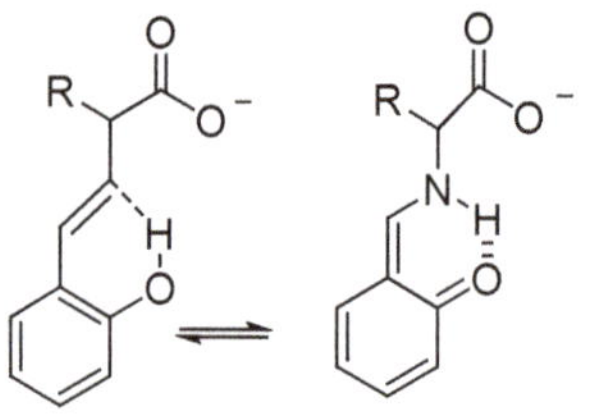

Figure 4. Ionic liquids based on amino acids.

In the reinvestigation of the diisopropylethylammonium formate, it was found that the 1:1 complex claimed by Anouti et al. [16] could not be reproduced. A more complex scheme was suggested (Figure 5) [17].

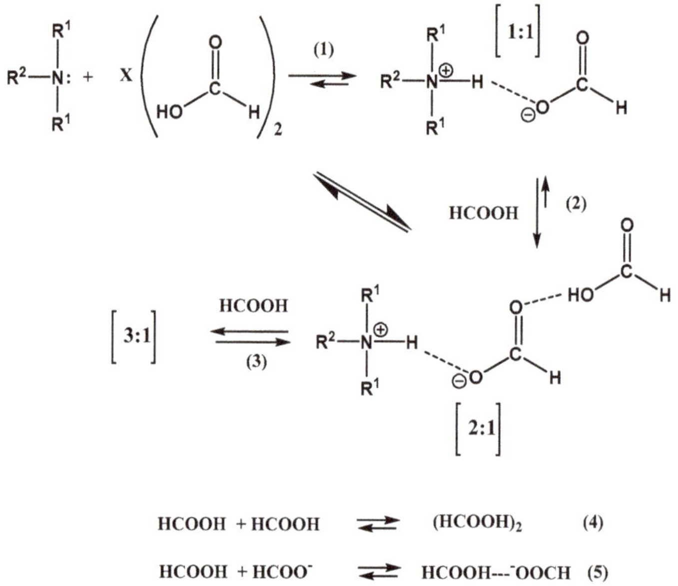

Figure 5. Suggested reaction scheme for the formation of the "ionic liquid". Reprinted with permission from Ref. [17]. Copyright 2016 American Chemical Society.

Primary deuterium isotope effects at the OH and NH$^+$ resonances at 243 K were measured as 2.24 and 0.34 ppm, respectively. These values dropped to 1.08 and 0.30 ppm at 193 K. The mere fact that the OH resonance could be observed and with that, the primary isotope effect, proves that this is not a 1:1 complex. A study of the acetic acid dimer gave a primary deuterium isotope effect of 0.3 ppm and that of the acetic acid-acetate dimer 0.6 ppm [18]. It is obvious, based on an analogy with that result, that the OH(D) primary isotope effects are partly equilibrium isotope effects (Section 4). On the other hand, the NH$^+$ is hydrogen bonded to the monomeric or to the dimeric acetate ion in the same fashion.

3.2. Ammonium Ions

Ammonium ions and alkyl ammonium ions play an important role in biology [19,20]. Attempts to mimic enzymes were carried out by Lehn et al. [21] An example is SC-24 (Figure 6). Other confinements have been investigated recently [22].

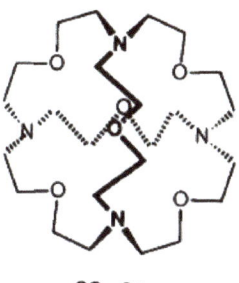

SC - 24

Figure 6. Structure of SC-24.

Arginine showed deuterium isotope effects at the η nitrogen, 0.307 ppm measured in a H_2O/D_2O mixture 1:1. Due to the technique used it is an average for the two Nη nitrogen obtained, which means that the measured isotope effect is close to one half of $^1\Delta N(D)$. The values served as a reference for non-interacting residues in lysozyme [9].

Platzer et al. [23] measured deuterium isotope effects at the side-chain nitrogen of protonated and non-protonated lysines and argines being part of a tripeptide, acetyl-Gly-X-Glyamide. For the side-chain nitrogen of lysine, the values were 1.05 ppm for ND_3^+ and ~1.9 ppm for ND_2. For arginines, the value for the ε nitrogen in the protonated state was 1.0 ppm and 1.4 ppm for the η nitrogen. In this case, it is not a pure one-bond isotope effect as long-range effects are present. In the case of lysines, the very different values between the protonated and non-protonated cases may be used to estimate the protonation state of lysines.

In Figure 7 is shown a plot of $^1\Delta N(D)_4$ vs. the heavy atom distances of halide ammonium ion salts, I^-, Br^-, Cl^- and F is shown$^-$. Marked with squares are data points for SC-24 and water. It is seen, that as the distance decreases, the one-bond deuterium isotope effect decreases. [24] A similar trend is found from theoretical calculations. [25] Furthermore, it is found that water is more effective than the halide ions. For SC-24, the one-bond deuterium isotope effects were found to be independent of the counter ions [26].

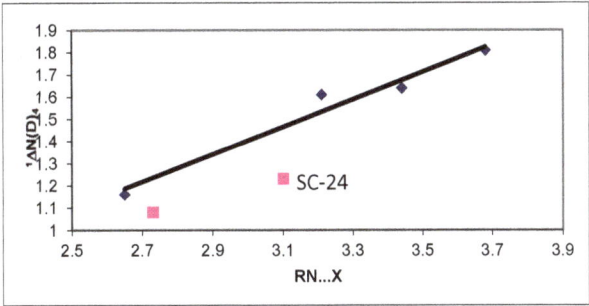

Figure 7. A plot of $^1\Delta N(D)_4$ vs. the heavy atom distances of halide ammonium ion salts. The second square is data for water. Data from Ref. [24].

The two-bond deuterium isotope effects for complexes with 18-crown-6, 18-crown-6(COOH)$_4$, dicylohexano 18-crown-6, kryptofix's 2.2.2, 2.2.1 and 5, SC-24 and the ionophore and nonactin, are negative. The two-bond deuterium isotope effects are roughly proportional to the NH chemical shifts. The higher the NH chemical shifts, the more negative the two-bond deuterium isotope effects. In other words, the more negative the stronger the hydrogen bond. For SC-24, 18-crown-6(COOH)$_4$ and nonactin, the two-bond deuterium isotope effects on ^1H chemical shifts are independent of the counter ion. For the others, the more negative values are usually found for Cl⁻ rather than for NO_3^- or I⁻ counter ions [27].

A plot of $^2\Delta H(D)$ vs. $^1\Delta N(D)$ shows very little correlation (Figure 8). It has been suggested that the ^{15}N chemical shifts and with that the one-bond isotope effects depend on orbital overlap with the counter ion, whereas the two-bond deuterium isotope effects on ^1H depend on electric field effects. In both cases, the isotope effect monitors ion pair formation.

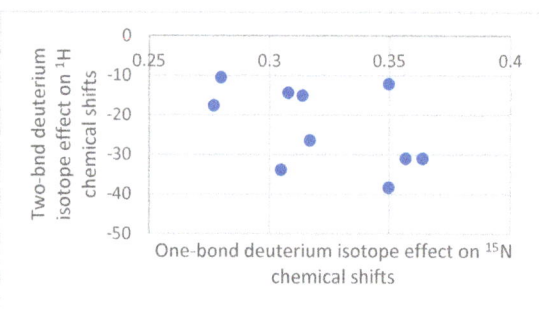

Figure 8. A plot of $^2\Delta H(D)$ vs. $^1\Delta N(D)$. Data from Ref. [26].

3.3. Enaminones and Similar Compounds

Data from the simple compounds such as those of Figure 9 may serve as reference points for more complex systems such as those of phenylene diamine derivatives of dehydracetic acid (Figure 10) [28]. The isotope effects in A are line with a non-tautomeric system, whereas those of B can only be explained by assuming a tautomeric equilibrium.

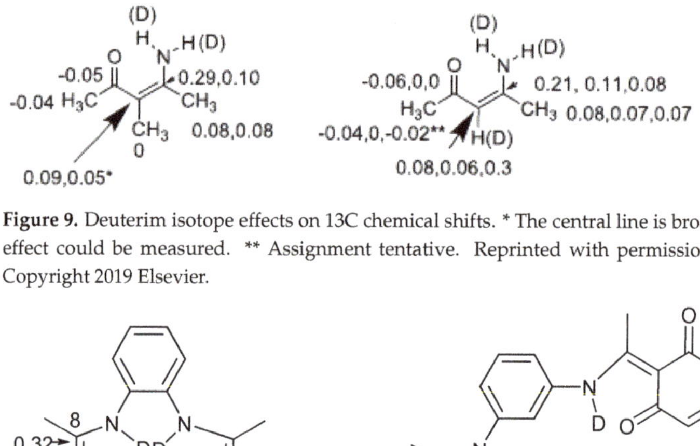

Figure 9. Deuterim isotope effects on 13C chemical shifts. * The central line is broad, so no isotope effect could be measured. ** Assignment tentative. Reprinted with permission from Ref. [29]. Copyright 2019 Elsevier.

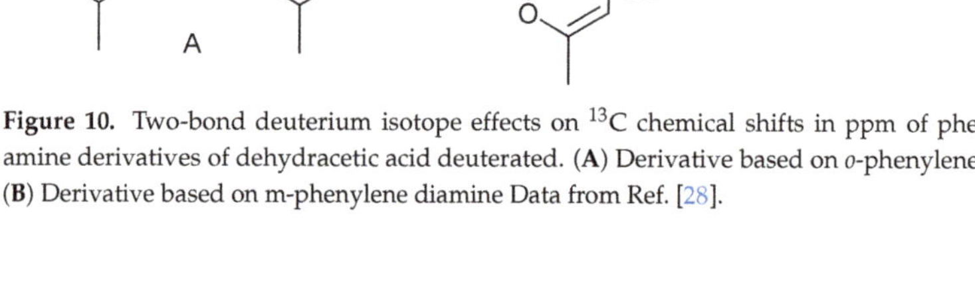

Figure 10. Two-bond deuterium isotope effects on ^{13}C chemical shifts in ppm of phenylenediamine derivatives of dehydracetic acid deuterated. (**A**) Derivative based on o-phenylene diamine; (**B**) Derivative based on m-phenylene diamine Data from Ref. [28].

Isotope effects have been measured in 1,4-dihydropyridines (Figure 11). The finding that the deuterium isotope effects on ^{13}C chemical shifts are very similar in derivatives A and B [30] seems to show that the potential hydrogen bonding in B is weak.

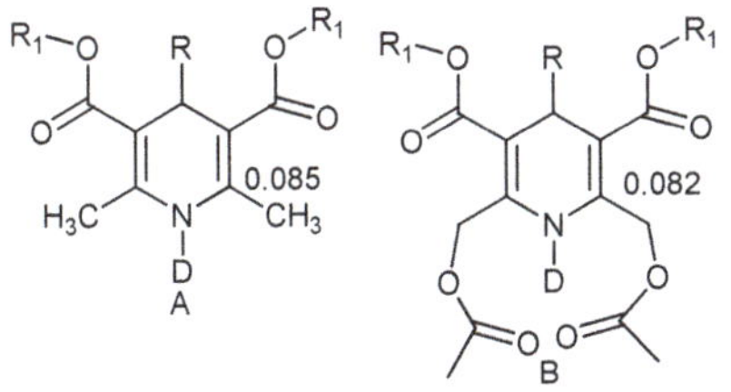

Figure 11. Two-bond deuterium isotope effects on ^{13}C chemical shifts of 1,4-dihydropyridines. (**A**) Without intra-molecualr hydrogen bond; (**B**) with intramolecular hydrogen bond From Ref. [30].

3.4. Dimers and Trimers

Tolstoy et al. in a very elegant way have used isotope effects on chemical shifts to determine the size of self-associated dimethylphosphinic, diphenylphosphoric acid, phenylphosphinic acid and bis(2,4,4-trimethylpentyl)phosphinic acid [31]. This has been extended to investigate heterodimers and heterotrimers of phosphinic and phosphoric acids (see Figure 12) [32]. Using the same technique, it could be proven that dimethylarsenic acid forms cyclic dimers in solution with two equivalent strong hydrogen bonds [33].

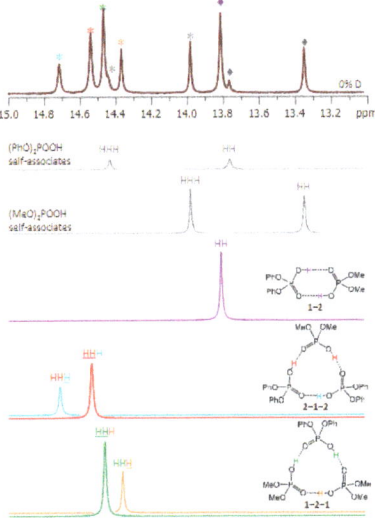

Figure 12. High-frequency part of the ^{1}H NMR spectrum of partially deuterated (OH/OD, 57%D) of a mixture of diphenylphosphoric and dimethylphospinic acid in CDF$_3$/CDF$_2$Cl at 100 K. Trimers are marked with asterisks and dimers with diamonds. Reprinted from Ref. [32].

3.5. Miscellaneous

The equilibrium in the system shown in Figure 13 was originally determined using deuterium isotope perturbation techniques [34] and later calculated [35]. Xu et al. [36] investigated the similar system, 2,4-dihydroxybenzaldehyde measuring integrals and found that the deuterium prefers the non-hydrogen-bonded bond OH-4. O´Leary [37] analyzed the 2,6-dihydroxybanzaldehyde system in terms of vibrations and found that the high-frequency modes resulted in a Keq less than one, whereas the low- and medium-frequency modes resulted in a Keq > 1.

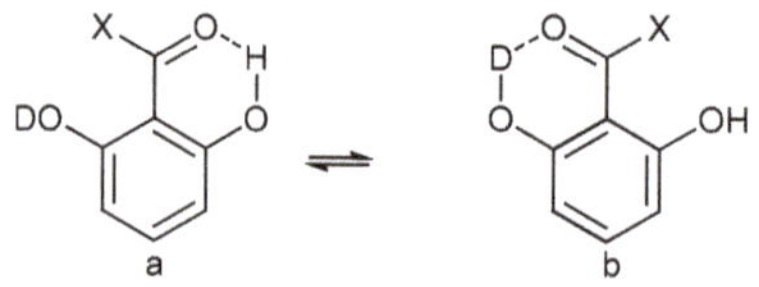

Figure 13. Equilibrium between monodeuterated 2,6-dihydroxyacylaromatics. X=H, CH$_3$ or OCH$_3$.

Deuterium isotope effects at C-2 at 1,1,1,3,3,3-hexafluoro-2-propanol-d$_2$ have been investigated in CDCl$_3$ and trimethylamine. The isotope effects were 0.364 and 0.341 ppm, respectively [38]. The authors ascribed the difference to complex formation, but the difference is very small and if anything in the wrong direction.

Schulz et al. [39] studied primary isotope effects in 1-N-TMPH-CH$_2$-2[HB(C$_6$F$_5$)$_2$]C$_6$H$_4$ (NHHB) (Figure 14). The molecule shows a strong dihydrogen bond. Deuteration of the NH proton led to a primary deuterium isotope effect of 0.56 ppm, whereas deuteration of the BH hydrogen did not lead to an isotope effect. The isotope effect of 0.56 ppm could indicate a double-well hydrogen bond potential [40]. However, the authors argued for a single-well potential, but they did not explain why the other effect was zero.

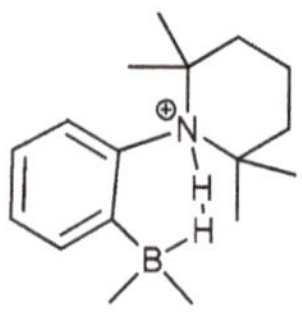

Figure 14. Structure of NHHB. The two hydrogens of the dihydrogen bond are selectively deuterated. From Ref. [39].

A xenon molecule in a deuterated hydrogen bond network of β-hydroquinones crystal shows an isotope effect of 2.4* ppm at 298 K and 2.6 ppm at 333 K. This effect is rather small considering the chemical shift range of Xe [41]. The effect is similar to that found in water/heavy water of 3.92* ppm. CH$_3$OD gives an isotope effect of the opposite sign.

3.6. Cooperativity

Hydrogen bonded systems with two hydrogen bond donors to the same acceptor e.g., 1,8-dihydroxyanthraquinones or the monoanion of 1,8,9-trihydroxyanthracene [42] can give rise to cooperativity. A second situation seen in Figure 12 is the trimers of phosphoric, phospinic acids and a third situation seen is the dimers of carboxylic acids (for carboxylic acids encapsulated see Section 4). For the monoanion of 1,8,9-trihydroxyanthracene the primary isotope effects were −0.2 ppm in both DMSO-d$_6$ and in 90%H$_2$O/10% DMSO-d$_6$. The degree of deuteration was 50% [42]. The finding of the same effect in those two solvents again showed that the hydrogen bond was strong enough not to be perturbed by the solvent. In the cyclic trimers both cooperative and anti-cooperative effects may be found. In the cooperative case, the X-D bond is weakened and the corresponding XdH bond is strengthened. The opposite is true for the anti-cooperative case [31]. In the cyclic trimers of phosphoric acids, cooperative effects are found [31]. In the case of hetero trimers of phosphinc and phosphoric acids, anti-cooperative effects are found probably due to steric factors [32]. For the trimers of dimethylarsinic acid, cooperativity is found [33].

4. Tautomerism

The use of isotope effects to investigate tautomerism has been treated in several reviews. [3–5] A key point is the observation in non-symmetrical systems of isotope effects on chemical shifts that they consist of both an intrinsic and an equilibrium contribution. A classic case is that of β-diketones, illustrated in Figure 15.

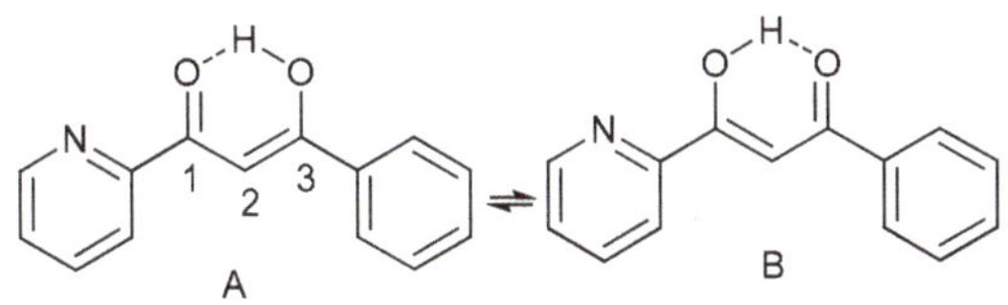

Figure 15. Tautomeric equilibrium illustrated by a β-diketone. The mole fraction of the B tautomer is x.

The equilibrium isotope effects can be formulated as seen in Equations (1)–(3).

$$^n\Delta X(D)_{int} = (1 - x)\ ^n\Delta X(D)_A + x\ ^n X(D)_B \tag{1}$$

$$^n\Delta X(D)_{eq} = (\delta X_B - \delta X_A)\ \Delta x \tag{2}$$

$$^n\Delta X(D)_{OBS} = {}^n\Delta X(D)_{int} + {}^n\Delta X(D)_{eq} \tag{3}$$

X could be ^{13}C, ^{1}H, ^{15}N, ^{19}F, etc. and x is the mole fraction of B. Δx is the change in the equilibrium upon deuteration.

This way, isotope effects on chemical shifts become a useful tool to establish whether or not tautomerism is present in cases in which this is not obvious. An example is 1,1′,1″-(2,4,6-trihydroxybenzene-1,3,5-triyl)triethanone (1,3,5-trihydroxy-2,4,6-triacetylbenzene). This was in CDCl$_3$ shown not to be tautomeric based on deuterium isotope effects on ^{13}C chemical shifts and DFT calculation of those [43]. This was further supported by low temperature studies showing that both the isotope effect and the OH chemical shifts were unchanged by lowering the temperature [44]. However, in ethanol this molecule was claimed to be tautomeric based on calculations [45]. This claim was investigated by the measurement of deuterium isotope effects on ^{13}C chemical shifts in a mixture of CDCl$_3$ and CH$_3$OH and CD$_3$OD, the latter in varying amounts. An extrapolation to 100% deuterium gives the isotope effects. A comparison of these with those in CDCl$_3$ are shown in Figure 16. No real differences are found showing that no tautomerism takes place in methanol and by analogy not in ethanol.

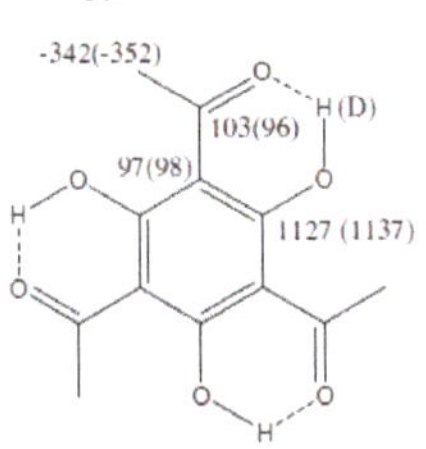

Figure 16. Deuterium isotope effects in CDCl$_3$ + CD$_3$OD and in brackets in CDCl$_3$. The values are the sums of deuteration at all OH sites. Reprinted with permission from Ref. [44]. Copyright 2014 Elsevier.

Mannich bases are compounds that may or may not be tautomeric depending on the substituents at the aromatic ring. This is demonstrated in derivatives of 2-hydroxy-3,4,5,6-tetrachlorobenzene (Figure 17) as well as other derivatives. Temperature may also play a role [46,47].

Other tautomeric examples based on usnic acid are seen below. Usnic acid has important biological applications. However, it is rather insoluble. An attempt to make it more soluble is to add a pegylated side-chain as shown in Figure 18. Measurements of deuterium isotope effects on ^{13}C chemical shifts showed that the equilibrium of the biologically important C ring is unperturbed [48].

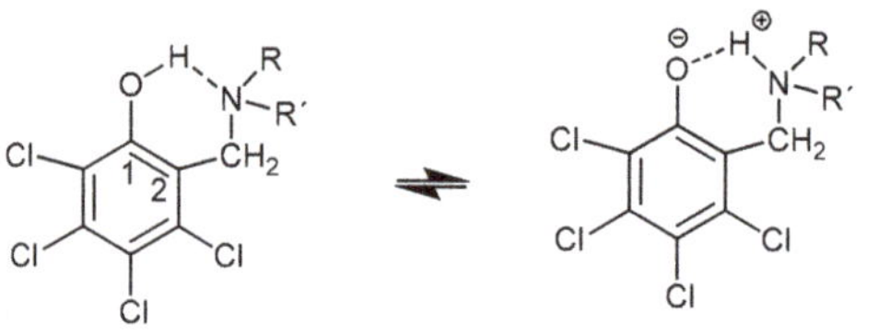

Figure 17. Tautomeric equilibrium of the Mannich base based on 2-hydroxy-3,4,5,6-tetrachlorobenzene.

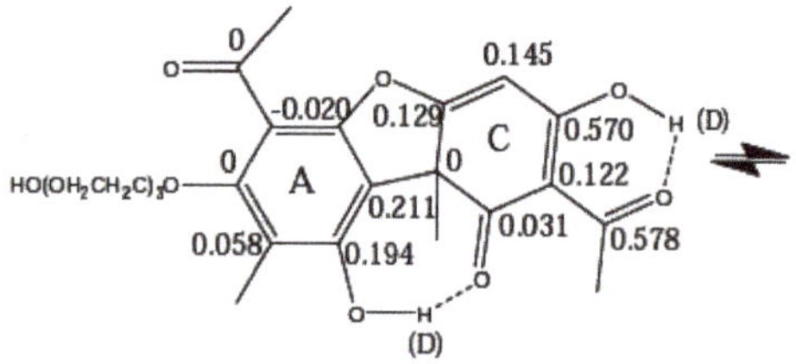

Figure 18. Deuterium isotope effects on ^{13}C chemical shifts of a pegylated usnic acid. From Ref. [48].

Another derivative is the Mannich base derived from usnic acid as shown in Figure 19 [49].

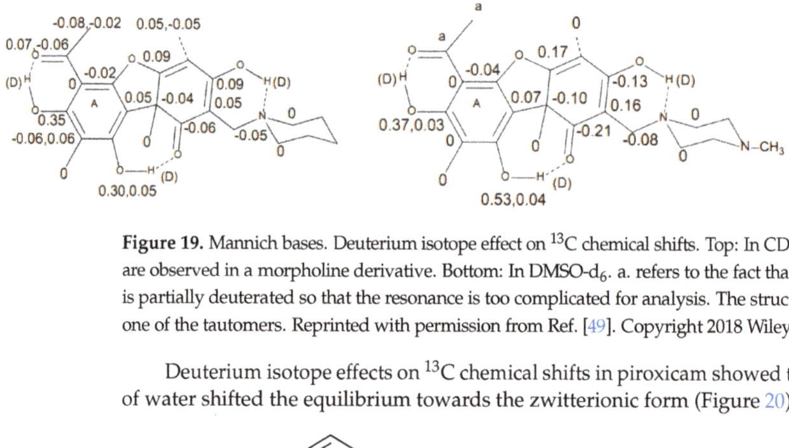

Figure 19. Mannich bases. Deuterium isotope effect on ^{13}C chemical shifts. Top: In CDCl$_3$. Similar effects are observed in a morpholine derivative. Bottom: In DMSO-d$_6$. a. refers to the fact that the methyl group is partially deuterated so that the resonance is too complicated for analysis. The structure shown is only one of the tautomers. Reprinted with permission from Ref. [49]. Copyright 2018 Wiley.

Deuterium isotope effects on ^{13}C chemical shifts in piroxicam showed that the addition of water shifted the equilibrium towards the zwitterionic form (Figure 20) [50].

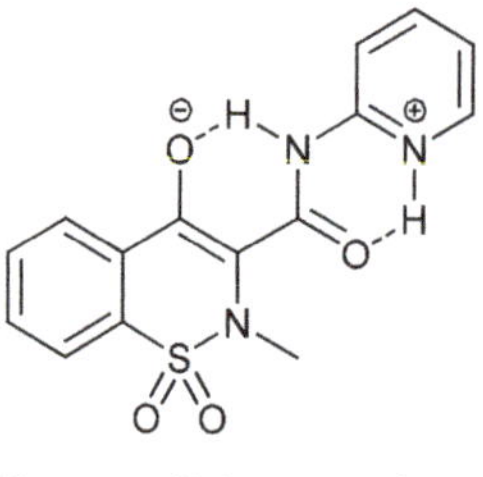

Figure 20. Zwitter ionic form of piroxicam.

Tautomerism may also occur in the solid state. An example is found in pyridoyl benzoyl β-diketones (Figure 21) [51]. In the liquid state, the two-bond deuterium isotope effects at C-1 and C-3 are slightly different reflecting that the equilibrium constant is slightly different from 1. In the solid state, the picture is very different (Figure 21). For **2** and **3** the effects are a clear sign of an equilibrium. However, for **1** a change in the crystal structure as a consequence of deuteration is suggested [51].

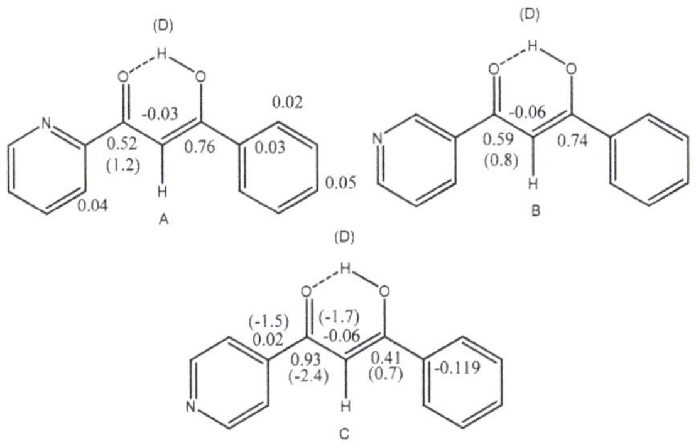

Figure 21. Deuterium isotope effects on ^{13}C chemical shifts in the liquid and solid state. Only one tautomer is shown (see Figure 15). Numbers in brackets are from the solid state. Deuteration may also take place at C-2. Those isotope effects are small and do not tell us about the equilibrium and are left out for clarity. See Ref. [51].

Changes in the crystal structure and the hydrogen bond structure is also discussed by Shi et al [52]. Ip et al. found in the triclinic phase a $^{1}\Delta$N(D) of -2.7^* ppm at 297 K. They concluded that deuteration of the XH proton of the complex between pentachlorophenol and 4-methylpyridine led a monoclinic structure with a weaker hydrogen bond than in the triclinic form [53].

Protonated proton sponges show both strong hydrogen bonds and tautomerism. The effect of the counter ion has been studied. In Figure 22, the counter ion is a proton-like sponge [54]. The effects in the proton sponge are similar to previous examples.

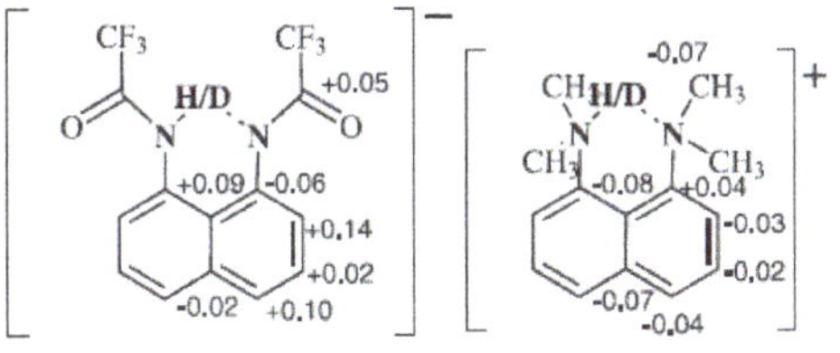

Figure 22. Deuterium isotope effects on ^{13}C chemical shifts of a proton sponge. Reprinted with permission from Ref. [54]. Copyright 2013 Wiley.

One of the questions in symmetric tautomeric systems with strong intramolecular hydrogen bonds is the position of the chelate proton. Two scenarios have been suggested: the chelate proton is jumping from one acceptor to the other or the proton is positioned at the center of the hydrogen bond. The difference being in a double-well potential or a single-well potential. For an early review see [55]. Based on calculations, Bogle and Singleton [56] proposed based that a coupling between a desymmetrizing mode and an anharmonic isotope-dependent mode could lead to isotope effects of the size found in, e.g., the phthalate monoanion (Figure 23A). In response to that, Perrin et al. studied ^{18}O-labelled difluoromaleimide [57] (Figure 23B). ^{19}F is a very chemical shift sensitive nucleus. It is thus very appropriate for the detection of small variations. The compound showed different chemical shifts for the two fluorines. This difference was related to a perturbation of the acidity of the carboxylic acid of the carboxylic acid due to the ^{18}O substitution and not a simple isotope effect, as the dianion did not show chemical shift differences.

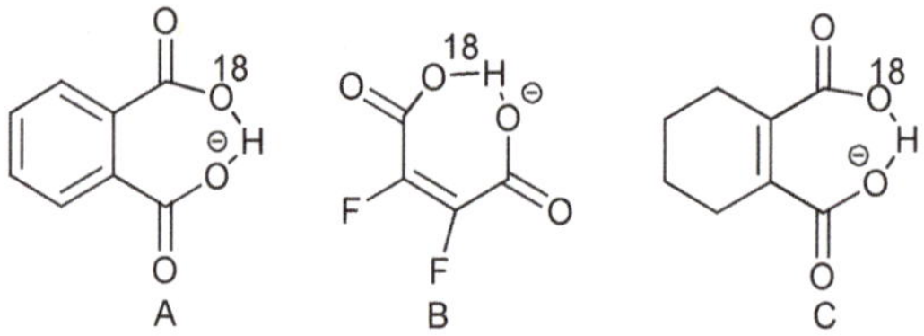

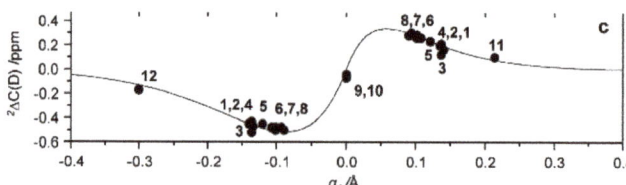

Figure 23. ^{18}O-labelled dicarboxylic acids. (**A**). phthalate monoanion; (**B**). difluoromaleimide; (**C**). cyclohexenedicarboxylic acid.

More recently, Perrin and Burke [58] found a temperature dependence of the C=O chemical shift of the ^{18}O-labelled carboxylic acid of ^{18}O-labelled cyclohexenedicarboxylic acid (Figure 23C). Based on this finding, they suggested an equilibrium.

In response to the paper by Bogle and Singleton, Perrin et al. [59] have studied ^{18}O-labelled 1,2-cyclohexendicarboxylate monoanion (Figure 23C) as well as the difluoromaleate anion (Figure 23B). In both cases they found a larger isotope effect at a lower temperature, which is against a desymmetrization, as this should become smaller at a lower temperature.

Limbach et al. [60] studied deuterated maleate and phthalate anions (Figure 23A) as well as a series of homoconjugated anions of carboxylic acid (deuteration at the OH proton). The primary isotope effects are plotted vs. the two-bond deuterium isotope effects on ^{13}C chemical shifts as seen in Figure 24. One can of course wonder why a distinction between a single-well and a double-well potential is so important, but this becomes clear when making plots such as the one in Figure 24. For maleate and phthalate, a single-well potential is assumed. Analysis of the data also allowed the construction of a rather complex correlation between $^2\Delta C(OD)$, q_1 and three fitting parameters.

Figure 24. Plot of two-bond deuterium isotope effects on ^{13}C chemical shifts of a series of primarily homoconjugated anions of carboxylic acids vs. q_1. Data points for **11** and **12** are heteroconjugated dimers. $q_1 = 0.5(r_{OH} - r_{HO})$. Reprinted with permission from Ref. [60]. Copyright 2012 American Chemical Society.

Succinic acid, *meso* and *rac*-succinic acid and methyl succinic acid with tetraalkylammonium ions as counter ions measured in CDF$_3$/CDF$_2$Cl at 300 to 120K showed double-well potential-based on isotope effects [61]. A plot of primary isotope effects vs. OH chemical shifts showed a large spread [62].

Carboxylic acid dimers have been studied trapped in a capsule. In case of partial deuteration, the encapsulation leads to slow exchange. The deuterium isotope effects on the OH chemical shift varies from 0.14 to 0.29* ppm for encapsulated acid vs. ~0.1* ppm for non-encapsulated acids. This can be related to the pressure of the encapsulation leading to a shorter O ... O distance and a stronger hydrogen bond. Isotope effects correlate with OH chemical shifts [63].

In Figure 25, a plot of primary deuterium isotope effects vs. OH chemical shifts for a series of primarily salicylates is shown [42]. This plot is very similar to plots in Refs. [3,62]. The change in the sign of the primary isotope effect was explained by Gunnarson et al. [40] finding a positive effect for weak double-well potentials and an increasing value as the anharmonicity increases. For those cases in which a single-well potential is the case, a negative value is found. As seen in Figure 25, the primary isotope effect for each compound is within 0.1 ppm in the different solvents showing that a solvent including water plays no special role for these hydrogen bonds.

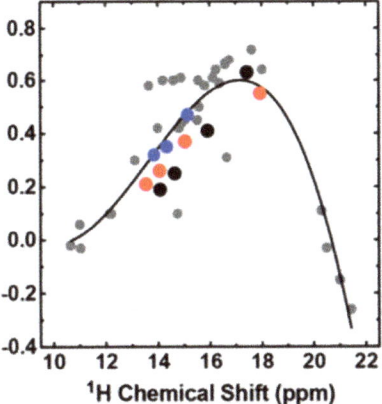

Figure 25. Plot primary deuterium isotope effects for a series of salicylates: 5-methylsalicylate, 4-methoxysalicylate, 5-formylsalicylate and 3,5-dinitrosalicylate. Chloroform (red), acetone (black) and water (blue) and for previously published O−H···O− hydrogen-bonded complexes in aprotic organic solvents (gray) [64] vs. OH chemical shifts. Reprinted with permission from Ref. [42]. Copyright 2015 American Chemical Society.

5. Hydrogen Bond Energy

Hydrogen bond energies and the less well-defined hydrogen bond strength is clearly very important for molecular properties [65]. One way of getting a qualitative idea is the monitor isotope effects in solvents of different polarity. Sigala et al. [42] found no difference in going to water (see Section 4) and also for the pegylated usnic acid the isotope effects were similar in CDCl$_3$ and in a mixture of water and DMSO [48].

Two-bond deuterium isotope effects on ^{13}C chemical shifts may also be related to electron densities at bond critical points [66] (Figure 26). The latter are related to hydrogen bond energies.

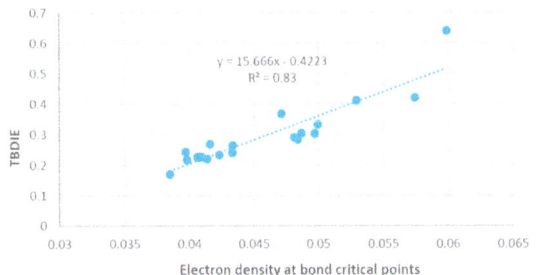

Figure 26. Two-bond deuterium isotope effects on ^{13}C chemical shifts vs. electron density at bond critical points. Reprinted from Ref. [67].

A quantitative approach was proposed by Reuben [68], who correlated two-bond deuterium isotope effects with hydrogen bond energies of intramolecular hydrogen bonds. Recently, this was tested in o-hydroxybenzaldehydes (Figure 27) [67]. The hydrogen bond energies were calculated by the hb and our method. This method was originally formulated by Cuma, Scheiner and Kar [69]. Another example is found in 5-acyl rhodanines and thiorhodanines with bulky acyl groups [70] again with a very good correlation.

Two-bond deuterium isotope effects are clearly a way of estimating hydrogen bond energies in cases in which the theoretical methods are less suited.

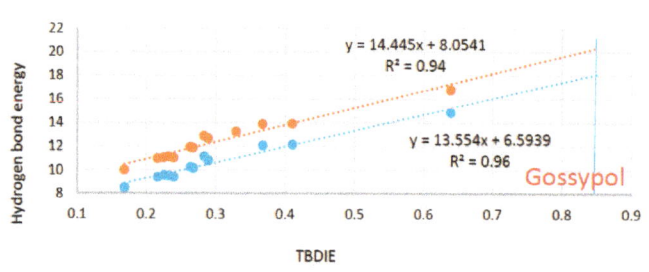

Figure 27. Plot of hydrogen bond energies vs. two-bond deuterium isotope effect on ^{13}C chemical shifts of *o*-hydroxyacyl aromatics. Plot of 'hb and out' hydrogen bond energies in kcal/mol. calculated either with MP2/6-311++G(d,p) or with B3LYP/6-311++G(d,p) vs. observed two-bond deuterium isotope effect (TBDIE) on ^{13}C chemical shifts in ppm. Top correlation line is B3LYP, bottom one is MP2. The observed TBDIE for gossypol is marked with the vertical blue line. Reprinted from Ref. [67].

6. Correlations

Isotope effects can also be correlated to other parameters related to, e.g., hydrogen bond strength and hydrogen bond energy.

Two-bond deuterium isotope effects on ^{13}C chemical shifts are proportional to XH chemical shift. For an example see Figure 28. Another example is given in Ref. [71]. See also Figures 24 and 25.

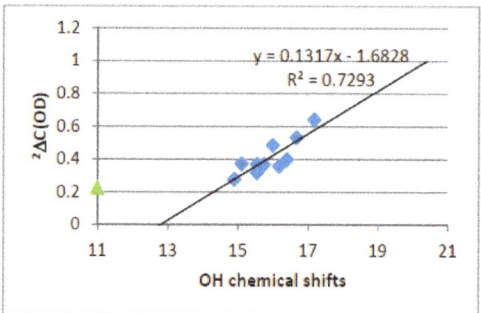

Figure 28. Plot of two-bond deuterium isotope effects vs. OH chemical shifts of 10-hydroybenzo[h]quinolones. Taken from Ref. [71].

7. Theoretical Calculations

A theory for the calculation of isotope effects on chemical shifts was presented by Jameson [72,73]. In a simplified way, deuterium isotope effects can be calculated by assuming that the XH bond is shortened upon deuteration. This approach has been described in a number of papers [62,71,72,74]. The shortening can be determined by the calculation of the hydrogen bond potential, but this is time consuming [43,75]. A simpler approach is to assume a reasonable value and as all the isotope effects of the molecule depend on the same shortening, a plot vs. experimental values will determine whether an intrinsic isotope effect is at hand. An example is shown in Figure 29.

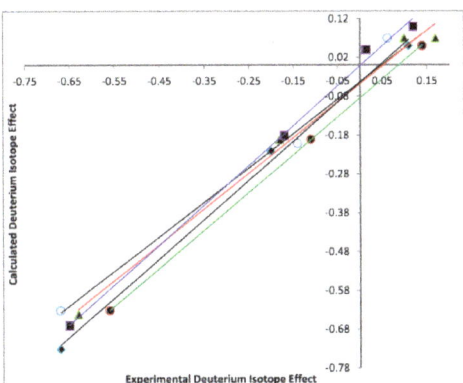

Figure 29. Plot of calculated vs. experimental deuterium isotope effects on ^{13}C chemical shifts. The investigated compounds are of X type with R' being methyl or phenyl and R being alkyl or aromatic. The calculations were of MP2/6-311+G(d2d,p) type. Reprinted with permission from Ref. [74]. Copyright 2018 Wiley.

An ab initio multi-component molecular orbital method on o-hydroxyacetophenones seems to overshoot the experimental values [76]. MC-MO calculations were also used for amino acid pairs [77]. A more recent example is a multicomponent hybrid density functional method combined with the polarizable continuum method [78]. This method was applied to picolinic N-oxide and led to decent predictions for the deuterium isotope effects on ^{13}C chemical shifts, whereas the primary deuterium isotope effects were less well predicted. In the case of deuterium isotope effects, this was also the case for acetylacetone. Gräfenstein has developed a difference dedicated second-order vibrational perturbation theory to calculate isotope effects [79]. This was applied to a series of o-hydroxybenzaldehydes [80]. Ab initio path integral molecular dynamics (PIMD) calculations showed a barrier-less proton transfer and a C$_2$V symmetry of the hydrogen bond. The calculated isotope effects were rather small [81]. An isocytosine dimer was studied at a low temperature. PIMD calculations were applied to the isocytosine base pair. For N3, the $^1\Delta N(D) = 0.88$ ppm [82]. The PIMD calculations gave somewhat too large values, but of the right order.

8. Conclusions

Isotope effects on chemical shifts cover a very broad range of hydrogen bonds ranging from the very weak over dihydrogen bonds to the very strong hydrogen bonds found in, e.g., dicarboxylic acid anions. Isotope effects on chemical shifts in small molecules provide a basis and an understanding for use in larger, e.g., biological molecules. One-bond deuterium isotope effects on ^{15}N chemical shifts and two-bond deuterium isotope effects on ^{1}H chemical shifts can be used to monitor the distance to nearby hydrogen bond acceptors and/or charges.

Isotope effects on chemical shifts are a strong tool in the investigation of tautomeric systems and can lift degeneracies in symmetrical systems as seen in the monanions of dicarboxylic acids. ^{18}O labelling of the latter is central in the discussion of single- vs. double-well potentials.

Two-bond deuterium isotope effects on ^{13}C chemical shifts are related to hydrogen bond strength and to hydrogen bond energies.

Funding: This research received no external funding.

Institutional Review Board Statement: Not applicable.

Informed Consent Statement: Not applicable.

Data Availability Statement: Not applicable.

Conflicts of Interest: The author declares no conflict of interest.

References

1. Chen, D.; Oezguen, N.; Urvi, P.; Ferguson, C.; Aannand, M.; Savidge, T.C. Regulation of protein-ligand binding affinity by hydrogen bond pairing. *Sci. Adv.* **2016**, *2*, e1501240. [CrossRef] [PubMed]
2. Liu, M.; Hansen, P.E.; Lin, X. Bromophenols in marine algae and their bioactivities. *Mar. Drugs* **2011**, *9*, 1273–1792. [CrossRef] [PubMed]
3. Sobczyk, L.; Obzrzud, M.; Filarowski, A. H/D Isotope Effects in Hydrogen Bonded Systems. *Molecules* **2013**, *18*, 4467–4476. [CrossRef] [PubMed]
4. Hansen, P.E. Isotope effects on chemical shifts as a tool in the study of tautomeric equilibria. In *Equilibria. Methods and Theories*; Antonov, L., Ed.; Wiley-VCH: Weinheim, Germany, 2014.
5. Hansen, P.E. Methods to distinguish tautomeric cases from static ones. In *Tautomerism: Ideas, Compounds, Applications*; Antonov, L., Ed.; Wiley-VCH: Weinheim, Germany, 2016.
6. Limbach, H.H.; Denisov, G.S.; Golubev, N.S. Hydrogen Bond Isotope Effects Studied by NMR. In *Isotope Effects in Chemistry and Biology*; Kohen, A., Limbach, H.H., Eds.; Taylor and Francis: Boca Raton, FL, USA, 2006.
7. Shahkhatuni, A.A.; Shahkhatuni, A.G.; Harutyunyan, A.S. Long range deuterium isotope effects on ^{13}C NMR chemical shifts of 2-alkanones in CD$_3$OD solutions of imidazolium acetate ionic liquids. *RSC Adv.* **2021**, *11*, 39051. [CrossRef]
8. Reuben, J. Isotopic Multiplets in the Carbon-13 NMR Spectra of Aniline Derivatives and Nucleosides with Partially Deuterated Amino Groups: Effects of Intra- and Intermolecular Hydrogen Bonding. *J. Am. Chem. Soc.* **1987**, *109*, 316–321. [CrossRef]
9. Mackenzie, H.W.; Hansen, D.F. A ^{13}C-detected ^{15}N double-quantum NMR experiment to probe arginine side-chain guanidinium 15N^{η} chemical shifts. *J. Biomol. NMR* **2017**, *69*, 123–132. [CrossRef]
10. Iwahara, J.; Jung, Y.-S.; Clore, G.M. Heteronuclear NMR Spectroscopy for Lysine NH$_3$ Groups in Proteins: Unique Effect of Water Exchange on ^{15}N Transverse Relaxation. *J. Am. Chem. Soc.* **2007**, *129*, 2971–2980. [CrossRef]
11. Kurutos, A.; Sauer, S.P.A.; Kamounah, F.S.; Hansen, P.E. Azo-hydrazone derived molecular switches: Synthesis and conformational investigation. *Magn. Reson. Chem.* **2021**, *59*, 1116–1125. [CrossRef]
12. Khrizman, A.R.; Cheng, H.Y.; Bottini, G.; Moyna, G. Observation of aliphatic C–HX hydrogen bonds in imidazolium ionic liquids. *Chem. Commun.* **2015**, *51*, 3193–3195. [CrossRef]
13. Bottini, G.; Moyna, G. Determining the relative strengths of aromatic and aliphatic C-H....X hydrogen bonds in imidazolium ionic liquids through measurements of H/D isotope effects on ^{19}F nuclear shielding. *Magn. Reson. Chem.* **2018**, *56*, 103–107. [CrossRef]
14. Remsing, R.C.; Wildin, J.L.; Rapp, A.L.; Moyna, G. Hydrogen Bonds in Ionic Liquids Revisited: $^{35/37}$Cl NMR Studies of Deuterium Isotope Effects in 1-*n*-Butyl-3-Methylimidazolium Chloride. *Chem. Phys. Lett. B* **2007**, *111*, 11619–11621. [CrossRef] [PubMed]
15. Ossowicz, P.; Janus, E.; Schroeder, G.; Rozwadowki, Z. Spectroscopic Studies of Amino Acid Ionic Liquid-Supported Schiff Bases. *Molecules* **2013**, *18*, 4986–5004. [CrossRef] [PubMed]
16. Anouti, M.; Caillo-Caravanier, M.; Le Floch, C.; Lemorant, D. Alkylammonium-based protic Ionic Liquids Part I: Preparation and Physicochemical Characterization. *J. Phys. Chem. B* **2008**, *112*, 9406–9411. [CrossRef] [PubMed]
17. Hansen, P.E.; Lund, T.; Krake, J.; Spanget-Larsen, J.; Hvidt, S. A Reinvestigation of the Ionic Liquid Diisopropylethylammonium formate by NMR and DFT Methods. *J. Phys. Chem. B.* **2016**, *120*, 11279–11286. [CrossRef] [PubMed]
18. Tolstoy, P.M.; Schah-Mohammedi, P.; Smirnov, S.N.; Golubev, N.S.; Denisov, G.S.; Limbach, H.-H. Characterization of fluxional Hydrogen-Bonded complexes of Acetic Acid and Acetate by NMR: Geometries and Isotope and Solvent Effects. *J. Am. Chem. Soc.* **2004**, *126*, 5621–5634. [CrossRef]
19. William, J.A.; Collinson, I. Ammonium Transporters: A molecular dual carriageway. *eLife* **2020**, *9*, e61148. [CrossRef]
20. Pflüger, T.; Hernández, C.F.; Lewe, P.; Frank, F.; Mertens, H.; Svergun, D.; Baumstark, M.W.; Lunin, V.Y.; Jetten, M.S.M.; Andrade, S.L.A. Signaling ammonium across membranes through an ammonium sensor histidine kinase. *Nat. Comm.* **2018**, *9*, 164. [CrossRef]
21. Graf, E.; Kintzinger, J.P.; Lehn, J.-M.; LeMoigne, J. Molecular Recognition. Selective Ammonium Cryptates of Synthetic Receptor Molecules Possessing a Tetrahedral Recognition Site. *J. Am. Chem. Soc.* **1982**, *104*, 1672–1678. [CrossRef]
22. Lambert, S.; Bartik, K.; Jabin, I. Specific Binding of Primary Ammonium Ions and Lysine-Containing Peptides in Protic Solvents by hexahomooxacalix[3]arenes. *J. Org. Chem.* **2020**, *85*, 10062–10071. [CrossRef]

23. Platzer, G.; Okon, M.; McIntosh, L.P. pH-dependent random coil ^{1}H, ^{13}C and ^{15}N chemical shifts of the ionizable amino acids: A guide for protein pKa measurements. *J. Biomol. NMR* **2014**, *60*, 109–129. [CrossRef]
24. Hansen, P.E. Deuterium Isotope Effects on ^{14,15}N Chemical Shifts of Ammonium Ions. A Solid State NMR Study. *Int. J. Inorg. Chem.* **2011**, 696497. [CrossRef]
25. Munch, M.; Hansen, A.E.; Hansen, P.E.; Bouman, T.D. Ab-Initio Calculations of Deuterium isotope Effects on hydrogen and Nitrogen nuclear Magnetic Shielding in the hydrated Ammonium Ion. *Acta Chem.Scand.* **1992**, *46*, 1065–1071. [CrossRef]
26. Hansen, P.E.; Hansen, A.E.; Lycka, A.; Buvari-Barcza, A. $^2\Delta$H(D) and $^1\Delta$N(D) Isotope Effects on Nuclear Shielding of Ammonium Ions in Complexes with Crown ethers and Cryptands. *Acta Chem. Scand.* **1993**, *47*, 777–788. [CrossRef]
27. Hansen, P.E.; Saeed, B.A. Ammonium Ions in a Confined Space. *J. Mol. Struct.* **2022**. submitted.
28. Jednacak, T.; Novak, P.; Urzarevic, K.; Bratos, I.; Mrkovic, J.; Cindric, M. Bioactive Phenylenediamine Derivatives of Dehydracetic Acid: Synthesis, Structural Characterization and Deuterium Isotope Effects. *Croat. Chem. Acta* **2011**, *84*, 203–209. [CrossRef]
29. Seyedkatouli, S.; Vakili, M.; Tayyari, S.F.; Hansen, P.E.; Kamounah, F.S. Molecular structure and intramolecular hydrogen bond strength of 3-methyl-4-amino-3-penten-2-one and its NMe and N-Ph substitutions by experimental and theoretical methods. *J. Mol. Struct.* **2019**, *1184*, 233–245. [CrossRef]
30. Petrova, M.; Muhamadejev, R.; Vigante, B.; Duburs, G.; Liepinsh, E. Intramolecular hydrogen bonds in 1,4-dihydropyridine derivatives. *R. Soc. Open Sci.* **2018**, *51*, 80088.
31. Mulloyarova, V.V.; Giba, I.S.; Kostin, M.A.; Denisov, G.S.; Shenderovich, I.G.; Tolstoy, P.M. Cyclic trimers of phosphinic acids in polar aprotic solvent:symmetry, chirality and H/D isotope effects on NMR chemical shifts. *Phys. Chem. Chem. Phys.* **2018**, *20*, 4901–4910. [CrossRef]
32. Mulloyarova, V.V.; Ustimchuck, D.O.; Filarowski, A.; Tolstoy, P.M. H/D Isotope Effects on ^{1}H-NMR Chemical Shifts in Cyclic Heterodimers and Heterotrimers of Phosphinic and Phosphoric Acids. *Molecules* **2020**, *25*, 1907. [CrossRef]
33. Mulloyarova, V.V.; Puzyk, A.M.; Efimova, A.A.; Antonov, S.S.; Evarestov, R.S.; Aliyarova, I.S.; Asfin, R.E.; Tolstoy, P.M. Solid-state and solution-state self-association of dimethylarsinic acid: IR, NMR and theoretical study. *J. Mol. Struct.* **2021**, *1234*, 130176. [CrossRef]
34. Hansen, P.E. Isotopic Perturbation of Equilibrium in 2,6-Dihydroxy Benzoyl Compounds. *Acta Chem. Scand.* **1988**, *B42*, 423–432. [CrossRef]
35. O´Leary, D.J.; Hickstein, D.D.; Hansen, B.K.V.; Hansen, P.E. Theoretic and NMR Studies of Deuterium Isotopic Perturbation of Hydrogen Bonding in Symmetrical Dihydroxy Compounds. *J. Org. Chem.* **2010**, *75*, 1331–1342. [CrossRef] [PubMed]
36. Xu, H.; Luo, T.; Wu, H.; Yuan, L.-B.; Zhao, S.-Q.; Liang, W.-J.; Zhong, S.-A.; Chen, Y.; Saunders, M.; Jiang, J.-Z.; et al. An equilibrium effect due to a strong hydrogen bond. *Chem. Phys. Lett.* **2018**, *713*, 117–120. [CrossRef]
37. O´Leary, D.J. Comment on "An equilibrium isotope effect due to a strong hydrogen bond". *Chem. Phys. Lett.* **2019**, *730*, 302–305. [CrossRef]
38. Pal, U.; Sen, S.; Maiti, N.C. Dα-H Carries Information of a Hydrogen Bond Involving the Geminal Hydroxyl Group: A Case Study with a Hydrogen-Bonded Complex of 1,1,1,3,3,3-Hexafluoro-2-propanol and Tertiary Amines. *J. Phys. Chem. A* **2014**, *118*, 1024–1030. [CrossRef]
39. Schulz, F.; Sumerin, V.; Heikkinen, S.; Pedersen, B.; Wang, C.; Atsumi, M.; Leskelä, M.; Repo, T.; Pyykkö, P.; Petry, W.; et al. Molecular Hydrogen Tweezers: Structure and Mechanisms by Neutron Diffraction, NMR, and Deuterium Labeling Studies in Solid and Solution. *J. Am. Chem. Soc.* **2011**, *133*, 20245–20257. [CrossRef]
40. Gunnarsson, G.; Wennerstöm, H.; Forsén, S. Proton and deuterium NMR of hydrogen bonds-relationship between isotope-effects and hydrogen-bond potential. *Chem. Phys. Lett.* **1976**, *38*, 96–99. [CrossRef]
41. Ilczyszyn, M.; Selent, M.; Ilczyszyn, M.M. Participation of Xenon Guest in Hydrogen Bond Network of β-Hydroquinone Crystal. *J. Phys. Chem. A* **2012**, *116*, 3206–3214. [CrossRef]
42. Sigala, P.A.; Ruben, E.A.; Liu, C.W.; Piccoli, P.M.B.; Hohenstein, E.G.; Martínez, T.J.; Schultz, A.J.; Herschlag, D. Determination of Hydrogen Bond Structure in Water versus Aprotic Environments to Test the Relationship between Length and Stability. *J. Am. Chem. Soc.* **2015**, *137*, 5730–5740. [CrossRef]
43. Abildgaard, J.; Bolvig, S.; Hansen, P.E. Unravelling the Electronic, Steric and Vibrational Contributions to Deuterium Isotope Effects on ^{13}C Chemcial Shifts by ab initio Model Calculations. Intramolecular Hydrogen bonded *o*-Hydroxy Acyl Aromatics. *J. Am. Chem. Soc.* **1998**, *12090*, 63–9069.
44. Hansen, P.E.; Kamounah, F.S.; Zhiryakova, D.; Manolova, Y.; Antonov, L. 1,1′,1″-(2,4,6-Trihydroxybenzene-1,3,5-triyl)triethanone non-tautomerism. *Tetrahedron Lett.* **2014**, *55*, 354–357. [CrossRef]
45. Serdiuk, I.E.; Wera, M.; Roshal, A.D.; Sowiński, P.; Zadykowicz, B.; Błażejowski, J. Tautomerism, structure and properties of 1,1′,1″-(2,4,6-trihydroxybenzene-1,3,5-triyl)triethanone. *Tetrahedron Lett.* **2011**, *52*, 2737–2740. [CrossRef]
46. Rospenk, M.; Koll, A.; Sobczyk, L. Proton transfer and secondary deuterium isotope effect in the C-13 NMR spectra of ortho-aminomethyl phenols. *Chem. Phys. Lett.* **1996**, *261*, 283–288. [CrossRef]
47. Hansen, P.E.; Spanget-Larsen, J. Structural studies on Mannich bases of 2-Hydroxy-3,4,5,6-tetrachlorobenzene. An UV, IR, NMR and DFT study. A mini-review. *J. Mol. Struct.* **2016**, *1119*, 235–239. [CrossRef]
48. Hansen, P.E.; Mortensen, J.; Kamounah, F.S. The importance of correct tautomeric structures for biological molecules. *JSM Chem.* **2015**, *3*, 1014–1019.

49. Nguyen, H.G.T.; Nguyen, V.N.; Kamounah, F.S.; Hansen, P.E. Structure of a new Usnic acid derivative from a deacylation reaction of in a Mannich reaction. NMR studies supported by theoretical calculations of NMR parameters. *Magn. Reson. Chem.* **2018**, *56*, 1094–1100. [CrossRef]
50. Ivanova, D.; Deneva, V.; Nedeltcheva, D.; Kamounah, F.S.; Gergov, G.; Hansen, P.E.; Kawauchi, S.; Antonov, L. Tautomeric transformations of Piroxicam in solution: A combined experimental and theoretical study. *RSC Adv.* **2015**, *5*, 31852–31860. [CrossRef]
51. Hansen, P.E.; Borisov, E.V.; Lindon, J.C. Determination of the Tautomeric Equilibria of Pyridoyl Benzoyl β-Diketones in the Liquid and Solid State through the use of Deuterium Isotope Effects on ^{1}H and ^{13}C NMR Chemical Shifts and Spin Coupling Constants. *Spectrochim. Acta* **2015**, *136*, 107–112. [CrossRef]
52. Shi, C.; Zhang, X.; Yu, C.-H.; Yao, Y.F.; Zhang, W. Geometric isotope effect of deuteration in a hydrogen-bonded host-guest crystal. *Nat. Comm.* **2018**, *9*, 481. [CrossRef]
53. Ip, B.C.K.; Shenderovich, I.G.; Tolstoy, P.M.; Frydel, J.; Denisov, G.S.; Buntkowsky, G.; Limbach, H.-H. NMR studies of solid Pentachlorophenol-4-Methylpyridine complexes Exhibiting Strong OHN Hydrogen Bonds: Geometric H/D Isotope Effects and Hydrogen Bond Coupling Cause Isotopic Polymorphism. *J. Phys. Chem. A* **2012**, *116*, 11370–11387. [CrossRef]
54. Pietrzak, M.; Grech, E.; Nowicka-Scheibe, J.; Hansen, P.E. Deuterium isotope effects on ^{13}C chemical shifts of negatively charged NH . . . N systems. *Magn. Reson. Chem.* **2013**, *51*, 683–688. [PubMed]
55. Perrin, C.L. Symmetry of hydrogen bonds in solution. *Pure Appl. Chem.* **2009**, *81*, 571–583. [CrossRef]
56. Bogle, X.S.; Singleton, D.A. Isotope-induced desymmetrization can mimic isotopic perturbation of equilibria. On the symmetry of bromonium ions and hydrogen bonds. *J. Am. Chem. Soc.* **2011**, *133*, 17172–17175. [CrossRef] [PubMed]
57. Perrin, C.L.; Kari, P.; Moore, C.; Rheingold, A.L. Hydrogen-bond symmetry in difluoromaleate monanion. *J. Am. Chem. Soc.* **2012**, *134*, 7766–7772. [CrossRef]
58. Perrin, C.L.; Burke, K.D. Variable-temperature study of hydrogen-bond symmetry in cyclohexene-1,2-dicarboxylate monoanion in chloroform-d. *J. Am. Chem. Soc.* **2014**, *136*, 4355–4362. [CrossRef]
59. Perrin, C.L.; Shrinidhi, A.; Burke, K.D. Isotopic-Perturbation NMR Study of Hydrogen-Bond Symmetry in solution: Temperature Dependence and Comparison of OHO and ODO Hydrogen bonds. *J. Am. Chem. Soc.* **2019**, *141*, 17278–17286. [CrossRef]
60. Guo, J.; Tolstoy, P.M.; Koeppe, B.; Golubev, N.S.; Denisov, G.S.; Smirnov, S.N.; Limbach, H.-H. Hydrogen Bond Geometries and Proton Tautomerism of Homoconjugated Anions of Carboxylic Acids Studied via H/D Isotope Effects on ^{13}C NMR Chemical Shifts. *J. Phys. Chem. A* **2012**, *116*, 11180–11188. [CrossRef]
61. Guo, J.; Tolstoy, P.M.; Koeppe, B.; Denisov, G.S.; Limbach, H.-H. NMR Study of Conformational Exchange and Double-Well Proton Potential in Intramolecular Hydrogen Bonds in Monoanions of Succinic Acid and Derivatives. *J. Phys. Chem. A* **2011**, *115*, 9828–9836. [CrossRef]
62. Hansen, P.E. Isotope effects on chemical shift in the study of intramolecular hydrogen bonds. *Molecules* **2015**, *20*, 2405–2424. [CrossRef]
63. Ajami, D.; Tolstoy, P.M.; Dube, H.; Odermatt, S.; Koeppe, B.; Guo, J.; Limbach, H.-H.; Rebek, J., Jr. Encapsulated Carboxylic Acid dimers with Compressed Hydrogen Bonds. *Angew. Chem. Int. Ed.* **2011**, *50*, 528–531. [CrossRef]
64. Hibbert, F.; Emsley, J. Hydrogen bonding and chemical reactivity. *J. Adv. Phys. Org. Chem.* **1990**, *26*, 255–379.
65. Hansen, P.E. A Spectroscopic Overview of Intramolecular NH . . . O,N,S Hydrogen Bonds. *Molecules* **2021**, *26*, 2409. [CrossRef] [PubMed]
66. Espinosa, E.; Molins, E.; Lecomte, C. Hydrogen bond strengths revealed by topological analyses of experimentally observed electron densities. *Chem. Phys. Lett.* **1998**, *285*, 170–173. [CrossRef]
67. Hansen, P.E.; Kamounah, F.S.; Saeed, B.A.; MacLachlan, M.J.; Spanget-Larsen, J. Intramolecular Hydrogen Bonds in Normal and Sterically Compressed o-Hydroxy Aromatic Aldehydes. Isotope Effects on Chemical Shifts and Hydrogen Bond Strength. *Molecules* **2019**, *24*, 4533. [CrossRef]
68. Reuben, J. Intramolecular hydrogen-bonding as reflected in the deuterium isotope effects on C-13 chemical-shifts-correlation with hydrogen bond energies. *J. Am. Chem. Soc.* **1986**, *108*, 1735–1738. [CrossRef]
69. Cuma, M.; Scheiner, S.; Kar, T. Competition between rotamerization and proton transfer in o-hydroxybenzaldehyde. *J. Am. Chem. Soc.* **1998**, *120*, 10497–10503. [CrossRef]
70. Elias, R.S.; Saeed, B.A.; Kamounah, F.S.; Duus, F.; Hansen, P.E. Strong Intramolecular Hydrogen Bonds and steric Effects. A NMR and Computational Study. *Magn. Reson. Chem.* **2020**, *58*, 154–162. [CrossRef]
71. Hansen, P.E.; Kamounah, F.S.; Gryko, D.T. Deuterium isotope effects on ^{13}C chemical shifts of 10-Hydroxybenzo[h]quinolones. *Molecules* **2013**, *18*, 4544–4560. [CrossRef]
72. Jameson, C.J. The Dynamic and Electronic Factors in Isotope Effects on NMR Parameters. In *Isotopes in the Physical and Biomedical Sciences. Isotopic Applications in NMR Studies*; Buncel, E., Jones, J.R., Eds.; Elsevier: Amsterdam, The Netherlands, 1991.
73. Jameson, C.J.; Osten, H.-J. The NMR isotope shift in polyatomic molecules. Estimation of the dynamic factors. *J. Chem. Phys.* **1984**, *81*, 4300–4305. [CrossRef]
74. Saeed, B.A.; Elias, R.S.; Kamounah, F.S.; Hansen, P.E. A NMR, MP2 and DFT Study of Thiophenoxyketenimines (o-ThioSchiff bases). *Magn. Reson. Chem.* **2018**, *56*, 172–182. [CrossRef]
75. Ikabata, Y.; Imamura, Y.; Nakai, N. Interpretation of Intermolecular Geometric isotope Effect in Hydrogen bonds: Nuclear Orbital plus Molecular orbital Study. *J. Phys. Chem A* **2011**, *115*, 1433–1439. [CrossRef] [PubMed]

76. Udagawa, T.; Ishimoto, T.; Tachikawa, M. Theoretical Study of H/D Isotope Effects on Nuclear Magnetic Shieldings Using an ab initio Multi-Component Molecular Orbital Method. *Molecules* **2013**, *18*, 5209–5220. [CrossRef] [PubMed]
77. Ullah, S.; Ishimoto, T.; Williamson, M.P.; Hansen, P.E. Ab Initio calculations of Deuterium isotope Effects on Chemical Shifts of Salt-bridged Lysines. *J. Phys. Chem. B* **2011**, *115*, 3208–3215. [CrossRef] [PubMed]
78. Kanematsu, Y.; Tachikawa, M. Development of multicomponent hybrid density functional theory with polarization continuum model for the analysis of nuclear quantum effect and solvent effect on NMR chemical shift. *J. Chem. Phys.* **2014**, *140*, 164111. [CrossRef] [PubMed]
79. Gräfenstein, J. Efficient calculation of NMR isotopic shifts: Difference-dedicated vibrational perturbation theory. *J. Chem. Phys.* **2019**, *151*, 244120. [CrossRef] [PubMed]
80. Gräfenstein, J. The Structure of the "Vibration Hole" around an Isotopic Substitution-Implications for the Calculation of Nuclear Magnetic Resonance (NMR) Isotopic shifts. *Molecules* **2020**, *25*, 2915. [CrossRef]
81. Kawashima, Y.; Tachikawa, M. Nuclear quantum effect on intramolecular hydrogen bond of hydrogen maleate anion: An ab initio path integral molecular dynamics study. *Chem. Phys. Lett.* **2013**, *571*, 23–27. [CrossRef]
82. Pohl, R.; Socha, O.; Slavicek, P.; Sala, M.; Hodgkinson, P.; Dracinsky, M. Proton transfer in guanine-cytosine base pair analogues studied by NMR spectroscopy and PIMD simulations. *Faraday Discuss.* **2018**, *212*, 331–344. [CrossRef]

Article

Simultaneous Estimation of Two Coupled Hydrogen Bond Geometries from Pairs of Entangled NMR Parameters: The Test Case of 4-Hydroxypyridine Anion

Elena Yu. Tupikina [1,*], Mark V. Sigalov [2] and Peter M. Tolstoy [1]

[1] Institute of Chemistry, St. Petersburg State University, 199034 St. Petersburg, Russia; peter.tolstoy@spbu.ru
[2] Department of Chemistry, Ben-Gurion University of the Negev, Beer-Sheva 84105, Israel; msigalov@bgu.ac.il
* Correspondence: e.tupikina@spbu.ru

Abstract: The computational method for estimating the geometry of two coupled hydrogen bonds with geometries close to linear using a pair of spectral NMR parameters was proposed. The method was developed based on the quantum-chemical investigation of 61 complexes with two hydrogen bonds formed by oxygen and nitrogen atoms of the 4-hydroxypyridine anion with OH groups of substituted methanols. The main idea of the method is as follows: from the NMR chemical shifts of nuclei of atoms forming the 4-hydroxylpyridine anion, we select such pairs, whose values can be used for simultaneous determination of the geometry of two hydrogen bonds, despite the fact that every NMR parameter is sensitive to the geometry of each of the hydrogen bonds. For these parameters, two-dimensional maps of dependencies of NMR chemical shifts on interatomic distances in two hydrogen bonds were constructed. It is shown that, in addition to chemical shifts of the nitrogen atom and quaternary carbon, which are experimentally difficult to obtain, chemical shifts of the carbons and protons of the CH groups can be used. The performance of the proposed method was evaluated computationally as well on three additional complexes with substituted alcohols. It was found that, for all considered cases, hydrogen bond geometries estimated using two-dimensional correlations differed from those directly calculated by quantum-chemical methods by not more than 0.04 Å.

Keywords: hydrogen bonds; NMR; structure determination

1. Introduction

Spectral NMR diagnostics is an important tool for studies of complexes with hydrogen bonds, as oftentimes it is the main way to obtain reliable information about hydrogen bond geometry and strength [1–7]. At the moment, there are a number of correlational equations linking the numerical value of spectral NMR parameters and the geometry/energy of a single hydrogen bond [8]. Some of them are well established and widely used, and their applicability was tested for multiple set of complexes, while others are applicable only to particular complexes and conditions. Among the parameters whose change upon the formation of a hydrogen-bonded complex is correlated with the strength of a X–H···Y hydrogen bond, one can distinguish two major groups. The parameters of the first group are characterized by a monotonous change along the proton transfer coordinate. As representatives of this group, one can name the chemical shifts of heavy nuclei δ_X and δ_Y and the spin–spin coupling constants $^1J_{XH}$ and $^1J_{HY}$ [9–13]. Parameters from the second group exhibit extremal values for the shortest (strongest) hydrogen bond. As an example, one could mention the chemical shift of the bridging proton δ_H or the spin–spin coupling constant $^{2h}J_{XY}$, whose change along the proton transfer coordinate can be characterized by a bell-shaped curve [3–5,14–17]. Parameters from the first group are preferable when solving the inverse spectral problem as their value unequivocally corresponds to a single geometry of a hydrogen bond. For parameters from the second group, there is an uncertainty that is caused by the fact that a single value of the parameter can be observed for two configurations of a hydrogen bond.

Despite the successful development of numerous methods of spectral diagnostics for estimating the geometry and strength of hydrogen bonds, there are some relevant issues that remain to be solved. For example, it is not known how the inverse spectral problem can be solved for systems with multiple mutually interacting hydrogen bonds, since the magnitude of each spectral parameter can be influenced by the presence of all hydrogen bonds simultaneously.

The aim of this computational work is to demonstrate the possibility of solving the inverse spectral problem for a system with two coupled hydrogen bonds, in particular, to find such pairs of spectral NMR variables that could be used for an unequivocal evaluation of geometries of a pair of mutually influencing hydrogen bonds. As a model system, we consider a 4-hydroxypyridine anion, which can form two hydrogen bonds as a hydrogen bond acceptor (from the oxygen and nitrogen side) with two substituted methanol molecules (Figure 1).

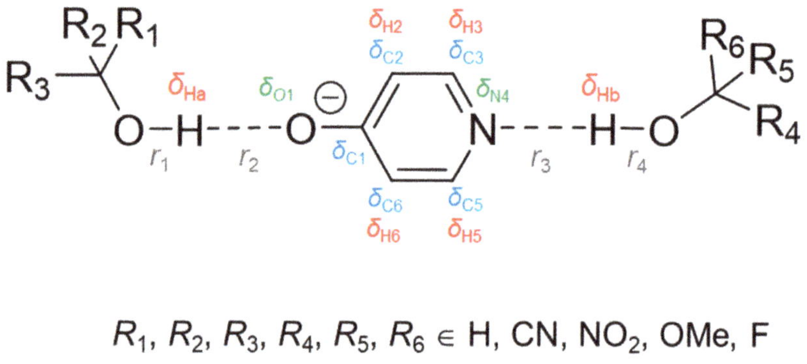

$R_1, R_2, R_3, R_4, R_5, R_6 \in H, CN, NO_2, OMe, F$

Figure 1. Schematic representation of investigated complexes with two hydrogen bonds formed by 4-hydroxypyridine anion as a hydrogen bond acceptor and two substituted methanols as donors. Geometric parameters (interatomic distances r_1, r_2, r_3 and r_4, grey) and spectral NMR parameters (chemical shifts of bridging protons δ_{Ha}, δ_{Hb}, red; atoms of 4-hydroxypyridine anion's ring: oxygen atom δ_{O1}, green; carbons $\delta_{C1}, \delta_{C2}, \delta_{C3}, \delta_{C5}, \delta_{C6}$, blue; hydrogens $\delta_{H2}, \delta_{H3}, \delta_{H5}, \delta_{H6}$, red, and nitrogen δ_{N4}, green) considered in this work are indicated.

Substitution was made by a subsequent replacement of hydrogen atoms by CN, NO_2, OMe or F groups. The resulting set of substituted methanols served as a source of proton donors of various strengths, regardless of their practical chemical stability. Table 1 shows the pattern of substitution used within each of the four sub-series of complexes (full list of substituents can be found in Table S1 in the Supporting Information). Together with the unsubstituted complex $R_1 = R_2 = R_3 = H$, $R_4 = R_5 = R_6 = H$, 61 complexes with two hydrogen bonds were considered in this work. In such systems, OHO and OHN hydrogen bonds can be of various strengths (from weak to moderate-strong, it is controlled by the choice of substituents). All hydrogen bonds are fairly linear (the range of OHO and OHN angles is 156°–179°, see Table S1) and are formed along the direction of the lone pair localization on proton-accepting O and N atoms. The central 4-hydroxypyridine anion can be negatively charged (OH···O⁻PyrN···HO), neutral (O⁻···HOPyrN···HO or OH···OPyrNH···O⁻) or even positively charged (O⁻···HOPyrNH⁺···O⁻), depending on proton positions in hydrogen bonds. Therefore, such a choice of model systems is suitable for the investigation of complexes with two mutually influencing hydrogen bonds (OHO and OHN) in a wide range of hydrogen bond geometries formed by neutral or charged species. Throughout the work, parameters associated with the OHO hydrogen bond will be denoted with an index "a" and those of OHN—with an index "b".

Table 1. List of substituents of 4-hydroxypyridine anion with two cyano-substituted methanol molecules. Pattern of substitution used for the selection of proton donors in complexes shown in Figure 1. X stands for one of the following substituents: CN, NO$_2$, OMe or F.

R_1	R_2	R_3	R_4	R_5	R_6
H	H	H	X	H	H
H	H	H	X	X	H
H	H	H	X	X	X
X	H	H	H	H	H
X	H	H	X	H	H
X	H	H	X	X	H
X	H	H	X	X	X
X	X	H	H	H	H
X	X	H	X	H	H
X	X	H	X	X	H
X	X	H	X	X	X
X	X	X	H	H	H
X	X	X	X	H	H
X	X	X	X	X	H
X	X	X	X	X	X

The proposed concept for simultaneous solution of the inverse spectral problem for both coupled hydrogen bonds is as follows. In order to determine the geometries of two hydrogen bonds, two spectral parameters are needed. A particular value of a chosen spectral parameter corresponds to multiple combinations of the two hydrogen bond geometries, forming an isoline on a distribution map of a spectral parameter as a function of two proton transfer coordinates (Figure 2, top). However, if one measures two spectral parameters, the intersection of two isolines (Figure 2, bottom) gives the geometry of the two hydrogen bonds in a coupled system. The case with a single intersection of two isolines is preferable and will be used in the following discussion as the criterion for a choice of spectral parameters for solving the inverse spectral problem.

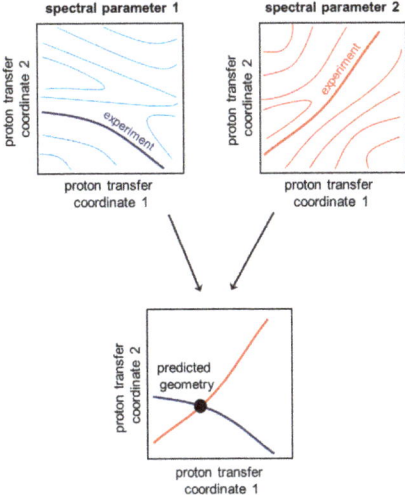

Figure 2. Schematic representation of the algorithm of inverse spectral problem solving. The isolines are drawn arbitrarily and do not correspond to any particular case.

2. Results and Discussion

The geometric parameters (interatomic distances r_1, r_2, r_3 and r_4) of hydrogen bonds in investigated complexes are collected in Table S1. The set of complexes covers a wide range of hydrogen bonds geometries—there are complexes without proton transfer in both hydrogen bonds (OH$\cdots$O$^-$PyrN$\cdots$HO), complexes with proton transfer in one of the hydrogen bonds (O$^-\cdots$HOPyrN$\cdots$HO or OH$\cdots$OPyrNH$\cdots$O$^-$) and complexes with two hydrogen bonds with proton transfer (O$^-\cdots$HOPyrNH$^+\cdots$O$^-$). The dependencies of the $q_{2a} = r_1 + r_2$ coordinate on $q_{1a} = 0.5 \cdot (r_1 - r_2)$ for the OHO hydrogen bond and the $q_{2b} = r_4 + r_3$ coordinate on $q_{1b} = 0.5 \cdot (r_4 - r_3)$ for the OHN hydrogen bond are shown in Figure 3. It is seen that the complexes cover a range of q_1 from ca. -0.4 to ca. 0.35 Å forming a parabolic curve.

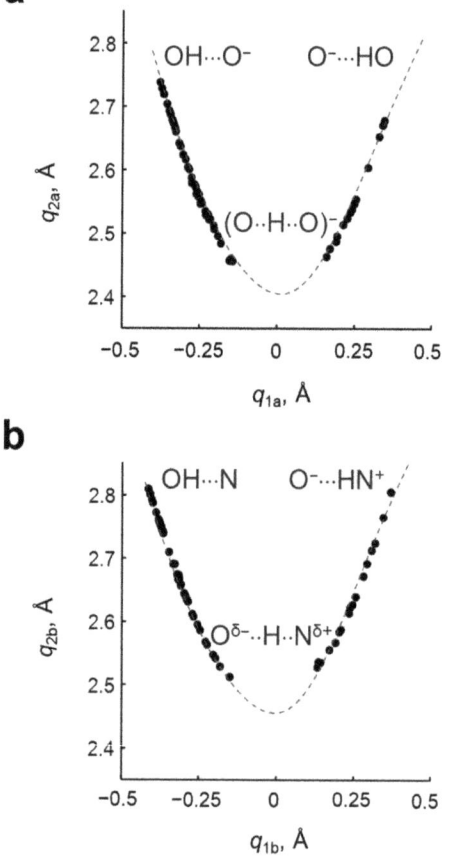

Figure 3. Dependencies of the (**a**) $q_{2a} = r_1 + r_2$ coordinate on $q_{1a} = 0.5 \cdot (r_1 - r_2)$ for OHO hydrogen bond; dependencies of the (**b**) $q_{2b} = r_4 + r_3$ coordinate on $q_{1b} = 0.5 \cdot (r_4 - r_3)$ for OHN hydrogen bond. Dashed lines are guides for the eye.

Despite the fact that OHO and OHN hydrogen bonds are divided in space by 4-hydroxypyridine anion, they "feel" the presence of each other through the electronic system of the complex as a whole, the phenomenon is called cooperativity. It manifests as the changes of the proton position in one hydrogen bond depending on the proton position in another hydrogen bond. For example, if one fixes the substituents R_1, R_2 and

R_3 in the proximity of the OHO hydrogen bond (see a set of points of the same shape and color in Figure 4), the variation of substituents in OHN hydrogen bond causes a change in q_{1a}. Complexes with a "tail-to-tail" (OH···O⁻PyrN···HO) and a "head-to-head" (O⁻···HOPyrNH⁺···O⁻) configuration are anti-cooperative (blue areas in Figure 4): i.e., the strengthening of one hydrogen bond causes the weakening of another hydrogen bond. For complexes with "head-to-tail geometry" (O⁻···HOPyrN···HO and OH···OPyrNH···O⁻) (green areas in Figure 4) cooperative effects are observed: i.e., the strengthening of one hydrogen bond causes the strengthening of another. Thus, the character of mutual influence of geometries of OHO and OHN hydrogen bonds in such a system can be either cooperative or anti-cooperative.

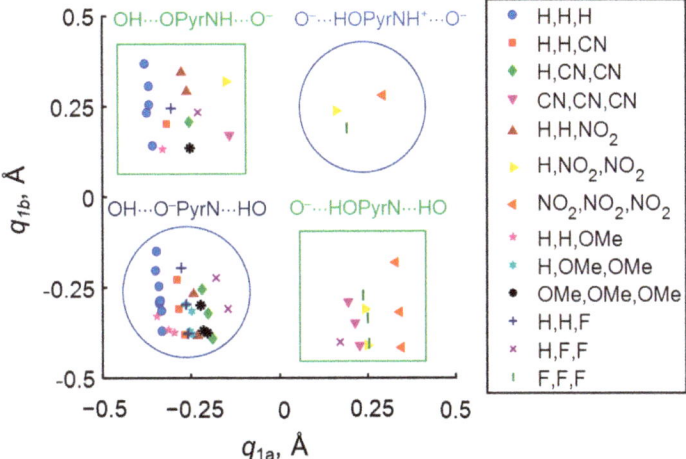

Figure 4. The dependence of proton position in the NH···O hydrogen bond q_{1b} on the proton position in the OH···O hydrogen bond q_{1a}. The shape and color of a marker indicate a series of complexes with the same set of substituents R_1, R_2 and R_3 (shown in legend) and varying set of substituents R_4, R_5 and R_6. Green areas correspond to cooperative hydrogen bonds and blue areas to anti-cooperative ones.

Among the NMR parameters of the 4-hydroxypyridine anion, which potentially are sensitive to the geometries of OH···O and OH···N hydrogen bonds, chemical shifts of the nuclei located in the proximity of hydrogen bonds, i.e., carbon (C1) and nitrogen (N4), are the first candidates. The changes of carbon and nitrogen chemical shifts upon complexation, $\Delta\delta_{C1}$ and $\Delta\delta_{N4}$, are shown in Figure 5.

In Figure 5, it is clearly seen that the C1 carbon chemical shift changes with the change of geometries of both hydrogen bonds—upon moving the bridging proton along OHO hydrogen bond (with fixed geometry of OHN hydrogen bond), $\Delta\delta_{C1}$ changes by up to 8 ppm, and moving it along the OHN hydrogen bond causes a change of $\Delta\delta_{C1}$ by up to 2.5 ppm. In other words, the steepness of the slope of the $\Delta\delta_{C1}(q_{1a}, q_{1b})$ surface is larger in the direction of q_{1a} than in the direction of q_{1b}. Contrarily, the nitrogen chemical shift is more sensitive to the geometry of the closest (OHN) hydrogen bond. The shape of the isolines of the distributions of $\Delta\delta_{C1}$ and $\Delta\delta_{N4}$ makes the unequivocal solving of the inverse spectral problem possible for most cases (except for $q_{1a} \approx -0.3$ Å due to the shape of the $\Delta\delta_{C1}$ isolines in this region caused by a presence of a hill with its maximal value). Together with the high sensitivity discussed above, this makes the pair $\Delta\delta_{C1}$ and $\Delta\delta_{N4}$ quite promising parameters. However, the experimental issues of measuring nitrogen chemical shift (low natural abundance of ^{15}N and line broadening due to quadrupolar interactions in ^{14}N NMR spectra) encourage us to discuss additional NMR parameters.

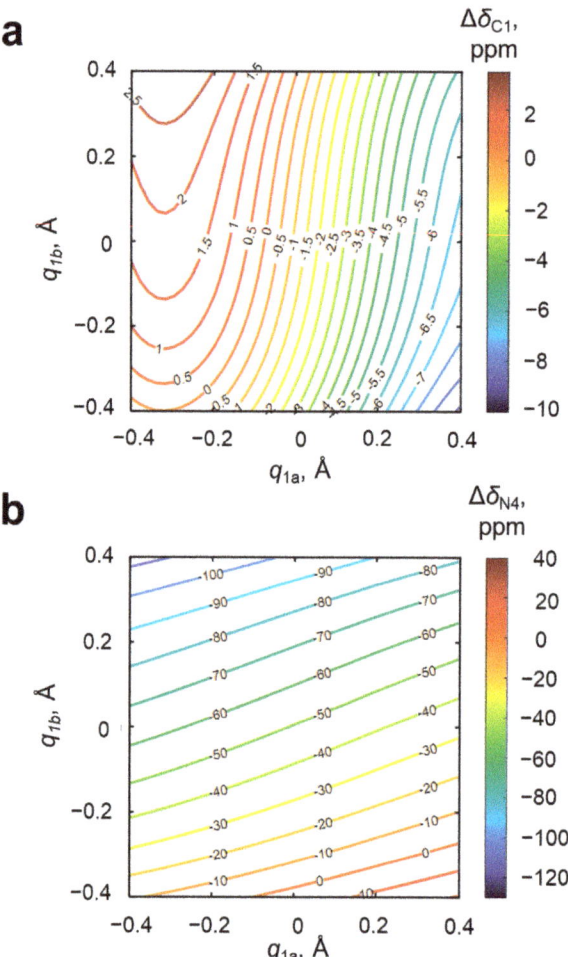

Figure 5. Distributions of the change upon complexation of chemical shift of (**a**) the C1 atom $\Delta\delta_{C1}$ and (**b**) the N4 atom $\Delta\delta_{N4}$ along q_1 coordinates for OHO (q_{1a}) and OHN (q_{1b}) hydrogen bonds. The coefficients a, b_1, c_1, d_1, b_2, c_2 and d_2 (Equation (1)) and R^2 are given in Table S2. Isolines are drawn with a step of 0.5 and 10 ppm, respectively.

The next pair of "promising" NMR parameters are the changes of chemical shifts of bridging protons in the OHO and OHN hydrogen bonds upon complexation, $\Delta\delta_{Ha}$ and $\Delta\delta_{Hb}$; their dependences on q_{1a} and q_{1b} are shown in Figure 6. Along the OHO hydrogen bond, $\Delta\delta_{Ha}$ increases from 8 to 17 ppm for $q_{1a} < 0$, reaches maximal value at $q_{1a} \approx 0$ and then decreases to 0 ppm. The change of the geometry of the OHN hydrogen bond (q_{1b}) does not significantly influence $\Delta\delta_{Ha}$. A similar situation is observed for $\Delta\delta_{2b}$: moving along the OHN hydrogen bond (q_{1b}) changes $\Delta\delta_{Hb}$ from 9 to 18 ppm for $q_{1b} \approx 0$ and then down to 0 ppm. It is clearly seen that both surfaces have a hill of maximal values of $\Delta\delta_H$ and slopes in the direction of one of q_1 axes (q_{1a} for $\Delta\delta_{Ha}$ and q_{1b} for $\Delta\delta_{Hb}$). It means that the magnitude of the bridging proton chemical shift in a given hydrogen bond is determined almost exclusively by the geometry and electronic features of the three atoms forming this bond. In other words, the chemical shifts of the two bridging protons are not coupled. For the purposes of solving the reverse spectral problem, this is an advantage, because the magnitude of each chemical shift can be used for the independent evaluation of hydrogen bond geometries.

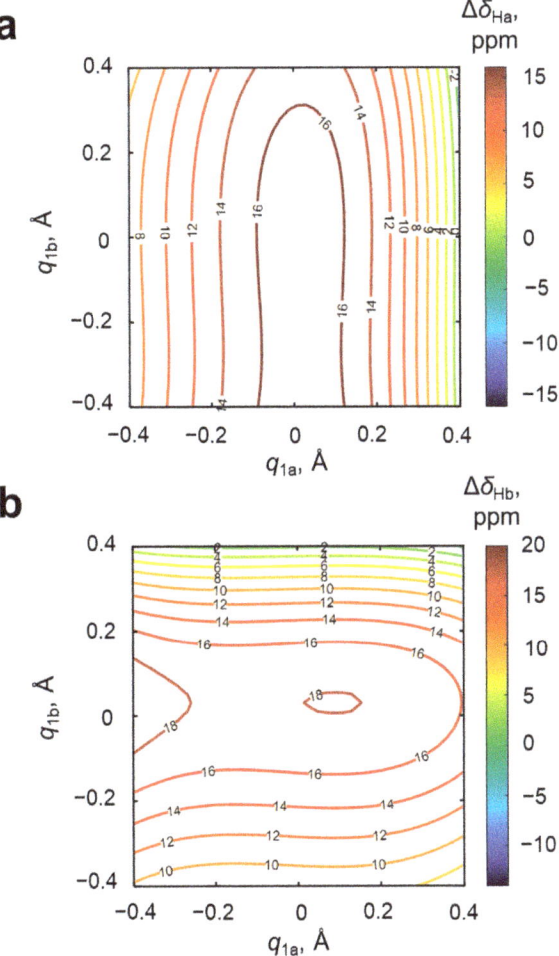

Figure 6. Distributions of the change upon complexation of chemical shift of bridging protons (**a**) in the OH···O hydrogen bond $\Delta\delta_{Ha}$ and (**b**) in the OH···N hydrogen bond $\Delta\delta_{Hb}$ along q_1 coordinates for OHO (q_{1a}) and OHN (q_{1b}) hydrogen bonds. The coefficients a, b_1, c_1, d_1, b_2, c_2 and d_2 (Equation (1)) and R^2 are given in Table S2. Isolines are drawn with a step of 2 ppm.

However, there are two principal problems with such an approach. The first problem was mentioned in introduction and is caused by the fact that a particular value of one of the proton chemical shifts ($\Delta\delta_{Ha}$ or $\Delta\delta_{Hb}$) corresponds to an isoline in distributions shown in Figure 6a,b that form a hairpin curve. Two hairpin isolines have four intersecting points (see Figure 7), i.e., each set of $\Delta\delta_{Ha}$ and $\Delta\delta_{Hb}$ values could correspond to four alternative hydrogen bond geometries. The second problem with using $\Delta\delta_{Ha}$ and $\Delta\delta_{Hb}$ for the evaluation of hydrogen bond geometries is due to the fact that, in the experimental spectrum, it is difficult to distinguish which signal will relate to the OHO and which to the OHN hydrogen bond. This issue demands the usage of additional NMR parameters, for example, chemical shifts of the nuclei of 4-hydroxypyridine.

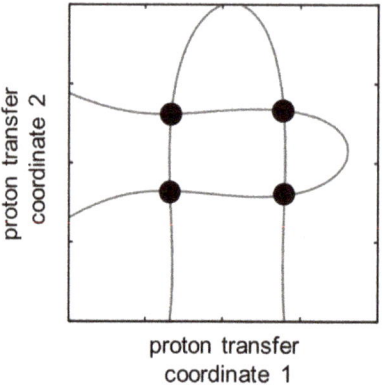

Figure 7. Schematic representation of the non-unequivocal inverse spectral problem solving. The black dots correspond to the four possible solutions. NMR parameters of the 4-hydroxypyridine anion, the chemical shift of carbon δ_{C2}, δ_{C3}, δ_{C5}, δ_{C6} and the hydrogen atoms δ_{H2}, δ_{H3}, δ_{H5}, δ_{H6}, were also analyzed for their sensitivity and applicability for the evaluation of the geometries of OHO and OHN hydrogen bonds. We found that the most promising parameters are arithmetically averaged chemical shifts of C2 and C6 atoms (δ_{C26}), C3 and C5 atoms (δ_{C35}) and H3 and H5 atoms (δ_{H35}). The distributions of changes of these parameters are shown in Figure 8. Both carbon chemical shifts, $\Delta\delta_{C26}$ and $\Delta\delta_{C35}$, are more sensitive to the closest hydrogen bond (OHO and OHN, respectively). The range of values of $\Delta\delta_{C26}$ was about 10 ppm, $\Delta\delta_{C35}$—15 ppm, which makes both of them suitable for the accurate evaluation of hydrogen bond geometries. The change of δ_{H35} within the change of the OHO and OHN hydrogen bonds geometries slightly exceeded 1 ppm. The topology of these three surfaces is such that almost all combinations of pair of isolines of $\Delta\delta_{C26}$, $\Delta\delta_{C35}$ and $\Delta\delta_{H35}$ have a single intersection point, thus making the solution of an inverse spectral problem unequivocal.

For an additional estimation of the mutual influence of OHO and OHN hydrogen bonds, calculations for two extra sets of systems with a single hydrogen bond were performed (4-hydroxypyridine anion with one substituted methanol from either the oxygen or nitrogen side, 10 complexes in each extra set). The results the of NMR calculations (δ_{C26} and δ_{C35}) are presented in Figures S1 and S2 of the Supporting Information. The difference plot between data shown in Figure 8 (system with two coupled hydrogen bonds) and Figures S1 and S2 (hypothetical system with no coupling between hydrogen bond modeled as a sum of systems with a single hydrogen bond) is shown in Figure S3. It is clearly seen that the cooperativity effects on the NMR parameters are substantial (± 2–4 ppm, making them non-negligible) and non-monotonous. The strongest effects are observed for the configuration $O^-\cdots HOPyrNH^+\cdots O^-$.

In order to test the approach proposed in this work, we performed additional calculations of structure and NMR parameters for three arbitrarily chosen complexes of 4-hydroxypyridine anion with CH_3CFHOH and CF_3CFHOH molecules: **1** ($CH_3CFHOH\cdots^-OPyrN\cdots HOCFHCF_3$), **2** ($CF_3CFHOH\cdots^-OPyrN\cdots HOCFHCF_3$) and **3** ($CF_3CFHOH\cdots^-OPyrN\cdots HOCFHCH_3$); the optimized structures of **1**–**3** are shown in Figure 9. The results of the comparison of the "predicted" geometry based on pairs of NMR chemical shifts and the "real" (calculated) geometries of the two hydrogen bonds are summarized in Table 2 and shown in Figures S4–S6 in the Supporting Information. It can be concluded that hydrogen bond geometries estimated using two-dimensional correlations differ from those directly calculated by quantum-chemical methods by not more than 0.04 Å. The only exception is complex **3** (CF_3CFHOH and CF_3CFHOH), for which the chemical shifts of carbons C2 and C6 differ significantly due to the presence of an additional hydrogen bond between the CH-group of the 4-hydroxypyridine anion and the fluorine atom of the substituted methanol (see Figure S7). In this case, using the chemical shift of carbon atom not involved in additional hydrogen bonding instead of the arithmetically averaged chemical shift $\Delta\delta_{C26}$ is more appropriate.

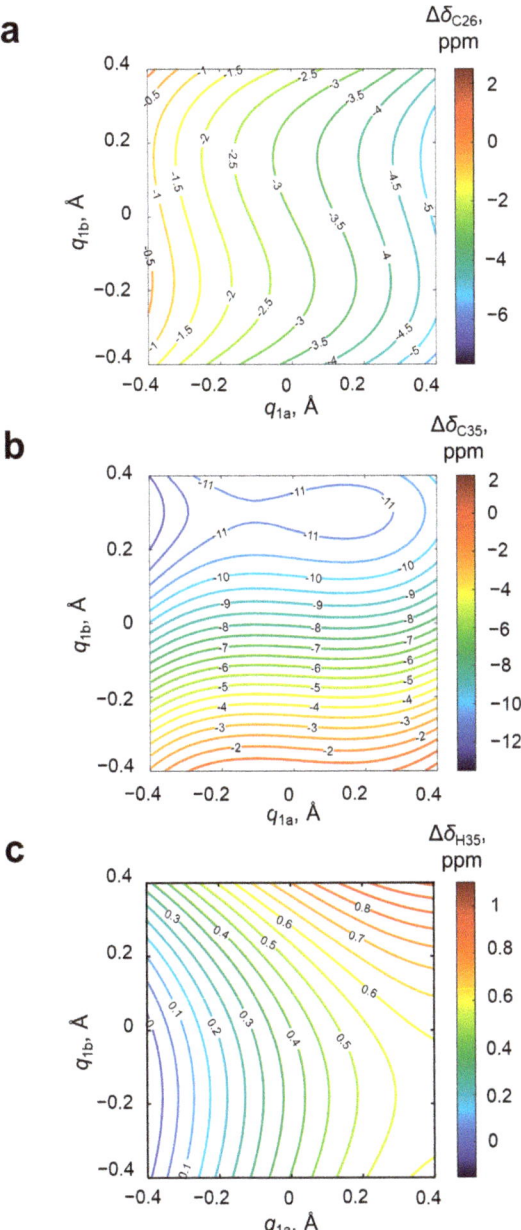

Figure 8. Distributions of the change upon complexation of average chemical shift of (**a**) C2 and C6 atoms $\Delta\delta_{C26}$, (**b**) C3 and C5 atoms $\Delta\delta_{C35}$ and (**c**) H3 and H5 atoms $\Delta\delta_{H35}$ along q_1 coordinates for OHO (q_{1a}) and OHN (q_{1b}) hydrogen bonds. The coefficients a, b_1, c_1, d_1, b_2, c_2 and d_2 (Equation (1)) and R^2 are given in Table S2. Isolines are drawn with a step of 0.5 ppm for (**a**,**b**) and 0.05 ppm for (**c**), respectively.

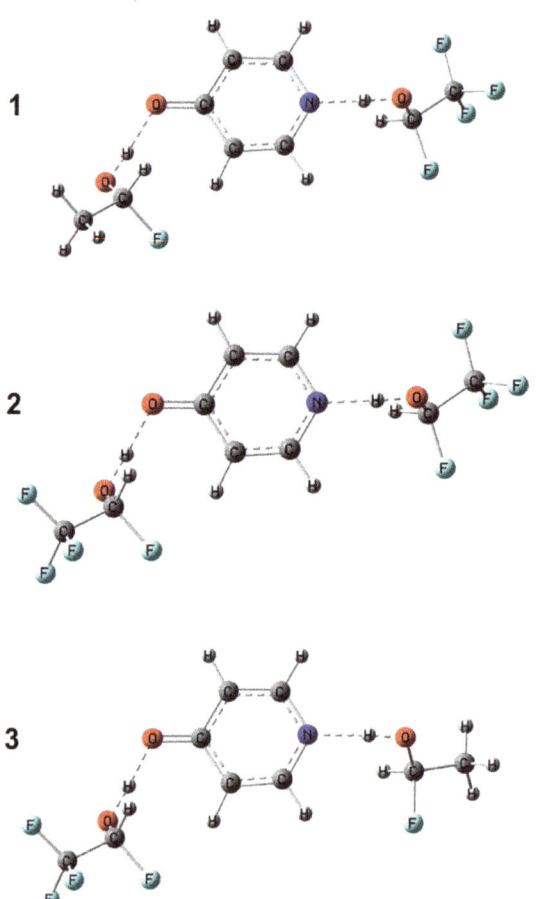

Figure 9. Structures of additional complexes with two hydrogen bonds formed by 4-hydroxypyridine anion as a hydrogen bond acceptor and two substituted methanols as donors used for testing the proposed approach. From top to bottom: **1** (proton donors CH$_3$CHFOH, CF$_3$CHFOH), **2** (proton donors CH$_3$CHFOH, CF$_3$CHFOH), **3** (proton donors CF$_3$CHFOH, CH$_3$CHFOH).

Table 2. Results of testing the proposed approach for three additional complexes, shown in Figure 8: **1** (proton donors CH$_3$CHFOH, CF$_3$CHFOH), **2** (proton donors CH$_3$CHFOH, CF$_3$CHFOH), **3** (proton donors CF$_3$CHFOH, CH$_3$CHFOH).

Complex (Proton Donors)	Calculated Geometry	Pair of Parameters for Prediction	Predicted Geometry
1 (CH$_3$CHFOH, CF$_3$CHFOH)	$q_{1a} = -0.27$ Å, $q_{1b} = -0.20$ Å	$\Delta\delta_{C1} = 1.2$ ppm, $\Delta\delta_{N4} = -37.0$ ppm	$q_{1a} = -0.29$ Å, $q_{1b} = -0.21$ Å
		$\Delta\delta_{C26} = -1.2$ ppm, $\Delta\delta_{C35} = -5.5$ ppm	$q_{1a} = -0.30$ Å, $q_{1b} = -0.17$ Å
		$\Delta\delta_{C35} = -5.5$ ppm, $\Delta\delta_{H35} = 0.03$ ppm	$q_{1a} = -0.32$ Å, $q_{1b} = -0.19$ Å

Table 2. Cont.

Complex (Proton Donors)	Calculated Geometry	Pair of Parameters for Prediction	Predicted Geometry
2 (CF$_3$CHFOH, CF$_3$CHFOH)	$q_{1a} = -0.19$ Å, $q_{1b} = -0.23$ Å	$\Delta\delta_{C1} = 0.6$ ppm, $\Delta\delta_{N4} = -29.4$ ppm	$q_{1a} = -0.20$ Å, $q_{1b} = -0.23$ Å
		$\Delta\delta_{C26} = -2.2$ ppm, $\Delta\delta_{C35} = -4.6$ ppm	$q_{1a} = -0.15$ Å, $q_{1b} = -0.20$ Å
		$\Delta\delta_{C35} = -4.6$ ppm, $\Delta\delta_{H35} = 0.22$ ppm	$q_{1a} = -0.18$ Å, $q_{1b} = -0.20$ Å
3 (CF$_3$CHFOH, CH$_3$CHFOH)	$q_{1\_a} = -0.17$ Å, $q_{1\_b} = -0.31$ Å	$\Delta\delta_{C1} = 0.0$ ppm, $\Delta\delta_{N4} = -14.8$ ppm	$q_{1a} = -0.20$ Å, $q_{1b} = -0.33$ Å
		$\Delta\delta_{C26} = -3.0$ ppm, $\Delta\delta_{C35} = -3.0$ ppm	$q_{1a} = -0.01$ Å, $q_{1b} = -0.28$ Å
		$\Delta\delta_{C35} = -3.0$ ppm, $\Delta\delta_{H35} = 0.24$ ppm	$q_{1a} = -0.16$ Å, $q_{1b} = -0.27$ Å

3. Computational Details

Geometry optimization was performed using second-order Moller–Plesset perturbation theory (MP2) [12,13] with Dunning' correlation-consistent polarized double-ζ basis set with diffuse functions aug-cc-pVDZ [14]. All calculated geometries were checked for the absence of imaginary frequencies. Chemical shieldings were calculated using DFT (B3LYP) with the augmented polarization consistent triple-ζ basis set aug-pcS-2, which is specially designed for the calculation of shieldings at the DFT level with a high accuracy [18].

Calculations were carried out using the Gaussian16 software. Computational resources were provided by the Computer Center of Saint Petersburg University Research Park (http://www.cc.spbu.ru/, accessed on 1 April 2020).

Visualization was performed in GaussView 6.0 and MATLAB 2021b software packages.

For the description of the geometries of OHO and OHN hydrogen bonds, the parameters q_1 and q_2 were used [2,16,17]: $q_1 = 0.5 \cdot (r_{OH} - r_{HY})$, $q_2 = r_{OH} + r_{HY}$, where Y = O, N. The meaning of q_1 and q_2 coordinates is pretty clear for linear hydrogen bonds—the q_1 coordinate is the shift of the hydrogen atom from the hydrogen bond center, the q_2 coordinate is the total length of the hydrogen bond (for strictly linear hydrogen bonds, q_2 is the distance between the heavy atoms O...O or O...N; it should be noted that, as hydrogen bonds deviate from linearity, q_2 loses this geometrical meaning, because in this case q_2 is slightly longer than the distance O...O or O...N). For the OHO hydrogen bond, the following pair of coordinates was used $q_{1a} = 0.5 \cdot (r_1 - r_2)$ and $q_{2a} = r_1 + r_2$ and for OHN, they were $q_{1b} = 0.5 \cdot (r_4 - r_3)$ and $q_{2b} = r_4 + r_3$, respectively.

The algorithm of construction for the distributions of spectral parameters ($\Delta\delta_{C1}$, $\Delta\delta_{N4}$ and others discussed in this work) along the q_{1a} and q_{1b} was as follows. For a set of complexes shown in Figure 1, NMR parameters were calculated. Let us denote the spectral parameter as f. The value of a change of a spectral parameter upon complexation is defined as $\Delta f = f_{complex} - f_{free}$ (where f_{free} is the value of a given spectral parameter for an isolated 4-hydroxypyridine anion or methanol, and $f_{complex}$ is the value of the same parameter within a hydrogen-bonded complex). Calculated Δf values were approximated as a function of q_{1a} and q_{1b} using the Curve Fitting Tool implemented in the MATLAB 2021b software package by a polynomial function of a third-degree along q_{1a} and q_{1b} without cross-terms:

$$\Delta f(q_{1a}, q_{2b}) = a + b_1 q_{1a} + c_1 q_{1a}^2 + d_1 q_{1a}^3 + b_2 q_{1b} + c_2 q_{1b}^2 + d_2 q_{1b}^3 \qquad (1)$$

The coefficients a, b_1, c_1, d_1, b_2, c_2 and d_2 for all discussed parameters and the coefficient of determination R^2 for each approximation are given in Table S2 in the Supporting Information. The resulting functions were plotted as contour plots, in which each isoline corresponds to the particular value of Δf and is marked by a color and an isovalue for clarity in the following figures. The polynomials were taken as one of the simplest forms that could describe the strongly non-monotonous behavior of spectral parameters. There is no deeper reason for this choice, and for the fitting purposes, other functions could be selected if the data points would allow doing so.

4. Conclusions

In this work, the possibility of solving the inverse spectral problem for a system with two coupled hydrogen bonds was demonstrated on the example of 61 complexes formed by 4-hydroxypyridine anion with two substituted methanols with OHO and OHN hydrogen bonds. The algorithm for using two-dimensional correlations is as follows: for measured or calculated change of two given spectral parameters, one should find isolines on their corresponding plots and overlap them (see Figure 2). The coordinates of the intersection point of the two isolines will give the proton positions in both hydrogen bonds. It was demonstrated that any pair of parameters $\Delta\delta_{N4}$, $\Delta\delta_{C1}$, $\Delta\delta_{C26}$, $\Delta\delta_{C35}$ and $\Delta\delta_{H35}$ is suitable for the evaluation of hydrogen bond geometries. However, $\Delta\delta_{C1}$, $\Delta\delta_{C26}$, $\Delta\delta_{C35}$ and $\Delta\delta_{H35}$ are more easily available from an experiment.

Thus far, we have only tested our approach computationally on three examples of complexes. The applicability of the method "in practice" awaits experimental verification by future researchers. Indeed, the proposed approach has obvious limitations. Firstly, such 2D maps are constructed for each compound individually (in our case for 4-hydroxypyridine anion). Secondly, the experimental application of the approach requires the measurements of signals that are not averaged out by fast (in the NMR time scale) molecular exchange—a condition that is not necessarily satisfied for intermolecular complexes. The coexistence of several types of molecular complexes in a solution at the same time can also complicate the solving of the reverse spectral problem. Thirdly, the accuracy of the 2D chemical shift maps worsens in the presence of additional non-covalent interactions between the proton donors and the central proton-accepting molecule. Fourthly, for other systems, multiple extrema in the 2D maps of spectral parameters can make determination of hydrogen bonds geometries non-unequivocal, as shown in Figure 7. An unfortunate combination of all four factors could, of course, render the proposed approach totally unreliable. Nevertheless, we have shown in principle that, for systems where these limitations are absent or can be neglected, the NMR spectral data can be sufficient to solve the two-dimensional reverse spectral problem even with a significant entanglement of spectral parameters.

The accuracy of geometry estimations with this approach for systems with two OHO and OHN hydrogen bonds (and without additional interactions) is in the range ± 0.04 Å.

Supplementary Materials: The following supporting information can be downloaded at: https://www.mdpi.com/article/10.3390/molecules27123923/s1, Table S1:Geometric parameters of OHO and OHN hydrogen bonds; Table S2: Coefficients a, b_1, c_1, d_1, b_2, c_2 and d_2 for each discussed spectral parameter, R^2 for each approximation; Figure S1:$\Delta\delta_{C35}$ for complexes of hydroxypyridine anion with a single substituted methanol molecule from an oxygen side (OHO hydrogen bond) $\Delta\delta_{C35a}$ and from a nitrogen site (OHN hydrogen bond) $\Delta\delta_{C35b}$. Distributions of the sum $\Delta\delta_{C35} = \Delta\delta_{C35a} + \Delta\delta_{C25b}$ along q_{1a} and q_{1b} hydrogen bonds; Figure S2: $\Delta\delta_{C26}$ for complexes of hydroxypyridine with a single substituted methanol molecule from an oxygen side (OHO hydrogen bond) $\Delta\delta_{C26a}$ and from a nitrogen site (OHN hydrogen bond) $\Delta\delta_{C26b}$. Distributions of the sum $\Delta\delta_{C26\_} = \Delta\delta_{C26a} + \Delta\delta_{C26b}$ along q_{1a} and q_{1b} hydrogen bonds; Figure S3: $\Delta\delta_{C26}$–$\Delta\delta_{C26}$, $\Delta\delta_{C35}$–$\Delta\delta_{C35\_}$ along q_{1a} and q_{1b}; Figure S4: Solution of the inverse spectral problem for $R_1 = H$, $R_2 = F$, $R_3 = CH_3$, $R_4 = H$, $R_5 = F$, $R_6 = CF_3$; Figure S5: Solution of the inverse spectral problem for $R_1 = H$, $R_2 = F$, $R_3 = CF_3$, $R_4 = H$, $R_5 = F$, $R_6 = CF_3$; Figure S6: Solution of the inverse spectral problem for $R_1 = H$, $R_2 = F$, $R_3 = CF_3$, $R_4 = H$, $R_5 = F$, $R_6 = CH_3$; Figure S7: Optimized geometry of complex 3 with CF_3CFOH and CH_3CFHOH as hydrogen bond donors.

Author Contributions: Conceptualization, P.M.T. and M.V.S.; methodology, E.Y.T.; validation, P.M.T. and M.V.S.; formal analysis, E.Y.T. and M.V.S.; investigation, E.Y.T. and M.V.S.; data curation, E.Y.T.; writing—original draft preparation, E.Y.T.; writing—review and editing, P.M.T. and M.V.S.; visualization, E.Y.T.; funding acquisition, P.M.T. All authors have read and agreed to the published version of the manuscript.

Funding: This research was funded by Russian Science Foundation grant number 18-13-00050.

Institutional Review Board Statement: Not applicable.

Informed Consent Statement: Not applicable.

Data Availability Statement: Data is contained within the article or Supplementary Material.

Acknowledgments: This study was supported by the Russian Science Foundation.

Conflicts of Interest: The authors declare no conflict of interest.

Sample Availability: Samples of the compounds are not available from the authors.

References

1. Arunan, E.; Desiraju, G.R.; Klein, R.A.; Sadlej, J.; Scheiner, S.; Alkorta, I.; Clary, D.C.; Crabtree, R.H.; Dannenber, J.J.; Hobza, P.; et al. Definition of the Hydrogen Bond (IUPAC Recommendations 2011). *Pure Appl. Chem.* **2011**, *83*, 1637–1641. [CrossRef]
2. Jeffrey, G.A.; Yeon, Y. The Correlation between Hydrogen-Bond Lengths and Proton Chemical Shifts in Crystals. *Acta Crystallogr. Sect. B Struct. Sci.* **1986**, *42*, 410–413. [CrossRef]
3. Lorente, P.; Shenderovich, I.G.; Golubev, N.S.; Denisov, G.S.; Buntkowsky, G.; Limbach, H.-H. 1H/15N NMR Chemical Shielding, Dipolar15N,2H Coupling and Hydrogen Bond Geometry Correlations in a Novel Series of Hydrogen-Bonded Acid-Base Complexes of Collidine with Carboxylic Acids. *Org. Magn. Reson.* **2001**, *39*, S18–S29. [CrossRef]
4. Benedict, H.; Shenderovich, I.G.; Malkina, O.L.; Malkin, V.G.; Denisov, G.S.; Golubev, N.S.; Limbach, H.H. Nuclear Scalar Spin-Spin Couplings and Geometries of Hydrogen Bonds. *J. Am. Chem. Soc.* **2000**, *122*, 1979–1988. [CrossRef]
5. Limbach, H.-H.; Tolstoy, P.M.; Pérez-Hernández, N.; Guo, J.; Shenderovich, I.G.; Denisov, G.S. OHO Hydrogen Bond Geometries and NMR Chemical Shifts: From Equilibrium Structures to Geometric H/D Isotope Effects, with Applications for Water, Protonated Water, and Compressed Ice. *Isr. J. Chem.* **2009**, *49*, 199–216. [CrossRef]
6. Tupikina, E.Y.; Denisov, G.S.; Melikova, S.M.; Kucherov, S.Y.; Tolstoy, P.M. New Look at the Badger-Bauer Rule: Correlations of Spectroscopic IR and NMR Parameters with Hydrogen Bond Energy and Geometry. FHF Complexes. *J. Mol. Struct.* **2018**, *1164*, 129–136. [CrossRef]
7. Giba, I.S.; Tolstoy, P.M. Self-Assembly of Hydrogen-Bonded Cage Tetramers of Phosphonic Acid. *Symmetry* **2021**, *13*, 258. [CrossRef]
8. Tolstoy, P.M.; Tupikina, E.Y. IR and NMR Spectral Diagnostics of Hydrogen Bond Energy and Geometry. In *Spectroscopy and Computation of Hydrogen-Bonded Systems*; Wojcik, M.J., Ozaki, Y., Eds.; Wiley: Hoboken, NJ, USA, 2022.
9. Ando, S.; Ando, I.; Shoji, A.; Ozaki, T. Intermolecular Hydrogen-Bonding Effect on Carbon-13 NMR Chemical Shifts of Glycine Residue Carbonyl Carbons of Peptides in the Solid State. *J. Am. Chem. Soc.* **2002**, *110*, 3380–3386. [CrossRef]
10. Guo, J.; Tolstoy, P.M.; Koeppe, B.; Golubev, N.S.; Denisov, G.S.; Smirnov, S.N.; Limbach, H.H. Hydrogen Bond Geometries and Proton Tautomerism of Homoconjugated Anions of Carboxylic Acids Studied via H/D Isotope Effects on 13C NMR Chemical Shifts. *J. Phys. Chem. A* **2012**, *116*, 11180–11188. [CrossRef] [PubMed]
11. Mulloyarova, V.V.; Giba, I.S.; Denisov, G.S.; Ostras', A.S.; Tolstoy, P.M. Conformational Mobility and Proton Transfer in Hydrogen-Bonded Dimers and Trimers of Phosphinic and Phosphoric Acids. *J. Phys. Chem. A* **2019**, *123*, 6761–6771. [CrossRef] [PubMed]
12. Tolstoy, P.M.; Smirnov, S.N.; Shenderovich, I.G.; Golubev, N.S.; Denisov, G.S.; Limbach, H.H. NMR Studies of Solid State—Solvent and H/D Isotope Effects on Hydrogen Bond Geometries of 1:1 Complexes of Collidine with Carboxylic Acids. *J. Mol. Struct.* **2004**, *700*, 19–27. [CrossRef]
13. Del Bene, J.E.; Bartlett, R.J.; Elguero, J. Interpreting 2hJ(F,N), 1hJ(H,N) and 1J(F,H) in the Hydrogen-Bonded FH–Collidine Complex. *Magn. Reson. Chem.* **2002**, *40*, 767–771. [CrossRef]
14. Harris, T.K.; Zhao, Q.; Mildvan, A.S. NMR Studies of Strong Hydrogen Bonds in Enzymes and in a Model Compound. *J. Mol. Struct.* **2000**, *552*, 97–109. [CrossRef]
15. Del Bene, J.E.; Alkorta, I.; Elguero, J. Spin-spin coupling across intramolecular N-H+-N hydrogen bonds in models for proton sponges: An ab initio investigation. *Magn. Reson. Chem.* **2008**, *46*, 457–463. [CrossRef] [PubMed]
16. Del Bene, J.E.; Elguero, J. Systematic Ab Initio Study of 15N 15N and 15N 1H Spin-Spin Coupling Constants across N-H +-N Hydrogen Bonds: Predicting N-N and N-H Coupling Constants and Relating Them to Hydrogen Bond Type. *J. Phys. Chem. A* **2006**, *110*, 7496–7502. [CrossRef] [PubMed]
17. Shenderovich, I.G.; Tolstoy, P.M.; Golubev, N.S.; Smirnov, S.N.; Denisov, G.S.; Limbach, H.H. Low-Temperature NMR Studies of the Structure and Dynamics of a Novel Series of Acid-Base Complexes of HF with Collidine Exhibiting Scalar Couplings across Hydrogen Bonds. *J. Am. Chem. Soc.* **2003**, *125*, 11710–11720. [CrossRef] [PubMed]
18. Jensen, F. Basis Set Convergence of Nuclear Magnetic Shielding Constants Calculated by Density Functional Methods. *J. Chem. Theory Comput.* **2008**, *4*, 719–727. [CrossRef] [PubMed]

Article

On the Coexistence of the Carbene···H-D Hydrogen Bond and Other Accompanying Interactions in Forty Dimers of N-Heterocyclic-Carbenes (I, IMe$_2$, I^{i}Pr$_2$, I^{t}Bu$_2$, IMes$_2$, IDipp$_2$, IAd$_2$; I = imidazol-2-ylidene) and Some Fundamental Proton Donors (HF, HCN, H$_2$O, MeOH, NH$_3$)

Mirosław Jabłoński

Faculty of Chemistry, Nicolaus Copernicus University in Toruń, ul. Gagarina 7, 87-100 Toruń, Poland; teojab@chem.umk.pl; Tel.: +48-056-611-4695

Citation: Jabłoński, M. On the Coexistence of the Carbene···H-D Hydrogen Bond and Other Accompanying Interactions in Forty Dimers of N-Heterocyclic-Carbenes (I, IMe$_2$, I^{i}Pr$_2$, I^{t}Bu$_2$, IMes$_2$, IDipp$_2$, IAd$_2$; I = imidazol-2-ylidene) and Some Fundamental Proton Donors (HF, HCN, H$_2$O, MeOH, NH$_3$). *Molecules* **2022**, *27*, 5712. https://doi.org/10.3390/molecules27175712

Academic Editor: Antonio Caballero

Received: 26 July 2022
Accepted: 29 August 2022
Published: 5 September 2022

Publisher's Note: MDPI stays neutral with regard to jurisdictional claims in published maps and institutional affiliations.

Copyright: © 2022 by the author. Licensee MDPI, Basel, Switzerland. This article is an open access article distributed under the terms and conditions of the Creative Commons Attribution (CC BY) license (https:// creativecommons.org/licenses/by/ 4.0/).

Abstract: The subject of research is forty dimers formed by imidazol-2-ylidene (I) or its derivative (IR$_2$) obtained by replacing the hydrogen atoms in both N-H bonds with larger important and popular substituents of increasing complexity (methyl = Me, iso-propyl = ^{i}Pr, tert-butyl = ^{t}Bu, phenyl = Ph, mesityl = Mes, 2,6-diisopropylphenyl = Dipp, 1-adamantyl = Ad) and fundamental proton donor (HD) molecules (HF, HCN, H$_2$O, MeOH, NH$_3$). While the main goal is to characterize the generally dominant C···H-D hydrogen bond engaging a carbene carbon atom, an equally important issue is the often omitted analysis of the role of accompanying secondary interactions. Despite the often completely different binding possibilities of the considered carbenes, and especially HD molecules, several general trends are found. Namely, for a given carbene, the dissociation energy values of the IR$_2$···HD dimers increase in the following order: NH$_3$ < H$_2$O < HCN ≤ MeOH ≪ HF. Importantly, it is found that, for a given HD molecule, IDipp$_2$ forms the strongest dimers. This is attributed to the multiplicity of various interactions accompanying the dominant C···H-D hydrogen bond. It is shown that substitution of hydrogen atoms in both N-H bonds of the imidazol-2-ylidene molecule by the investigated groups leads to stronger dimers with HF, HCN, H$_2$O or MeOH. The presented results should contribute to increasing the knowledge about the carbene chemistry and the role of intermolecular interactions, including secondary ones.

Keywords: carbene; N-heterocyclic carbene; NHC; imidazol-2-ylidene; hydrogen bond; intermolecular interaction; secondary interaction; organometallic chemistry; DFT

1. Introduction

In carbenes [1–80], the carbon atom is sp^2-hybridized. However, unlike the vast majority of organic compounds, in carbenes, it is merely divalent, consequently forming only one (R=C) or at most two (R$_1$R$_2$C) covalent bonds. Thus, two valence electrons remain available, giving the possibility of two spin states, singlet or triplet [5,6,8,70]. In the spin singlet state, both electrons are paired to form a lone electron pair. Additionally, the carbene carbon atom has an unfilled *p*-orbital perpendicular to the sp^2 hybrids. While the triplet state is also possible in some carbenes, a number of factors are known to increase singlet stability. The close presence of a strongly electronegative atom, such as, e.g., N, is important here, which stabilizes the singlet state through both the inductive σ-electron withdrawing effect and the π-electron charge donation from the lone electron pair of this atom to the vacant *p*-orbital of the carbene carbon atom [5,6,8,18,70]. In the case of cyclic carbenes, another factor stabilizing the singlet state is the π-electron delocalization within the entire ring, which may be related to its aromaticity, as is the case, for example, in the imidazol-2-ylidene molecule [38]. Another factor that stabilizes the singlet state of a carbene is a low value of the R$_1$–C–R$_2$ angle [3–5]. Naturally, this condition is obviously present in cyclic carbenes. While all of the conditions mentioned so far exist in imidazol-2-ylidene, they need not necessarily occur simultaneously in other N-heterocyclic carbenes (NHCs) [61]. In their case, an important stabilizing factor is the size of the substituents attached to both nitrogen atoms, as larger substituents prevent carbene dimerization [1].

The presence of a lone electron pair on the carbene carbon atom makes carbenes good Lewis bases, showing strongly nucleophilic properties. Carbenes are mainly known for their association with transition metal atoms (see, e.g., refs. [11,30,31,36,55,63]), making them extremely useful compounds in organic, organometallic and materials chemistry as well as homogeneous catalysis [21,30,33,57,61,62,69]. In addition to bonds with metals, however, carbenes quite willingly also form other intermolecular connections, such as hydrogen bonds [9,10,16,19,38,51,56,58], lithium bonds [39,45,46,62,69], beryllium bonds [17,53,54,62,69,73,79], magnesium bonds [14,22,40,41,62,69,73,79], triel bonds [13,20,23,27,69], tetrel bonds [42,65,66,69], pnictogen bonds [47,48,50,69], chalcogen bonds [52,69], halogen bonds [12,15,43,59,69] (in particular to iodine [12,15]), and aerogen bonds [67]. Moreover, in addition to these possibilities resulting from the presence of a lone electron pair, the presence of an empty p orbital in singlet carbenes also gives them electrophilic properties that seem to be much less studied [7,24,29,68,71,74,75].

Considering the fundamental role of hydrogen bonds [81–96], it is somewhat surprising that hydrogen bonds involving carbenes are studied only very sporadically [9,10,16,19,38,51,56,58]. Obviously, due to considerable methodological limitations, the first theoretical reports (in 1983 by Pople et al. [9] and then in 1986 by Pople [10]) concerned a simple H$_2$C···HF dimer. Then, in 1996, Alkorta and Elguero investigated dimers between H$_2$C or F$_2$C and a few simple proton donors [19]. More recently, Jabłoński and Palusiak studied hydrogen bonds between carbenes CF$_2$, CCl$_2$ and imidazol-2-ylidene and such proton donors as H$_2$O and HCF$_3$ [38]. In a wonderful extensive work on the application of theoretical methods in the study of carbene chemistry, Gerbig and Ley [56] also mentioned the imidazol-2-ylidene···HCF$_3$ dimer in which there is C···H-C hydrogen bonding and the accompanying N-H···F (both described earlier in ref. [38]). Samanta et al. [58] showed the possibility of the formation of either C···H-N or C···H-O hydrogen bonds between the simple heterocyclic derivative of 1,3-di(methyl)imidazol-2-ylidene (IMe$_2$) and MeNH$_2$ or, in particular, MeOH, respectively. In the latter case, the NHC acts as an esterification catalyst activating the alcohol molecule. On the other hand, from the experimental point of view, hydrogen bonding with carbene was first announced by Arduengo et al. [16] in 1995 with the report essentially relating to the C–H–C bridge in a bis(carbene)–proton complex formed by 1,3-di(2,4,6-trimethylphenyl)-imidazol-2-ylidene (IMes$_2$). Much later, in 2011, it was shown that IMes$_2$ and 1,3-di(2,6-diisopropylphenyl)imidazolidin-2-ylidene (SI^{i}Pr$_2$) can form a C···H-O type hydrogen bond to 1-hydroxy-2,2,6,6-tetramethyl-piperidine (TEMPO-H) [51]. In addition to crystallographic and NMR studies, results of theoretical calculations were also reported.

As shown above, the theoretical studies for hydrogen bonding involving carbenes have generally been down to very small carbenes, at most imidazol-2-ylidene (I) or 1,3-di(methyl)imidazol-2-ylidene (IMe$_2$). However, there are no theoretical studies in which the imidazol-2-ylidene derivative would contain even larger substituents on both nitrogen atoms. Moreover, the influence of the presence of these substituents on the possible interactions accompanying the leading C···H-D hydrogen bond has not been investigated yet. This article aims to fill this gap. Namely, this article examines the hydrogen bonds between imidazol-2-ylidene (I) and its seven popular derivatives containing gradually more bulky substituents (methyl = Me, iso-propyl = ^{i}Pr, tert-butyl = ^{t}Bu, phenyl = Ph, mesityl = Mes, 2,6-diisopropylphenyl = Dipp, 1-adamantyl = Ad) on both nitrogen atoms and five fundamental proton donor molecules (HF, HCN, H$_2$O, MeOH, NH$_3$). The possible combinations form forty dimers, which can be briefly designated as IR$_2$···HD, where R is one of the substituents mentioned previously. The general scheme of the dimers in question is shown in Figure 1.

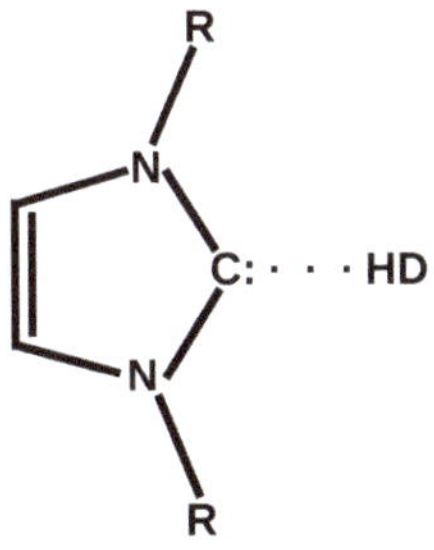

Figure 1. General scheme of the IR₂···HD dimers (R = H, Me, ^{i}Pr,^{t}Bu, Ph, Mes, Dipp, Ad). The colon on the carbene carbon atom represents a lone electron pair.

It should be emphasized that the C···H-D hydrogen bonds studied here engage the carbene species in their singlet spin states and thus a lone electron pair on the carbene carbon atom (Figure 1) and are therefore considerably distant from hydrogen bonds involving radicals [97,98]. In addition to the description of C···HD hydrogen bonds, an equally important goal is to analyze the possibility of the emergence of various types of accompanying interactions and their impact on the structure of the obtained dimers. This concerns the important issue of the coexistence of various types of interactions in molecular systems. Recently, the author of the present paper has shown [80] that various types of secondary interactions have a significant effect on the mutual orientation of the ZnX₂ molecular plane relative to the imidazol-2-ylidene ring in various types of IR₂ carbenes. This result is important because the possible torsion of the planes is usually attributed to the steric effects resulting from the presence of bulky R substituents.

2. Theoretical Methods

Geometries of all the systems were fully optimized on the ωB97X-D/6-311++G(d,p) level of theory, i.e., using the ωB97X-D range-separated gradient- and dispersion-corrected hybrid exchange-correlation functional [99] of density functional theory [100–102] and the 6-311++G(d,p) basis set being of the triple-zeta type and containing both polarization and diffuse functions on all atoms [103–107]. It is worth noting that ωB97X-D was one of the best functionals out of 200 tested [108]. There were no imaginary frequencies showing that equilibrium structures were obtained each time. Both the geometry optimization and the frequency analysis were performed using the Gaussian 16 (Revision C.01) program [109].

Values of the electron density at the bond critical points (bcp) [110–112] of the C···H-D hydrogen bonds and other accompanying interactions were computed using the AIMAll program [113]. Indeed, the C···H-D hydrogen bond should also be investigated through quantum chemical topology, which, over the years, was shown to be an efficient theoretical approach [114,115].

The dissociation energy was calculated as the difference between the total energy of a dimer and the sum of total energies of isolated subsystems in their own fully optimized structures. Total energies were corrected for the zero-point vibrational energies (ZPVE). Dissociation energies are given as positive values. In order to determine the binding energy (in fact, the interaction energy [116]) of the individual interaction of interest, the formula suggested recently by Emamian et al. [116] was used:

$$E_b [\text{kcal/mol}] = -223.08 \cdot \rho_{bcp} [\text{a.u.}] + 0.7423 \quad (1)$$

This formula is based on the electron density value determined at the bond critical point of this interaction, which is the parameter that, as has been shown [116], is best correlated with binding energy among many wave function-based HB descriptors. As a result, Emamian et al. proposed using this equation for a quick estimate of the energy of hydrogen-bond-forming networks. It is worth emphasizing that the values of the dissociation energies and the (sum) of the binding energies determined by Equation (1) are not comparable, since the former is a global quantity relating to the entire dimer, while the latter is a local parameter relating to an individual interaction. Besides, the dissociation energy takes into account deformation energies and ZPVEs.

3. Results and Discussion

3.1. Hydrogen Bonds to Carbenes Observed in the Solid State

It should be noted that despite the high reactivity of carbene compounds, hydrogen bonds with the participation of the carbene carbon atom as a proton acceptor quite often occur in the crystals. Of course, due to the abundance of C-H bonds, the most common hydrogen bond of this type is C···H-C, but C···H-N or C···H-O can also be found in the Cambridge Structural Database [117]. Some more interesting examples are shown in Figure 2.

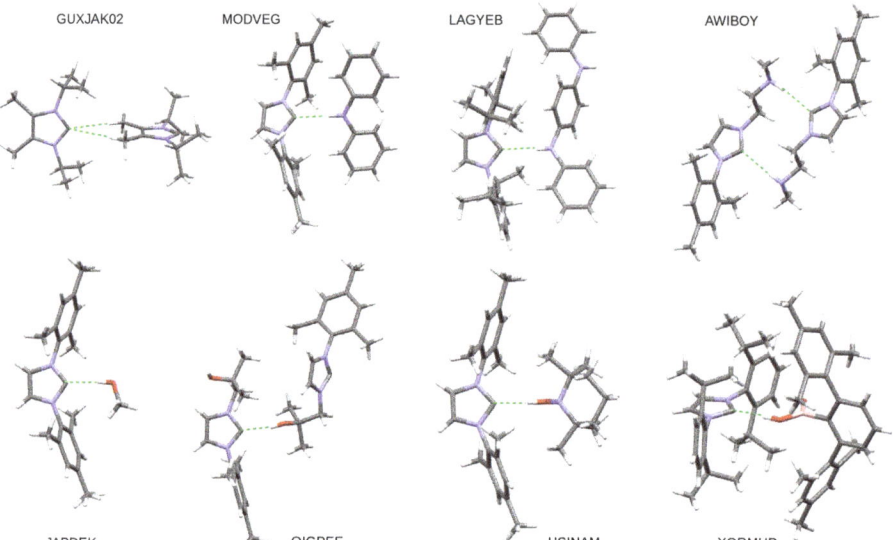

Figure 2. Examples of the occurrence of hydrogen bonds (marked with a dashed green line) involving a carbene carbon atom in crystallographic structures. The atoms are labeled as follows: carbon—gray, hydrogen—white, nitrogen—blue, oxygen—red, and boron—pink.

There are two C···H-C type hydrogen bonds in the crystal form of 1,3-diisopropyl-4,5-dimethylimidazol-2-ylidene (GUXJAK02) [118]. Both the dimer between 1,3-dimesityl-imidazol-2-ylidene and diphenylamine (MODVEG) [119] and between 1,3-bis[2,6-diisopropylphenyl]imidazol-2-ylidene and N^1,N^4-diphenylbenzene-1,4-diamine (LAGYEB) [120] have one C···H-N hydrogen bond, while the dimer of N-methyl-2-(3-phenyl-2,3-dihydro-1H-imidazol-1-yl)ethanamine (AWIBOY) [121] has two. 1,3-dimesitylimidazol-2-ylidene in methanol (JAPDEK) [122], on the other hand, is the simplest example of the occurrence of a C···H-O type hydrogen bond and was reported as the first X-ray structure of a carbene–alkohol hydrogen-bonded complex. The -OH group donors are somewhat more complex in QIGPEE (i.e., the dimer of 1-(2-hydroxy-2-methylpropyl)-3-(2,4,6-trimethylphenyl)-imidazol-2-ylidene) [123], USINAM (i.e., the dimer between 1,3-dimesitylimidazol-2-ylidene and 2,2,6,6-tetramethylpiperidin-1-ol) [51], and XORMUP (the dimer between 1,3-bis[2,6-bis(propan-2-yl)phenyl]-imidazol-2-ylidene and (2,4,6,2″,4″,6″-hexamethyl[1,1′:3′,1″-terphenyl]-2′-yl)boronic acid) [124]. Interestingly, in the last system, there is the -B(OH)$_2$

group, of which, however, only one -O-H bond is involved in the formation of the C···H-O bridge with the carbene molecule.

The examples presented show clearly that the carbene carbon can actually engage in hydrogen bonding quite easily by acting as a proton acceptor. It can also be seen that these carbenes have large substituents (most often in the form of mesityl groups) in the 1 and 3 positions of the imidazol-2-ylidene parent molecule. Therefore, the study of these types of systems using theoretical methods is also very important [80]. However, the presence of bulky substituents significantly increases the size of the systems and thus significantly increases the computational cost of theoretical research. Moreover, their presence increases the probability of the occurrence of interactions other than C···H-D, which additionally complicates the research. Probably for these reasons the amount of theoretical studies on this type of systems is small. As stated in the Introduction, the purpose of this article is to fill this gap to some extent. The investigated systems will be described in the next subsection.

3.2. Investigated Systems

The subject of the research in this article is the hydrogen bonds between imidazol-2-ylidene (I) and its seven derivatives having a pair of R substituents (methyl = Me, iso-propyl = ^{i}Pr, tert-butyl = ^{t}Bu, phenyl = Ph, mesityl = Mes, 2,6-diisopropylphenyl = Dipp, 1-adamantyl = Ad) attached to both nitrogens and five popular and important H-D proton donors (HF, HCN, H_2O, MeOH, NH_3). Dimers (complexes) having this bond will be denoted as $IR_2 \cdots HD$ for simplicity. Apart from the description of C···H-D hydrogen bonds in these systems, an important topic will be the possible presence of other accompanying bonds as they can sometimes significantly influence the structure of a complex [80]. Values of some selected parameters characterizing the C···H-D hydrogen bond or the proton-donor molecule in the $IR_2 \cdots HD$ dimers are listed in Table 1. Because imidazol-2-ylidene is somewhat different in terms of its binding capacity, the imidazol-2-ylidene complexes will be discussed first.

Table 1. Values of selected parameters characterizing the C···H-D hydrogen bond or the proton-donor molecule in the $IR_2 \cdots HD$ dimers (I = imidazol-2-ylidene; R = Me, ^{i}Pr, ^{t}Bu, Ph, Mes, Dipp, Ad; HD = HF, HCN, H_2O, MeOH, NH_3): D_0–dissociation energy (in kcal/mol), $\rho_{C\cdots H}$–the electron density at the bond critical point of C···H-D (in a.u.), E_b–binding energy of the C···H-D hydrogen bond obtained by Equation (1) (in kcal/mol), $d_{C\cdots H}$–length of the C···H distance (in Å), α_{CHD}–the C-H-D angle (in degrees), Δd_{HD}–elongation of the H-D proton-donor bond (in Å) relative to isolated HD (HF–0.917 Å, HCN–1.068 Å, H_2O–0.958 Å, MeOH–0.957 Å, and NH_3–1.013 Å).

HD	Property	Carbene							
		I	IMe$_2$	I^{i}Pr$_2$	I^{t}Bu$_2$	IPh$_2$	IMes$_2$	IDipp$_2$	IAd$_2$
HF	D_0	16.1	18.1	19.0	18.3	17.5	18.9	19.9	19.2
	$\rho_{C\cdots H}$	0.065	0.074	0.077	0.075	0.073	0.076	0.079	0.076
	E_b	−13.7	−15.7	−16.3	−15.9	−15.5	−16.1	−16.8	−16.3
	$d_{C\cdots H}$	1.676	1.626	1.612	1.626	1.632	1.614	1.599	1.618
	α_{CHD}	180.0	180.0	180.0	179.4	171.5	169.8	180.0	180.0
	Δd_{HD}	0.061	0.075	0.080	0.084	0.075	0.080	0.083	0.088
HCN	D_0	8.6	9.4	9.9	10.0	9.8	11.0	11.4	11.3
	$\rho_{C\cdots H}$	0.025	0.027	0.028	0.029	0.027	0.029	0.030	0.030
	E_b	−4.8	−5.3	−5.5	−5.8	−5.2	−5.6	−5.9	−6.0
	$d_{C\cdots H}$	2.138	2.105	2.087	2.071	2.103	2.070	2.058	2.055
	α_{CHD}	180.0	180.0	180.0	180.0	173.7	167.2	179.7	180.0
	Δd_{HD}	0.029	0.033	0.035	0.038	0.031	0.034	0.035	0.040

Table 1. Cont.

HD	Property				Carbene				
		I	IMe$_2$	I^{i}Pr$_2$	I^{t}Bu$_2$	IPh$_2$	IMes$_2$	IDipp$_2$	IAd$_2$
H$_2$O	D_0	8.4	9.0	9.5	10.3	9.7	10.7	11.2	9.9
	$\rho_{C\cdots H}$	0.027	0.034	0.035	0.040	0.035	0.036	0.033	0.033
	E_b	−5.3	−6.8	−7.0	−8.3	−7.1	−7.3	−6.7	−6.6
	$d_{C\cdots H}$	2.035	1.968	1.959	1.901	1.954	1.930	1.976	1.991
	α_{CHD}	140.4	163.8	166.4	174.4	166.6	163.4	168.5	172.9
	Δd_{HD}	0.019	0.025	0.026	0.033	0.026	0.028	0.023	0.027
MeOH	D_0	9.3	10.4	11.2	12.2	11.5	13.1	14.4	12.2
	$\rho_{C\cdots H}$	0.028	0.036	0.037	0.041	0.038	0.037	0.038	0.038
	E_b	−5.5	−7.2	−7.4	−8.4	−7.6	−7.6	−7.7	−7.6
	$d_{C\cdots H}$	2.035	1.956	1.945	1.898	1.927	1.923	1.927	1.931
	α_{CHD}	140.8	162.1	163.6	173.9	172.3	165.2	173.2	176.5
	Δd_{HD}	0.018	0.025	0.026	0.032	0.026	0.026	0.026	0.030
NH$_3$	D_0	7.2	5.1	5.3	4.5	5.2	6.7	7.3	5.4
	$\rho_{C\cdots H}$	0.011	0.018	0.019	0.018	0.019	0.020	0.018	0.018
	E_b	−1.6	−3.3	−3.4	−3.2	−3.5	−3.7	−3.2	−3.2
	$d_{C\cdots H}$	2.521	2.280	2.276	2.314	2.271	2.222	2.291	2.307
	α_{CHD}	123.1	157.0	158.2	173.6	165.9	158.5	161.1	174.4
	Δd_{HD}	0.007	0.011	0.011	0.011	0.010	0.012	0.009	0.011

3.3. Imidazol-2-Ylidene Complexes

Compared to all other molecules, imidazol-2-ylidene is distinguished by the presence of highly polar N-H bonds (Figure 3), which are good proton donors and therefore can eagerly form hydrogen bonds as long as there is an atom with good proton-acceptor properties, such as O, N or a halogen atom, in their presence.

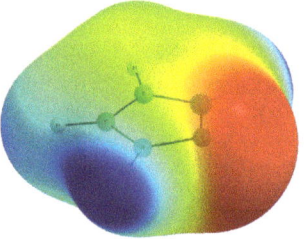

Figure 3. Map of the electrostatic potential (in a.u.) of imidazol-2-ylidene: −0.06, red; −0.03, yellow; 0.00, green; 0.03, cyan; 0.06, blue.

The stable dimers between imidazol-2-ylidene and HF, HCN, H$_2$O, MeOH or NH$_3$ are shown in Figure 4.

In the case of HF and HCN, the obtained hydrogen bonds are linear, the former of which is very strong, which is expressed by a large value of the determined dissociation (16.1 kcal/mol) and binding (−13.7 kcal/mol) energies, a short distance C$\cdots$H (1.676 Å) and a fairly high value of the electron density determined at the critical point (ρ_{bcp}) of this bond (0.065 a.u.). In addition, the formation of the C$\cdots$H-F hydrogen bond leads to a significant elongation of the H-F bond (+0.061 Å) and a very large red-shift of its stretching vibration frequency (−1317 cm^{-1}). On the other hand, the interaction in the I$\cdots$HCN dimer is almost half as weak (8.6 kcal/mol), which is also related to the much longer C$\cdots$H distance (2.138 Å) and the much lower ρ_{BCP} value (0.025 a.u.). The elongation of the C-H bond in HCN and the red-shift of its stretching vibration frequency are only +0.029 Å and −397 cm^{-1}, respectively.

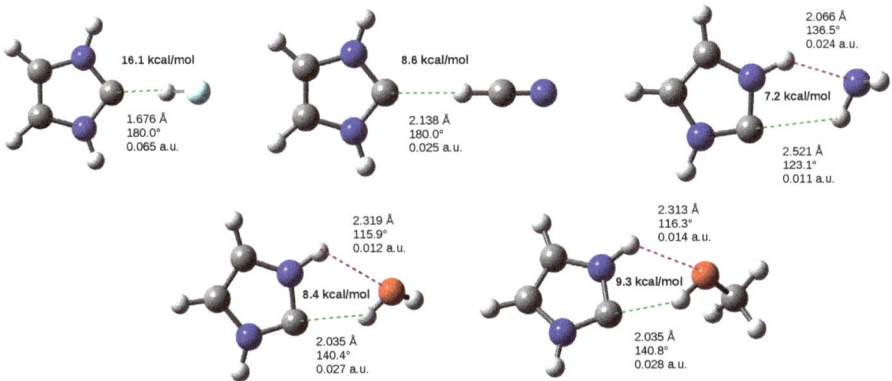

Figure 4. Imidazol-2-ylidene dimers with HF, HCN, NH₃, H₂O and MeOH. The values of the dissociation energy are given in bold, while the three values next to the interaction shown refer to the length of the hydrogen bond, its angle and the value of the electron density at the bond critical point.

As can be seen in Figure 4, unlike I···HF and I···HCN, in the remaining systems, i.e., I···NH₃, I···H₂O and I···MeOH, the C···H-D hydrogen bond is not linear due to the presence of another hydrogen bond. The presence of this accompanying hydrogen bond results not only from the presence of a strongly polar N-H bond in the imidazol-2-ylidene molecule, but above all from the presence of the D atom having an easily accessible lone electron pair. This possibility leads to the formation of the N-H···O hydrogen bond in the dimers I···H₂O and I···MeOH and N-H···N in the I···NH₃ dimer. Importantly, this situation, i.e., the simultaneous presence of two strong hydrogen bonds, means that the determined values of the dissociation energy do not represent strengths of the C···HD hydrogen bonds only, because, as representing intermolecular interactions globally, they should rather be assigned to both the hydrogen bonds, i.e., C···H-D and the accompanying one. Moreover, it is not easy to extract energy values for individual bonds. Nevertheless, as already mentioned in the Theoretical Methods section, Emamian et al. [116] have recently shown that among many wave function-based HB descriptors (including those based on QTAIM), the electron density determined at the bond critical point of HB best correlates with the binding (in fact, interaction) energy of this HB, giving Equation (1). As a result, they proposed using this equation for a quick estimate of the energy of hydrogen bonds forming networks [116]. Using this proposal, in the case of the I···NH₃ dimer, the value is −4.6 kcal/mol for the N-H···N hydrogen bond and only −1.7 kcal/mol for C···H-N. Thus, the N-H···N bond is definitely stronger than N-H···C, which is also expressed in a definitely shorter distance H···N (2.066 Å) than H···C (2.521 Å). However, in the case of I···H₂O and I···MeOH dimers, the O-H···C bond is definitely stronger (−5.3 and −5.5 kcal/mol, respectively) than the accompanying N-H···O bond (−1.9 and −2.4 kcal/mol, respectively). This is also reflected in shorter C···H distance (2.035 Å) than H···O (2.319 and 2.313 Å, respectively).

The dimers I···NH₃, I···H₂O and I···MeOH are good simple examples clearly showing that imidazol-2-ylidene willingly forms accompanying hydrogen bonds by the strongly polar N-H bond. The formation of such bonds is particularly easy when the D atom has lone electron pairs readily available, as is the case, for example, of N and O atoms. The formation of these bonds is prevented after replacing the H atoms in both N-H bonds with non-polar R substituents. However, in addition to the frequently mentioned steric effects, this type of modification of the imidazol-2-ylidene molecule may lead to other accompanying secondary interactions, which is discussed in the other subsections.

3.4. The IR$_2$···HD Dimers

3.4.1. The IR$_2$···HF and IR$_2$···HCN Dimers

Due to the fairly large similarity in the values of the C-H-D angle, i.e., the local geometry of the C···H-D hydrogen bond, in the HF and HCN dimers, it is convenient to consider these dimers together. The fully optimized structures of the dimers IR$_2$···HF and IR$_2$···HCN are shown in Figure 5 along with the obtained values of dissociation energy, distance C···H, angle C-H-D and value of the electron density at the bond critical point of the C···H hydrogen bond.

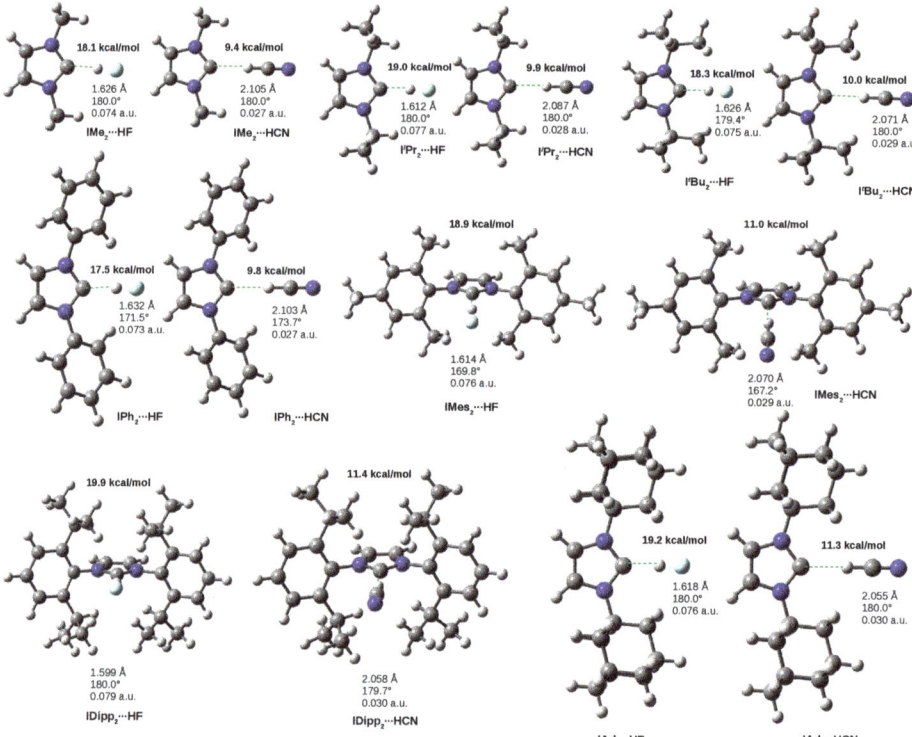

Figure 5. The IR$_2$···HF and IR$_2$···HCN (R = Me, ^{i}Pr, ^{t}Bu, Ph, Mes, Dipp, Ad) dimers. The values of the dissociation energy are given in bold, while the three values refer to the length of the hydrogen bond, the C-H-D angle and the value of the electron density at the bond critical point of the C···H-D hydrogen bond.

First, let us note that, as with the I···HF and I···HCN dimers (Figure 4), the HF dimers are much stronger than their HCN counterparts. The dissociation energies are in the 17.5–19.9 kcal/mol range while the HCN dimer values are only 9.4–11.4 kcal/mol. In the former case, the strongest complex is IDipp$_2$···HF, and the weakest is IPh$_2$···HF, while in the latter, the strongest complex is also the one involving IDipp$_2$, but the weakest dimer is formed by IMe$_2$ instead. Much greater strength of hydrogen bonds C···H in IR$_2$···HF dimers than IR$_2$···HCN is also visible in much smaller distances H···C (1.599–1.632 Å vs 2.055–2.105 Å), much higher values of ρ_{bcp} (0.073–0.079 a.u. vs 0.027–0.030 a.u.), much greater extensions of the H-D bond (0.075–0.088 Å vs 0.031–0.040 Å) and much higher values of the ν_{HD} stretching vibration frequency red-shift (from −1591 up to −1814 cm^{-1} vs. from −462 up to −548 cm^{-1}). It is worth noting that although the dimers involving Ad (i.e.,

IAd$_2 \cdots$ HF and IAd$_2 \cdots$ HCN) are not the strongest (although the difference compared to their counterparts involving Dipp is small, especially in the case of the HCN dimer), the effect of extending the proton-donor bond and red-shift values are in these systems greatest. Therefore, it can be concluded that together with IDipp$_2$, IAd$_2$ also forms strong C$\cdots$H hydrogen bonds, which have the greatest impact on the characteristics of HF and HCN proton-donor molecules. It is also worth noting that for both HF and HCN, the dissociation energies obtained for the dimers shown in Figure 5 are clearly greater than the energies obtained for I$\cdots$HF and I$\cdots$HCN (16.1 and 8.6 kcal/mol, respectively, as shown in Table 1 and Figure 4), and thus the substitution of hydrogen atoms in both N-H bonds in imidazol-2-ylidene even by methyl groups increases the strength of the dimer. This can most likely be explained by the weak inductive effect of these groups, which increases the charge on lone electron pairs.

It may be asked why the systems with IDipp$_2$ and IAd$_2$ are characterized by the highest values of dissociation energies (of course also the shortest H$\cdots$C distances and the highest ρ_{bcp}, Δd_{HD} and red-shift values). Very often, the increase in dimer strength is due to the presence of additional competing interactions, the presence of which is often suggested by the significant non-linearity of the dominant hydrogen bond, as was the case with the dimers I$\cdots$H$_2$O, I$\cdots$MeOH and I$\cdots$NH$_3$ shown in Figure 4 and discussed earlier. However, the reference to the C-H-D angle value can be very deceptive because these values for both IDipp$_2$ and IAd$_2$ are exactly 180°, so the C$\cdots$H-D hydrogen bonds in the dimers of these carbenes with HF and HCN are linear. What is more, interestingly, carbenes IPh$_2$, and especially IMes$_2$, are characterized by a distinct nonlinearity of C$\cdots$H-D hydrogen bonds. In these cases, the non-linearity may indeed result from the presence of other competing interactions which affect the alignment of the proton-donor molecule and thus also the geometry of the C$\cdots$H-D hydrogen bond.

In order to search for such competitive interactions, the determination of a molecular graph defined by QTAIM [110–112] may be a particularly helpful tool. Figure 6 presents molecular graphs of several selected IR$_2 \cdots$HF and IR$_2 \cdots$HCN dimers.

As can be clearly seen, in each of the examples, the molecular graph shows the presence of accompanying interatomic interactions, which are indicated by color-coded arrows. IPh$_2 \cdots$HF is a simple example in which the molecular graph suggests the presence of two accompanying C-H$\cdots$F type hydrogen bonds. A similar situation occurs in the slightly larger IMes$_2 \cdots$HF, where the proton-donating C-H bond is derived from one of the methyl groups in the mesityl group. Both these bonds should be very weak, which is suggested by low electron density values (ca. 0.008 a.u.). Using Equation (1) gives the value of -1.0 kcal/mol. In contrast, in the IMes$_2 \cdots$HCN dimer, a similar pair of bond paths indicates the presence of CH$\cdots$C-type hydrogen bonds, which should be even weaker (ρ_{bcp} amounts to ca. 0.0034 a.u. only, and therefore, the bonding effect is negligible) than C-H$\cdots$F. The IAd$_2 \cdots$HF dimer is a very simple example of a system featuring a C$\cdots$F tetrel bonding pair. This time the binding energy value is significant (ca. -1.4 kcal/mol), suggesting that the pair of these interactions may contribute significantly to the overall binding effect of the dimer.

The presence of 2,6-diisopropylphenyl (Dipp) substituent allows a particularly large number of accompanying interactions [80]. Apart from pairs of hydrogen bonds of the type C$\cdots$H-F (or C$\cdots$H-C in IDipp$_2 \cdots$HCN), both these dimers, i.e., IDipp$_2 \cdots$HF and IDipp$_2 \cdots$HCN, experience the presence of bond paths between the hydrogen atom of a C-H bond and the carbene carbon atom. Therefore, this result suggests that apart from the dominant C$\cdots$H-D hydrogen bond (D = F or C), the carbene carbon atom engages in two additional C$\cdots$H-C hydrogen bonds, which, however, should be very weak (ca. -0.9 and -1.1 kcal/mol in IDipp$_2 \cdots$HF and IDipp$_2 \cdots$HCN, respectively). In the case of IDipp$_2 \cdots$HCN, the molecular graph also suggests the presence of a pair of *intra*molecular hydrogen bonds of the C-H$\cdots$N type, as nitrogen atoms belong to the imidazol-2-ylidene ring. Since the values of ρ_{bcp} are significant (ca. 0.016 a.u.), the use of Equation (1) yields the value of ca. -2.8 kcal/mol. Apart from the aforementioned accompanying hydrogen bonds, the molecular graphs of dimers IDipp$_2 \cdots$ HF and IDipp$_2 \cdots$ HCN also show the presence of numerous C-H$\cdots$H-C interactions, which, however, seems to be a fairly common feature in the case of crowded systems with many C-H bonds [80,125–127].

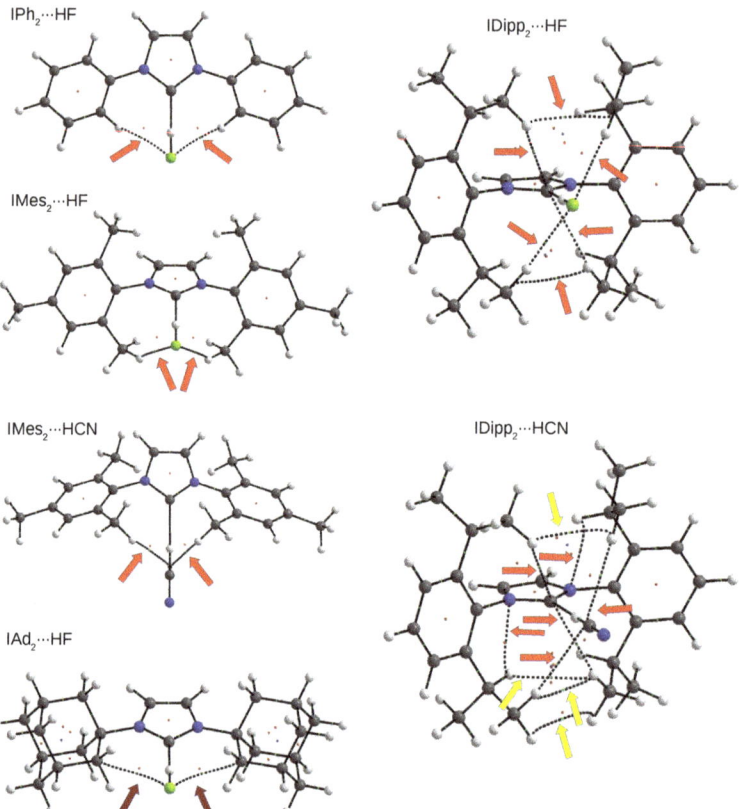

Figure 6. Molecular graphs of selected IR$_2 \cdots$ HF and IR$_2 \cdots$ HCN dimers. Arrows show the presence of bond paths for some accompanying interatomic interactions (hydrogen bonds—red, tetrel bonds—brown, and C-H$\cdots$H-C contacts—yellow). Large balls represent atoms (hydrogen—white, carbon—gray, nitrogen—blue, and fluorine—light green) and small balls represent critical points (bond critical points—light green, ring critical points—red, and cage critical points—blue).

As shown (Figure 6), molecular graphs suggest the presence of many different interactions, including intermolecular ones. Their role in the overall binding of the system and their influence on the overall structure is not easy to quantify, although there are many descriptors for individual, i.e., local, interactions [116]. One such descriptor is the electron density value determined at the bond critical point of a given interaction, which, as already mentioned, best correlates with the binding energy [116]. In the case of IR$_2 \cdots$ HF and IR$_2 \cdots$ HCN dimers, the estimates based on Equation (1) show that these interactions should be much weaker than the dominant C$\cdots$H-D hydrogen bond. Nevertheless, also taking into account their large number, it is highly likely that their presence can influence the overall structure of the dimer. In particular, their presence can significantly influence the angle values as they are generally associated with small force constants.

3.4.2. The IR$_2$···H$_2$O and IR$_2$···MeOH Dimers

The significant non-linearity of the hydrogen bonds in the dimers I···H$_2$O and I···MeOH already indicates the presence of additional intermolecular interactions, and indeed, as shown in Figure 4 and discussed previously, there is an additional hydrogen bond of the N-H···O type in both of these dimers. The structures of these dimers also prove that lone electron pairs on oxygen are readily available, and therefore, both H$_2$O and MeOH will be willing to engage in the formation of accompanying hydrogen bonds. The resulting IR$_2$···H$_2$O and IR$_2$···MeOH dimer structures are shown in Figure 7.

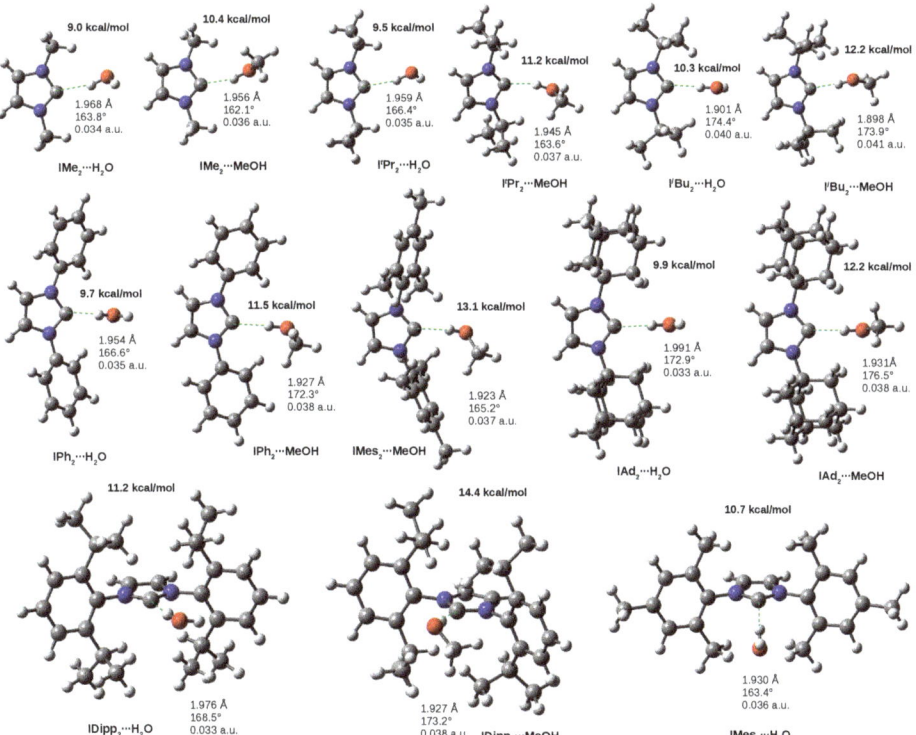

Figure 7. The IR$_2$···H$_2$O and IR$_2$···MeOH (R = Me, ^{i}Pr, ^{t}Bu, Ph, Mes, Dipp, Ad) dimers. The values of the dissociation energy are given in bold, while the three values refer to the length of the hydrogen bond, the C-H-D angle and the value of the electron density at the bond critical point of the C···H-D hydrogen bond.

Among the dimers with water, IDipp$_2$···H$_2$O and IMes$_2$···H$_2$O are characterized by the highest dissociation energy value (11.2 and 10.7 kcal/mol, respectively), and the lowest is for IMe$_2$···H$_2$O (9.0 kcal/mol). It is similar in the case of dimers with MeOH: IDipp$_2$···MeOH (14.4 kcal/mol), IMes$_2$···MeOH (13.1 kcal/mol), IMe$_2$···MeOH (10.4 kcal/mol). As was the case with dimers involving HF or HCN, the presence of any R substituent clearly increases the value of the dissociation energy. Clearly (Table 1, Figure 7), for a given carbene IR$_2$, MeOH forms a stronger complex than H$_2$O. This is also reflected in the shorter distances C···H. For example, for IPh$_2$ the distances are 1.927 and 1.954 Å, respectively, and for IAd$_2$, they are 1.931 and 1.991 Å, respectively. The dissociation energies for H$_2$O are rather similar and generally slightly lower than those for HCN. It is worth noting that in the case of dimers with H$_2$O and MeOH, the highest (i.e., most negative) values of E_b, i.e., the binding energy obtained by Equation (1), have been obtained for the I^{t}Bu$_2$ carbene, characterized by the formation of an almost linear (ca. 174°) C···H-O bond. On the contrary, the smallest values of E_b have been obtained for IMe$_2$ (−6.8 and −7.2 kcal/mol for H$_2$O and MeOH, respectively). However, they are much larger than in the unsubstituted imidazol-2-ylidene (−5.3 and −5.5 kcal/mol, respectively).

Taking into account that the molecular graphs obtained for the IR$_2$ ··· HF and IR$_2$ ··· HCN dimers (Figure 6) in many cases indicated the presence of two, four or even more additional coexisting interactions, it is now worth analyzing the molecular graphs for IR$_2$ ··· H$_2$O and IR$_2$ ··· MeOH dimers. The most interesting examples are shown in Figure 8.

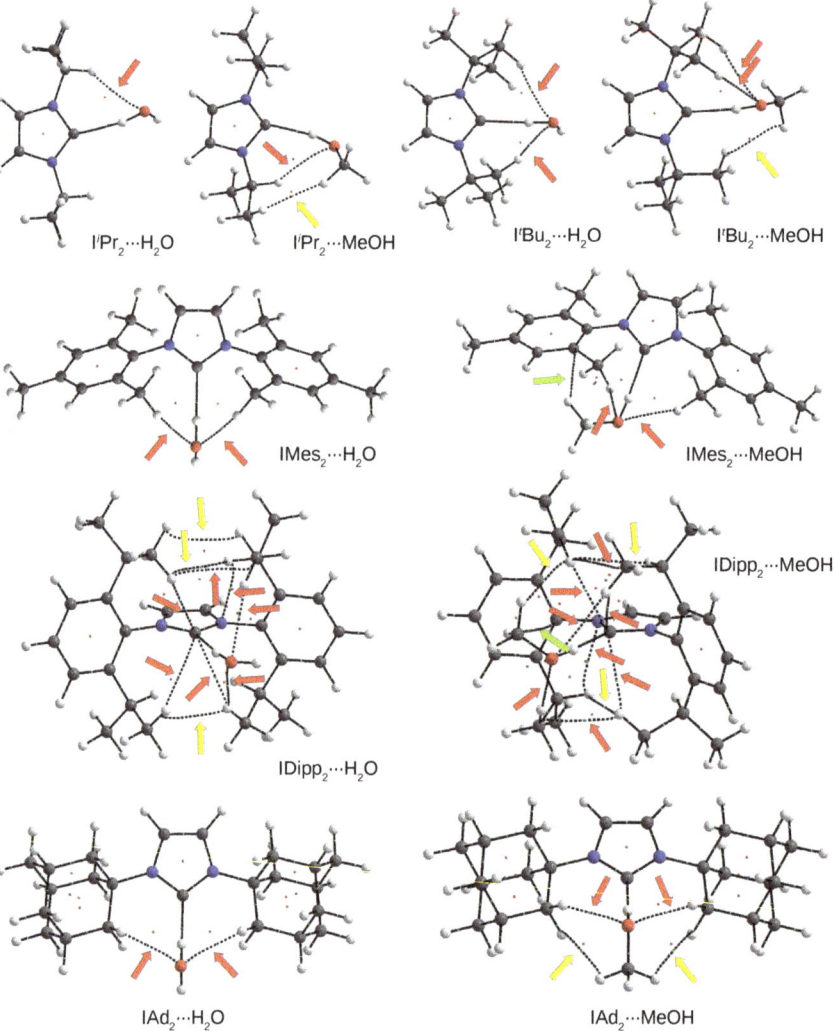

Figure 8. Molecular graphs of selected IR$_2$ ··· H$_2$O and IR$_2$ ··· MeOH dimers. Arrows show the presence of bond paths for some accompanying interatomic interactions (hydrogen bonds—red, C-H···H-C contacts—yellow, and C-H···π hydrogen bonds—light green). Large balls represent atoms (hydrogen—white, carbon—gray, nitrogen—blue, and oxygen—red) and small balls represent critical points (bond critical points—light green, ring critical points—red, and cage critical points—blue).

In the dimers shown in Figure 8, generally one (e.g., I^{i}Pr$_2$···H$_2$O) or two (e.g., I^{t}Bu$_2$···H$_2$O) bond paths are present for the accompanying C-H···O hydrogen bonds (indicated by red arrows). It is clear, however, that for some carbenes, the MeOH dimers contain additional bond paths for C-H···H-C contacts (indicated by yellow arrows). Moreover, the presence of a methyl group in MeOH allows in some cases (see the dimers with IMes$_2$ and IDipp$_2$) an additional (i.e., compared with H$_2$O) C-H···π-type hydrogen bond (green arrow). As was the case with the HF and HCN dimers, IDipp$_2$ produces the most intricate molecular graph, suggesting the existence of many interactions accompanying the main C···H-O hydrogen bond involving the carbene carbon atom. In addition to three C-H···H-C interactions, numerous accompanying hydrogen bonds are present. Taking this into account, this fact may explain the exceptionally high value of dissociation energy in dimers with IDipp$_2$, and especially in IDipp$_2$···MeOH (Table 1 and Figure 7). Nevertheless, all these accompanying interactions should be much weaker than the leading C···H-O hydrogen bond involving the carbene carbon atom. For example, C-H···O hydrogen bonds are generally in the range of -0.6 to -1.5 kcal/mol (e.g., -0.8 kcal/mol in I^{t}Bu$_2$···H$_2$O, -1.0 kcal/mol in I^{t}Bu$_2$···MeOH, -1.1 kcal/mol in I^{i}Pr$_2$···H$_2$O and I^{i}Pr$_2$···MeOH, and -1.4 kcal/mol in IMes$_2$···H$_2$O), however they are slightly stronger in IAd$_2$···H$_2$O (-1.6 kcal/mol) and especially in IAd$_2$···MeOH (-1.9 kcal/mol). The C-H···π bonds in IMes$_2$···MeOH and IDipp$_2$···MeOH are much weaker (-0.6 kcal/mol). On the other hand, the energy of C-H···H-C interactions is negligible, below -0.5 kcal/mol (e.g., -0.4 kcal/mol in IAd$_2$···MeOH). On the other hand, the *intra*molecular C-H···N hydrogen bonds in IDipp$_2$···H$_2$O and IDipp$_2$···MeOH are significantly strong (-2.8 kcal/mol). All these energy values should be compared with the energies of the dominant C···H-O hydrogen bond, which in the case of IR$_2$···H$_2$O and IR$_2$···MeOH dimers range from -6.6 to -8.4 kcal/mol (Table 1).

3.4.3. The IR$_2$···NH$_3$ Dimers

The values of dissociation (D_0) and binding (E_b) energies listed in Table 1 show that the dimers with NH$_3$ as well as the C···H-N hydrogen bonds in them should be the weakest. Namely, the D_0 values are from 4.5 kcal/mol in I^{t}Bu$_2$···NH$_3$ to 7.3 kcal/mol in IDipp$_2$···NH$_3$. The E_b values form a fairly narrow range from -3.2 kcal/mol (not including I···NH$_3$ with a much lower value of -1.6 kcal/mol) to -3.7 for IDipp$_2$···NH$_3$. Such small values of D_0 and E_b may result from the fact that the N-H bond in ammonia is a worse proton donor than the O-H bond in water or methanol. The C-H-D angle (α_{CHD}) values show that in none of the dimers, the C···H-N hydrogen bond is linear (although in I^{t}Bu$_2$···NH$_3$ and IAd$_2$···NH$_3$ this angle is ca. 174°); quite the opposite, the deviation from linearity is significant (157°–166°), although much smaller than in I···NH$_3$ (123°). Such large deviations from linearity may indicate the presence of additional coexisting interactions. The structures of the IR$_2$···NH$_3$ dimers are shown in Figure 9, while their molecular graphs can be found in Figure 10.

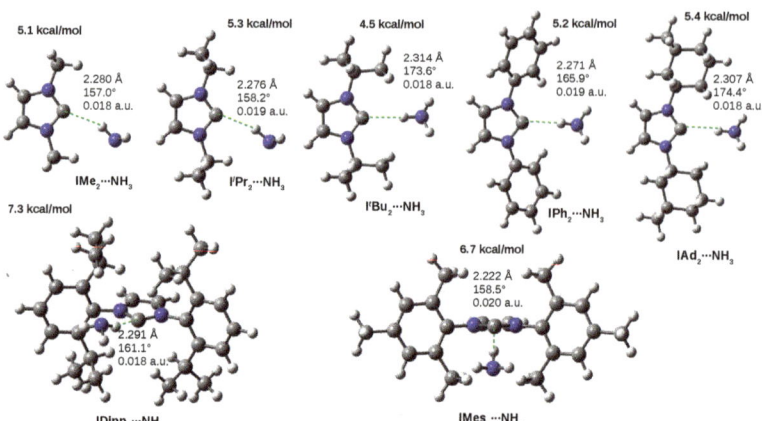

Figure 9. The IR$_2$···NH$_3$ (R = Me, ^{i}Pr, ^{t}Bu, Ph, Mes, Dipp, and Ad) dimers. The values of the dissociation energy are given in bold, while the three values refer to the length of the hydrogen bond, the C-H-D angle and the value of the electron density at the bond critical point of the C···H-D hydrogen bond.

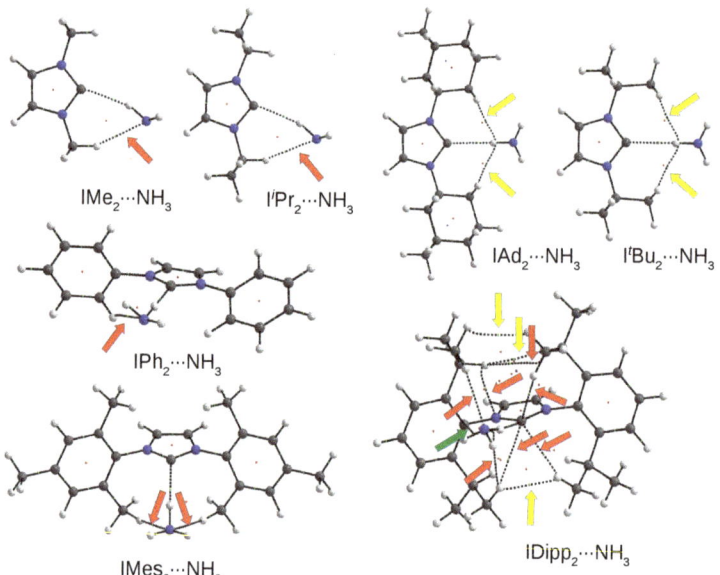

Figure 10. Molecular graphs of the IR$_2$···NH$_3$ dimers. Arrows show the presence of bond paths for some accompanying interatomic interactions (hydrogen bonds—red, C-H···H-C or C-H···H-N contacts—yellow, and N···π contact—dark green). Large balls represent atoms (hydrogen—white, carbon—gray, and nitrogen—blue) and small balls represent critical points (bond critical points—light green, ring critical points—red, and cage critical points—blue).

A careful comparison of the molecular graphs obtained for the IR$_2$···H$_2$O and IR$_2$···MeOH dimers (Figure 8) with the molecular graphs obtained for the IR$_2$···NH$_3$ dimers (Figure 10) shows some impoverishment in the second case. In addition, there is a visible change in the nature of some accompanying interactions in some systems with the same IR$_2$ carbene. For example, the molecular graph of the I^{i}Pr$_2$···NH$_3$ dimer contains only one bond path for the interaction (C-H···N) accompanying the dominant C···H-N hydrogen bond, while I^{i}Pr$_2$···MeOH in addition to the additional C-H···O also contains a bond path for C-H···H-C. The I^{t}Bu$_2$···H$_2$O dimer contains two bond paths for C-H···O hydrogen bonds, whereas I^{t}Bu$_2$···NH$_3$ also has two additional bond paths, but for C-H···H-C interactions. The situation is similar for the molecular graph involving Ad; in the case of dimer with H$_2$O, there are two bond paths for C-H···O hydrogen bonds, whereas in the case of IAd$_2$···NH$_3$, there are two bond paths for C-H···H-N. Keeping in mind that C-H···H-C interactions are weaker than C-H···O/N hydrogen bonds, together with a smaller number of accompanying interactions, this finding may explain the weaker strength of complexes with NH$_3$. Similarly, it can be seen that IDipp$_2$ creates many bond paths, of which as many as three (except three for N-H bonds, of course) lead to the nitrogen atom of the ammonia molecule. Two of them represent very weak C-H···N hydrogen bonds (−0.1 and −1.3 kcal/mol), while the other, interestingly, determines the N···π contact (−0.5 kcal/mol). Of course, all these interactions are much weaker than the dominant C···H-N hydrogen bond (−3.2 kcal/mol). Nevertheless, their multitude makes the dissociation energy for IDipp$_2$···NH$_3$ clearly the highest (7.3 kcal/mol) among all the IR$_2$···NH$_3$ dimers (see Table 1 or Figure 9). Of course, it should be noted that a similar dissociation energy value was obtained for a simple I···NH$_3$ dimer (7.2 kcal/mol); however, as already shown (Figure 4), such a large value in this case results from the presence of a strong N-H···N hydrogen bond (−4.6 kcal/mol), which overwhelms the weaker C···H-N bond (−1.6 kcal/mol).

3.5. Relationship between Dimer Strength and the C···H Distance

As has been shown in the previous subsections, many of the dimers considered here have some secondary interactions accompanying the leading C···H-D hydrogen bond. The strength of at least some of them may be considerable. Therefore, one should not expect a good overall (i.e., for all the IR$_2$···HD dimers) correlation between the dissociation energy, D_0, and the distance C···H. Moreover, good correlations are also not to be expected even in cases where individual proton-donor molecules are considered. Of course, the exception here is HF, and perhaps HCN, which more often forms linear or nearly linear C···H hydrogen bonds (Table 1). This is confirmed in Figure 11 (left), where it is shown that the R^2 (coefficient of determination) values for the linear correlation between D_0 and $d_{C···H}$ are very poor for NH$_3$ (even with the rejection of the clearly outlier point for I···NH$_3$), H$_2$O and MeOH (0.217, 0.319, 0.550, respectively) and much better for HCN (0.886) and especially for HF (0.930).

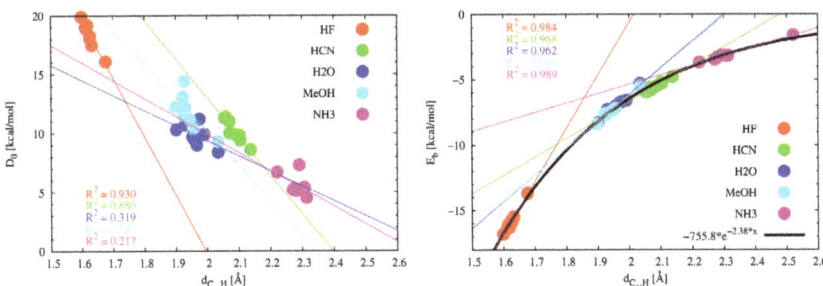

Figure 11. The dependence of either the dissociation energy (**left**) or the binding energy according to Equation (1) (**right**) on the distance C···H in the IR$_2$···HD dimers.

This result clearly shows that in the case of more complex dimers, i.e., having more significant intermolecular contacts, one should not expect a good relationship between the dimer dissociation energy and the length of any single intermolecular contact. Rather, a good correlation can be expected in the case of a clearly local parameter, which is the binding (interaction) energy determined from Equation (1), and therefore based on the electron density value at the critical point of the C···H-D hydrogen bond. Indeed, as shown in Figure 11 (right), for HF, MeOH and NH_3, the linear correlations between E_b and $d_{C...H}$ are very good ($R^2 > 0.98$) and only slightly worse for HCN and H_2O (0.968 and 0.962, respectively). Importantly, for all the $IR_2 \cdots HD$ dimers, an excellent exponential relationship, $E_b = -755.8\, e^{-2.38\, d_{C...H}}$, has been found (black curve in Figure 11). This result is not unexpected as a good common exponential relationship between ρ_{bcp} and the hydrogen bond length has been found for both different X and Y atoms in the X-H···Y hydrogen bond [128–131].

4. Conclusions

In this article, forty dimers formed by imidazol-2-ylidene or its derivatives, in which the hydrogen atoms of both N-H bonds were replaced by important and popular substituents of increasing complexity (methyl = Me, iso-propyl = ^{i}Pr, tert-butyl = ^{t}Bu, phenyl = Ph, mesityl = Mes, 2,6-diisopropylphenyl = Dipp, 1-adamantyl = Ad), and five fundamental proton donor (HD) molecules (HF, HCN, H_2O, MeOH, NH_3), have been studied. Although the most important goal was to describe the hydrogen bond formed by the carbene carbon atom, C···H-D, an equally important issue, which is, in the author's opinion, insufficiently described in the literature, was the possibility of the formation of various accompanying interactions and their influence on the structure of the considered dimers. Despite different interaction abilities represented by carbenes and especially HD molecules, the following general trends have been found, in addition to many specific results:

- For a given carbene, dissociation energies of the $IR_2 \cdots HD$ dimers increase in the following order: $NH_3 < H_2O < HCN \leq MeOH \ll HF$.
- For a given HD molecule (HF, HCN, H_2O, MeOH, or NH_3), $IDipp_2$, i.e., 1,3-bis[2,6-diisopropylphenyl]imidazol-2-ylidene, has been found to form the strongest dimers. This has been attributed to the multiplicity of various interactions accompanying the dominant C···H-D hydrogen bond.
- The substitution of hydrogen atoms in both N-H bonds of the imidazol-2-ylidene molecule by Me, ^{i}Pr, ^{t}Bu, Ph, Mes, Dipp or Ad groups leads to stronger dimers with HF, HCN, H_2O or MeOH.

The article clearly shows that various types of secondary interactions, thus often omitted in the analysis, can have a significant impact on the structure of a molecular system and its strength. The results should improve our understanding of carbene chemistry and the role of intermolecular interactions.

Funding: This research received no external funding.

Data Availability Statement: Data available from the author on reasonable request.

Conflicts of Interest: The authors declare no conflict of interest.

Abbreviations

The following abbreviations are used in this article:

NHC	N-heterocyclic carbene
I	imidazol-2-ylidene
IR_2	R-substituted imidazol-2-ylidene
Me	methyl group
^{i}Pr	isopropyl group
^{t}Bu	tert-butyl group
Ph	phenyl group
Mes	mesityl group
Dipp	2,6-diisopropylphenyl group
Ad	adamantyl group
QTAIM	quantum theory of atoms in molecules

References

1. Wanzlick, H.W. Aspects of Nucleophilic Carbene Chemistry. *Angew. Chem. Int. Ed. Engl.* **1962**, *1*, 75–80. [CrossRef]
2. Kirmse, W. *Carbene Chemistry*; Academic Press: Cambridge, MA, USA, 1964.
3. Hoffmann, R.; Zeiss, G.D.; Van Dine, G.W. The Electronic Structure of Methylenes. *J. Am. Chem. Soc.* **1968**, *90*, 1485–1499. [CrossRef]
4. Gleiter, R.; Hoffmann, R. On Stabilizing a Singlet Methylene. *J. Am. Chem. Soc.* **1968**, *90*, 5457–5460. [CrossRef]
5. Baird, N.C.; Taylor, K.F. Multiplicity of the Ground State and Magnitude of the T_1–S_0 gap in Substituted Carbenes. *J. Am. Chem. Soc.* **1978**, *100*, 1333–1338. [CrossRef]
6. Harrison, J.F.; Liedtke, R.C.; Liebman, J.F. The Multiplicity of Substituted Acyclic Carbenes and Related Molecules. *J. Am. Chem. Soc.* **1979**, *101*, 7162–7168. [CrossRef]
7. Schoeller, W.W. Electrophilicity and nucleophilicity in singlet carbenes. II. Electrophilic selectivity. *Tetrahedron Lett.* **1980**, *21*, 1509–1510. [CrossRef]
8. Mueller, P.H.; Rondan, N.G.; Houk, K.N.; Harrison, J.F.; Hooper, D.; Willen, B.H.; Liebman, J.F. Carbene Singlet–Triplet Gaps. Linear Correlations with Substituent π Donation. *J. Am. Chem. Soc.* **1981**, *103*, 5049–5052. [CrossRef]
9. Pople, J.A.; Raghavachari, K.; Frisch, M.J.; Binkley, J.S.; Schleyer, P.v.R. Comprehensive Theoretical Study of Isomers and Rearrangement Barriers of Even-Electron Polyatomic Molecules $H_m ABH_n$ (A, B = C, N, O, and F). *J. Am. Chem. Soc.* **1983**, *105*, 6389–6398. [CrossRef]
10. Pople, J.A. A theoretical search for the methylenefluoronium ylide. *Chem. Phys. Lett.* **1986**, *132*, 144–146. [CrossRef]
11. Schubert, U. *Advances in Metal Carbene Chemistry*; Springer: Dordrecht, The Netherlands, 1989.
12. Arduengo, A.J., III; Kline, M.; Calabrese, J.C.; Davidson, F. Synthesis of a Reverse Ylide from a Nucleophilic Carbene. *J. Am. Chem. Soc.* **1991**, *113*, 9704–9705. [CrossRef]
13. Arduengo, A.J., III; Dias, H.V.R.; Calabrese, J.C.; Davidson, F. A Stable Carbene-Alane Adduct. *J. Am. Chem. Soc.* **1992**, *114*, 9724–9725. [CrossRef]
14. Arduengo, A.J., III; Rasika Dias, H.V.; Davidson, F.; Harlow, R.L. Carbene adducts of magnesium and zinc. *J. Organomet. Chem.* **1993**, *462*, 13–18. [CrossRef]
15. Kuhn, N.; Kratz, T.; Henkel, G. A Stable Carbene Iodine Adduct: Secondary Bonding in 1,3-Diethyl-2-iodo-4,5-dimethylimidazolium Iodide. *J. Chem. Soc. Chem. Commun.* **1993**, 1778–1779. [CrossRef]
16. Arduengo, A.J., III; Gamper, S.F.; Tamm, M.; Calabrese, J.C.; Davidson, F.; Craig, H.A. A Bis(carbene)–Proton Complex: Structure of a C–H–C Hydrogen Bond. *J. Am. Chem. Soc.* **1995**, *117*, 572–573. [CrossRef]
17. Herrmann, W.A.; Runte, O.; Artus, G. Synthesis and structure of an ionic beryllium–"carbene" complex. *J. Organomet. Chem.* **1995**, *501*, C1–C4. [CrossRef]
18. Boehme, C.; Frenking, G. Electronic Structure of Stable Carbenes, Silylenes, and Germylenes. *J. Am. Chem. Soc.* **1996**, *118*, 2039–2046. [CrossRef]
19. Alkorta, I.; Elguero, J. Carbenes and Silylenes as Hydrogen Bond Acceptors. *J. Phys. Chem.* **1996**, *100*, 19367–19370. [CrossRef]
20. Li, X.-W.; Su, J.; Robinson, G.H. Syntheses and molecular structure of organo-group 13 metal carbene complexes. *Chem. Commun.* **1996**, 2683–2684. [CrossRef]
21. Herrmann, W.A.; Köcher, C. N-Heterocyclic Carbenes. *Angew. Chem. Int. Ed. Engl.* **1997**, *36*, 2162–2187. [CrossRef]
22. Arduengo, A.J., III; Davidson, F.; Krafczyk, R.; Marshall, W.J.; Tamm, M. Adducts of Carbenes with Group II and XII Metallocenes. *Organometallics* **1998**, *17*, 3375–3382. [CrossRef]
23. Hibbs, D.E.; Hursthouse, M.B.; Jones, C.; Smithies, N.A. Synthesis, crystal and molecular structure of the first indium trihydride complex, [InH3CN(Pri)C2Me2N(Pri)]. *Chem. Commun.* **1998**, 869–870. [CrossRef]
24. Goumri-Magnet, S.; Polishchuck, O.; Gornitzka, H.; Marsden, C.J.; Baceiredo, A.; Bertrand, G. The Electrophilic Behavior of Stable Phosphanylcarbenes Towards Phosphorus Lone Pairs. *Angew. Chem. Int. Ed.* **1999**, *38*, 3727–3729. [CrossRef]
25. Bourissou, D.; Guerret, O.; Gabbaï, F.P.; Bertrand, G. Stable Carbenes. *Chem. Rev.* **2000**, *100*, 39–91. [CrossRef] [PubMed]
26. Bertrande, G. *Carbene Chemistry: From Fleeting Intermediates to Powerful Reagents*; FontisMedia S.A.: Lausanne, Switzerland; Marcel Dekker, Inc.: New York, NY, USA, 2002.
27. Merceron, N.; Miqueu, K.; Baceiredo, A.; Bertrand, G. Stable (Amino)(phosphino)carbenes: Difunctional Molecules. *J. Am. Chem. Soc.* **2002**, *124*, 6806–6807. [CrossRef] [PubMed]
28. Wang, D.; Wurst, K.; Buchmeiser, M.R. N-heterocyclic carbene complexes of Zn(II): Synthesis, X-ray structures and reactivity. *J. Organomet. Chem.* **2004**, *689*, 2123–2130. [CrossRef]
29. Moss, R.A.; Platz, M.S.; Jones, M., Jr. (Eds.) *Reactive Intermediate Chemistry*; John Wiley & Sons, Inc.: Hoboken, NJ, USA, 2005.
30. Scott, N.M.; Nolan, S.P. Stabilization of Organometallic Species Achieved by the Use of N-Heterocyclic Carbene (NHC) Ligands. *Eur. J. Inorg. Chem.* **2005**, 1815–1828. [CrossRef]
31. Frenking, G.; Solà, M.; Vyboishchikov, S.F. Chemical bonding in transition metal carbene complexes. *J. Organomet. Chem.* **2005**, *690*, 6178–6204. [CrossRef]
32. Garrison, J.C.; Youngs, W.J. Ag(I) N-Heterocyclic Carbene Complexes: Synthesis, Structure, and Application. *Chem. Rev.* **2005**, *105*, 3978–4008. [CrossRef]
33. Nolan, S.P. *N-Heterocyclic Carbenes in Synthesis*; Wiley-VCH: Weinheim, Germany, 2006.

34. Carey, F.A.; Sundberg, R.J. *Carbenes, Part B: Reactions and Synthesis. Advanced Organic Chemistry*; Springer: New York, NY, USA, 2007.
35. Díez-González, S.; Nolan, S.P. Stereoelectronic parameters associated with N-heterocyclic carbene (NHC) ligands: A quest for understanding. *Coord. Chem. Rev.* **2007**, *251*, 874–883. [CrossRef]
36. Jacobsen, H.; Correa, A.; Poater, A.; Costabile, C.; Cavallo, L. Understanding the M–(NHC) (NHC = N-heterocyclic carbene) bond. *Coord. Chem. Rev.* **2009**, *253*, 687–703. [CrossRef]
37. de Frémont, P.; Marion, N.; Nolan, S.P. Carbenes: Synthesis, properties, and organometallic chemistry. *Coord. Chem. Rev.* **2009**, *253*, 862–892. [CrossRef]
38. Jabłoński, M.; Palusiak, M. Divalent carbon atom as the proton acceptor in hydrogen bonding. *Phys. Chem. Chem. Phys.* **2009**, *11*, 5711–5719. [CrossRef] [PubMed]
39. Li, Q.; Wang, H.; Liu, Z.; Li, W.; Cheng, J.; Gong, B.; Sun, J. Ab Initio Study of Lithium-Bonded Complexes with Carbene as an Electron Donor. *J. Phys. Chem. A* **2009**, *113*, 14156–14160. [CrossRef] [PubMed]
40. Arrowsmith, M.; Hill, M.S.; MacDougall, D.J.; Mahon, M.F. A Hydride-Rich Magnesium Cluster. *Angew. Chem. Int. Ed.* **2009**, *48*, 4013–4016. [CrossRef] [PubMed]
41. Arnold, P.L.; Casely, I.J.; Turner, Z.R.; Bellabarba, R.; Tooze, R.B. Magnesium and zinc complexes of functionalised, saturated N-heterocyclic carbene ligands: Carbene lability and functionalisation, and lactide polymerisation catalysis. *Dalton Trans.* **2009**, *35*, 7236–7247. [CrossRef] [PubMed]
42. Wang, Y.; Robinson, G.H. Unique homonuclear multiple bonding in main group compounds. *Chem. Commun.* **2009**, 5201–5213. [CrossRef] [PubMed]
43. Li, Q.; Wang, Y.; Liu, Z.; Li, W.; Cheng, J.; Gong, B.; Sun, J. An unconventional halogen bond with carbene as an electron donor: An ab initio study. *Chem. Phys. Lett.* **2009**, *469*, 48–51. [CrossRef]
44. Hindi, K.M.; Panzner, M.J.; Tessier, C.A.; Cannon, C.L.; Youngs, W.J. The Medicinal Applications of Imidazolium Carbene–Metal Complexes. *Chem. Rev.* **2009**, *109*, 3859–3884. [CrossRef]
45. Wang, Y.; Xie, Y.; Abraham, M.Y.; Wei, P.; Schaefer, H.F., III; Schleyer, P.v.R.; Robinson, G.H. A Viable Anionic N-Heterocyclic Dicarbene. *J. Am. Chem. Soc.* **2010**, *132*, 14370–14372. [CrossRef]
46. Li, Z.-F.; Yang, S.; Li, H.-X. Theoretical prediction characters of unconventional weak bond with carbene as electron donors and Li-Y (Y = OH, H, F, NC and CN) as electron acceptors. *J. Mol. Struct. THEOCHEM* **2010**, *952*, 56–60.
47. Wang, Y.; Xie, Y.; Abraham, M.Y.; Gilliard, R.J., Jr.; Wei, P.; Schaefer, H.F., III; Schleyer, P.v.R.; Robinson, G.H. Carbene-Stabilized Parent Phosphinidene. *Organometallics* **2010**, *29*, 4778–4780. [CrossRef]
48. Abraham, M.Y.; Wang, Y.; Xie, Y.; Wei, P.; Schaefer, H.F., III; Schleyer, P.v.R.; Robinson, G.H. Carbene Stabilization of Diarsenic: From Hypervalency to Allotropy. *Chem. Eur. J.* **2010**, *16*, 432–435. [CrossRef]
49. Mercs, L.; Albrecht, M. Beyond catalysis: N-Heterocyclic Carbene Complexes as Components for Medicinal, Luminescent, and Functional Materials Applications. *Chem. Soc. Rev.* **2010**, *39*, 1903–1912. [CrossRef] [PubMed]
50. Patel, D.S.; Bharatam, P.V. Divalent N(I) Compounds with Two Lone Pairs on Nitrogen. *J. Phys. Chem. A* **2011**, *115*, 7645–7655. [CrossRef] [PubMed]
51. Giffin, N.A.; Makramalla, M.; Hendsbee, A.D.; Robertson, K.N.; Sherren, C.; Pye, C.C.; Masuda, J.D.; Clyburne, J.A.C. Anhydrous TEMPO-H: Reactions of a good hydrogen atom donor with low-valent carbon centres. *Org. Biomol. Chem.* **2011**, *9*, 3672–3680. [CrossRef] [PubMed]
52. Zhao, Q.; Feng, D.; Sun, Y.; Hao, J.; Cai, Z. Theoretical Investigations on the Weak Nonbonded C=S···CH$_2$ Interactions: Chalcogen-Bonded Complexes With Singlet Carbene as an Electron Donor. *Int. J. Quant. Chem.* **2011**, *111*, 3881–3887. [CrossRef]
53. Gilliard, R.J., Jr.; Abraham, M.Y.; Wang, Y.; Wei, P.; Xie, Y.; Quillian, B.; Schaefer, H.F., III; Schleyer, P.v.R.; Robinson, G.H. Carbene-Stabilized Beryllium Borohydride. *J. Am. Chem. Soc.* **2012**, *134*, 9953–9955. [CrossRef]
54. Arrowsmith, M.; Hill, M.S.; Kociok-Köhn, G.; MacDougall, D.J.; Mahon, M.F. Beryllium-Induced C–N Bond Activation and Ring Opening of an N-Heterocyclic Carbene. *Angew. Chem. Int. Ed.* **2012**, *51*, 2098–2100. [CrossRef]
55. Moss, R.A.; Doyle, M.P. *Contemporary Carbene Chemistry*; John Wiley & Sons, Inc.: Hoboken, NJ, USA, 2013.
56. Gerbig, D.; Ley, D. Computational methods for contemporary carbene chemistry. *WIREs Comput. Mol. Sci.* **2013**, *3*, 242–272. [CrossRef]
57. Nelson, D.J.; Nolan, S.P. Quantifying and understanding the electronic properties of N-heterocyclic carbenes. *Chem. Soc. Rev.* **2013**, *42*, 6723–6753. [CrossRef]
58. Samanta, R.C.; De Sarkar, S.; Fröhlich, R.; Grimme, S.; Studer, A. N-Heterocyclic carbene (NHC) catalyzed chemoselective acylation of alcohols in the presence of amines with various acylating reagents. *Chem. Sci.* **2013**, *4*, 2177–2184. [CrossRef]
59. Esrafili, M.D.; Mohammadirad, N. Insights into the strength and nature of carbene···halogen bond interactions: A theoretical perspective. *J. Mol. Model.* **2013**, *19*, 2559–2566. [CrossRef] [PubMed]
60. Budagumpi, S.; Endud, S. Group XII Metal–N-Heterocyclic Carbene Complexes: Synthesis, Structural Diversity, Intramolecular Interactions, and Applications. *Organometallics* **2013**, *32*, 1537–1562. [CrossRef]
61. Hopkinson, M.N.; Richter, C.; Schedler, M.; Glorius, F. An overview of N-heterocyclic carbenes. *Nature* **2014**, *510*, 485–496. [CrossRef] [PubMed]
62. Bellemin-Laponnaz, S.; Dagorne, S. Group 1 and 2 and Early Transition Metal Complexes Bearing N-Heterocyclic Carbene Ligands: Coordination Chemistry, Reactivity, and Applications. *Chem. Rev.* **2014**, *114*, 8747–8774. [CrossRef]

63. Visbal, R.; Concepción Gimeno, M. N-Heterocyclic Carbene Metal Complexes: Photoluminescence and Applications. *Chem. Soc. Rev.* **2014**, *43*, 3551–3574. [CrossRef]
64. Santoro, O.; Nahra, F.; Cordes, D.B.; Slawin, A.M.Z.; Nolan, S.P.; Cazin, C.S.J. Synthesis, characterization and catalytic activity of stable [(NHC)H][ZnXY$_2$] (NHC = N-Heterocyclic carbene, X, Y = Cl, Br) species. *J. Mol. Catal.* **2016**, *423*, 85–91. [CrossRef]
65. Del Bene, J.E.; Alkorta, I.; Elguero, J. Carbon–Carbon Bonding between Nitrogen Heterocyclic Carbenes and CO$_2$. *J. Phys. Chem. A* **2017**, *121*, 8136–8146. [CrossRef]
66. Liu, M.; Li, Q.; Li, W.; Cheng, J. Carbene tetrel-bonded complexes. *Struct. Chem.* **2017**, *28*, 823–831. [CrossRef]
67. Esrafili, M.D.; Sabouri, A. Carbene–aerogen bonds: An ab initio study. *Mol. Phys.* **2017**, *115*, 971–980. [CrossRef]
68. Moss, R.A.; Wang, L.; Cang, H.; Krogh-Jespersen, K. Extremely reactive carbenes: Electrophiles and nucleophiles. *J. Phys. Org. Chem.* **2017**, *30*, e3555. [CrossRef]
69. Nesterov, V.; Reiter, D.; Bag, P.; Frisch, P.; Holzner, R.; Porzelt, A.; Inoue, S. NHCs in Main Group Chemistry. *Chem. Rev.* **2018**, *118*, 9678–9842. [CrossRef] [PubMed]
70. Alkorta, I.; Elguero, J. A LFER analysis of the singlet-triplet gap in a series of sixty-six carbenes. *Chem. Phys. Lett.* **2018**, *691*, 33–36. [CrossRef]
71. Jabłoński, M. The first theoretical proof of the existence of a hydride-carbene bond. *Chem. Phys. Lett.* **2018**, *710*, 78–83. [CrossRef]
72. Dagorne, S. Recent Developments on *N*-Heterocyclic Carbene Supported Zinc Complexes: Synthesis and Use in Catalysis. *Synthesis* **2018**, *50*, 3662–3670. [CrossRef]
73. Walley, J.E.; Wong, Y.-O.; Freeman, L.A.; Dickie, D.A.; Gilliard, R.J., Jr. N-Heterocyclic Carbene-Supported Aryl- and Alk- oxides of Beryllium and Magnesium. *Catalysts* **2019**, *9*, 934. [CrossRef]
74. Jabłoński, M. In search for a hydride-carbene bond. *J. Phys. Org. Chem.* **2019**, *32*, e3949. [CrossRef]
75. Yourdkhani, S.; Jabłoński, M. Physical nature of silane$\cdots$carbene dimers revealed by state-of-the-art ab initio calculations. *J. Comput. Chem.* **2019**, *40*, 2643–2652. [CrossRef]
76. Procter, R.J.; Uzelac, M.; Cid, J.; Rushworth, P.J.; Ingleson, M.J. Low-Coordinate NHC–Zinc Hydride Complexes Catalyze Alkyne C–H Borylation and Hydroboration Using Pinacolborane. *ACS Catal.* **2019**, *9*, 5760–5771. [CrossRef]
77. Dzieszkowski, K.; Barańska, I.; Mroczyńska, K.; Słotwiński, M.; Rafiński, Z. Organocatalytic Name Reactions Enabled by NHCs. *Materials* **2020**, *13*, 3574. [CrossRef]
78. Specklin, D.; Fliedel, C.; Dagorne, S. Recent Representative Advances on the Synthesis and Reactivity of N-Heterocyclic-Carbene-Supported Zinc Complexes. *Chem. Rec.* **2021**, *21*, 1130–1143. [CrossRef] [PubMed]
79. Jabłoński, M. Study of Beryllium, Magnesium, and Spodium Bonds to Carbenes and Carbodiphosphoranes. *Molecules* **2021**, *26*, 2275. [CrossRef]
80. Jabłoński, M. Theoretical Study of N-Heterocyclic-Carbene–ZnX$_2$ (X = H, Me, Et) Complexes. *Materials* **2021**, *14*, 6147. [CrossRef]
81. Pauling, L. *The Nature of the Chemical Bond*; Cornell University Press: New York, NY, USA, 1960.
82. Pimentel, G.C.; McClellan, A.L. *The Hydrogen Bond*; W.H. Freeman & Co.: San Francisco, CA, USA, 1960.
83. Hamilton, W.C.; Ibers, J.A. *Hydrogen Bonding in Solids*; W. A. Benjamin: New York, NY, USA, 1968.
84. Vinogradov, S.N.; Linnell, R.H. *Hydrogen Bonding*; Van Nostrand-Reinhold: Princeton, NJ, USA, 1971.
85. Schuster, P.; Zundel, G.; Sandorfy, C. (Eds.) *The Hydrogen Bond. Recent Developments in Theory and Experiments*; North Holland: Amsterdam, The Netherlands, 1976; Volumes I–III.
86. Schuster, P. *Intermolecular Interactions: From Diatomics to Biopolymers*; Pullman, B., Ed.; John Wiley: New York, NY, USA, 1978.
87. Hobza, P.; Zahradník, R. *Weak Intermolecular Interactions in Chemistry and Biology*; Academia: Prague, Czech Republic, 1980.
88. Jeffrey, G.A.; Saenger, W. *Hydrogen Bonding in Biological Structures*; Springer: Berlin/Heidelberg, Germany, 1991.
89. Hadži, D. (Ed.) *Theoretical Treatments of Hydrogen Bonding*; John Wiley: Chichester, UK, 1997.
90. Jeffrey, G.A. *An Introduction to Hydrogen Bonding*; Oxford University Press: New York, NY, USA, 1997.
91. Scheiner, S. (Ed.) *Molecular Interactions. From van der Waals to Strongly Bound Complexes*; Wiley: Chichester, UK, 1997.
92. Scheiner, S. *Hydrogen Bonding: A Theoretical Perspective*; Oxford University Press: New York, NY, USA, 1997.
93. Desiraju, G.R.; Steiner, T. *The Weak Hydrogen Bond in Structural Chemistry and Biology*; Oxford University Press: New York, NY, USA, 1999.
94. Grabowski, S.J. (Ed.) *Hydrogen Bonding—New Insights*; Springer: Dordrecht, The Netherlands, 2006.
95. Maréchal, Y. *The Hydrogen Bond and the Water Molecule*; Elsevier: Amsterdam, The Netherlands, 2007.
96. Gilli, G.; Gilli, P. *The Nature of the Hydrogen Bond. Outline of a Comprehensive Hydrogen Bond Theory*; Oxford University Press: Oxford, UK, 2009.
97. Scheiner, S. Evaluation of DFT Methods to Study Reactions of Benzene With OH Radical. *Int. J. Quant. Chem.* **2012**, *112*, 1879–1886. [CrossRef]
98. Li, Q.-Z.; Li, H.-B. Hydrogen Bonds Involving Radical Species. In *Noncovalent Forces*; Scheiner, S., Ed.; Springer: Cham, Switzerland, 2015.
99. Chai, J.-D.; Head-Gordon, M. Long-range corrected hybrid density functionals with damped atom–atom dispersion correction. *Phys. Chem. Chem. Phys.* **2008**, *10*, 6615–6620. [CrossRef]
100. Hohenberg, P.; Kohn, W. Inhomogeneous Electron Gas. *Phys. Rev.* **1964**, *136*, B864–B871. [CrossRef]
101. Kohn, W.; Sham, L.J. Self-Consistent Equations Including Exchange and Correlation Effects. *Phys. Rev.* **1965**, *140*, A1133–A1138. [CrossRef]

102. Parr, R.G.; Yang, W. *Density-Functional Theory of Atoms and Molecules*; Oxford University Press: New York, NY, USA, 1989.
103. Krishnan, R.; Binkley, J.S.; Seeger, R.; Pople, J.A. Self-consistent molecular orbital methods. XX. A basis set for correlated wave functions. *J. Chem. Phys.* **1980**, *72*, 650–654. [CrossRef]
104. McLean, A.D.; Chandler, G.S. Contracted Gaussian basis sets for molecular calculations. I. second row atoms, Z=11–18. *J. Chem. Phys.* **1980**, *72*, 5639–5648. [CrossRef]
105. Curtiss, L.A.; McGrath, M.P.; Blandeau, J.-P.; Davis, N.E.; Binning, R.C., Jr.; Radom, L. Extension of Gaussian-2 theory to molecules containing third-row atoms Ga–Kr. *J. Chem. Phys.* **1995**, *103*, 6104–6113. [CrossRef]
106. Frisch, M.J.; Pople, J.A.; Binkley, J.S. Self-consistent molecular orbital methods 25. Supplementary functions for Gaussian basis sets. *J. Chem. Phys.* **1984**, *80*, 3265–3269. [CrossRef]
107. Clark, T.; Chandrasekhar, J.; Spitznagel, G.W.; Schleyer, P.v.R. Efficient Diffuse Function-Augmented Basis Sets for Anion Calculations. III. The 3-21+G Basis Set for First-Row Elements, Li–F. *J. Comput. Chem.* **1983**, *4*, 294–301. [CrossRef]
108. Mardirossian, N.; Head-Gordon, M. Thirty years of density functional theory in computational chemistry: An overview and extensive assessment of 200 density functionals. *Mol. Phys.* **2017**, *19*, 2315–2372. [CrossRef]
109. Frisch, M.J.; Trucks, G.W.; Schlegel, H.B.; Scuseria, G.E.; Robb, M.A.; Cheeseman, J.R.; Scalmani, G.; Barone, V.; Petersson, G.A.; Nakatsuji, H.; et al. *Gaussian 16, Revision C.01*; Gaussian, Inc.: Wallingford, CT, USA, 2019.
110. Bader, R.F.W. *Atoms in Molecules: A Quantum Theory*; Oxford University Press: New York, NY, USA, 1990.
111. Popelier, P.L.A. *Atoms in Molecules. An Introduction*; Longman: Singapore, 2000.
112. Matta, C.F.; Boyd, R.J. *The Quantum Theory of Atoms in Molecules*; Wiley-VCH: Weinheim, Germany, 2007.
113. Keith, T.A. AIMAll (Version 15.05.18); TK Gristmill Software: Overland Park, KS, USA, 2015. Available online: aim.tkgristmill.com (accessed on 26 July 2022).
114. Gatti, C. Chemical bonding in crystals: New directions. *Z. Kristallogr.* **2005**, *220*, 399–457. [CrossRef]
115. Gatti, C.; Macchi, P. (Eds.) *A Guided Tour Through Modern Charge-Density Analysis*; Springer: Dordrecht, The Netherlands, 2011.
116. Emamian, S.; Lu, T.; Kruse, H.; Emamian, H. Exploring Nature and Predicting Strength of Hydrogen Bonds: A Correlation Analysis Between Atoms-in-Molecules Descriptors, Binding Energies, and Energy Components of Symmetry-Adapted Perturbation Theory. *J. Comput. Chem.* **2019**, *40*, 2868–2881. [CrossRef]
117. Groom, C.R.; Bruno, I.J.; Lightfoot M.P.; Ward, S.C. The Cambridge Structural Database. *Acta Cryst.* **2016**, *B72*, 171–179. [CrossRef]
118. Bläser, D.; Boese, R.; Göhner, M.; Herrmann, F.; Kuhn, N.; Ströbele, M. 2,3-Dihydro-1,3,4,5-tetraisopropylimidazol-2-yliden / 2,3-Dihydro-1,3,4,5-tetraisopropylimidazol-2-ylidene. *Z. Naturforsch. B* **2014**, *69*, 71–76. [CrossRef]
119. Cowan, J.A.; Clyburne, J.A.C.; Davidson, M.G.; Harris, R.L.W.; Howard, J.A.K.; Küpper, P.; Leech, M.A.; Richards, S.P. On the Interaction between N-Heterocyclic Carbenes and Organic Acids: Structural Authentication of the First N–H···C Hydrogen Bond and Remarkably Short C–H···O Interactions. *Angew. Chem., Int. Ed.* **2002**, *41*, 1432–1434. [CrossRef]
120. Kinney, Z.J.; Rheingold, A.L.; Protasiewicz, J.D. Preferential N–H···:C< hydrogen bonding involving ditopic NH-containing systems and N-heterocyclic carbenes. *RSC Adv.* **2020**, *10*, 42164–42171.
121. Li, C.-Y.; Kuo, Y.-Y.; Tsai, J.-H.; Yap, G.P.A.; Ong, T.-G. Amine-Linked N-Heterocyclic Carbenes: The Importance of an Pendant Free-Amine Auxiliary in Assisting the Catalytic Reaction. *Chem. Asian J.* **2011**, *6*, 1520–1524. [CrossRef]
122. Movassaghi, M.; Schmidt, M.A. N-Heterocyclic Carbene-Catalyzed Amidation of Unactivated Esters with Amino Alcohols. *Org. Lett.* **2005**, *7*, 2453–2456. [CrossRef] [PubMed]
123. Srivastava, R.; Moneuse, R.; Petit, J.; Pavard, P.-A.; Dardun, V.; Rivat, M.; Schiltz, P.; Solari, M.; Jeanneau, E.; Veyre, L.; et al. Early/Late Heterobimetallic Tantalum/Rhodium Species Assembled Through a Novel Bifunctional NHC-OH Ligand. *Chem. Eur. J.* **2018**, *24*, 4361–4370. [CrossRef]
124. Guo, R.; Huang, X.; Zhao, M.; Lei, Y.; Ke, Z.; Kong, L. Bifurcated Hydrogen-Bond-Stabilized Boron Analogues of Carboxylic Acids. *Inorg. Chem.* **2019**, *58*, 13370–13375. [CrossRef] [PubMed]
125. Jabłoński, M. Does the Presence of a Bond Path Really Mean Interatomic Stabilization? The Case of the Ng@Superphane (Ng = He, Ne, Ar, and Kr) Endohedral Complexes. *Symmetry* **2021**, *13*, 2241. [CrossRef]
126. Jabłoński, M. Endo- and exohedral complexes of superphane with cations. *J. Comput. Chem.* **2022**, *43*, 1120–1133. [CrossRef] [PubMed]
127. Jabłoński, M. The physical nature of the ultrashort spike–ring interaction in iron maiden molecules. *J. Comput. Chem.* **2022**, *43*, 1206–1220. [CrossRef] [PubMed]
128. Mallinson, P.R.; Smith, G.T.; Wilson, C.C.; Grech, E.; Wozniak, K. From Weak Interactions to Covalent Bonds: A Continuum in the Complexes of 1,8-Bis(dimethylamino)naphthalene. *J. Am. Chem. Soc.* **2003**, *125*, 4259–4270. [CrossRef]
129. Dominiak, P.M.; Makal, A.; Mallinson, P.R.; Trzcinska, K.; Eilmes, J.; Grech, E.; Chruszcz, M.; Minor, W.; Woźniak, K. Continua of Interactions between Pairs of Atoms in Molecular Crystals. *Chem. Eur. J.* **2006**, *12*, 1941–1949. [CrossRef]
130. Jabłoński, M.; Palusiak, M. Basis Set and Method Dependence in Atoms in Molecules Calculations. *J. Phys. Chem. A* **2010**, *114*, 2240–2244. [CrossRef]
131. Jabłoński, M.; Solà, M. Influence of Confinement on Hydrogen Bond Energy. The Case of the FH···NCH Dimer. *J. Phys. Chem. A* **2010**, *114*, 10253–10260. [CrossRef]

Article

The Role of Hydrogen Bonds in Interactions between [PdCl$_4$]$^{2-}$ Dianions in Crystal

Rafał Wysokiński [1,*], Wiktor Zierkiewicz [1,*], Mariusz Michalczyk [1,*], Thierry Maris [2] and Steve Scheiner [3]

[1] Faculty of Chemistry, Wrocław University of Science and Technology, Wybrzeże Wyspiańskiego 27, 50-370 Wrocław, Poland
[2] Département de Chimie, Université de Montréal, Montréal, QC H3C 3J7, Canada; thierry.maris@umontreal.ca
[3] Department of Chemistry and Biochemistry, Utah State University, Logan, UT 84322-0300, USA; steve.scheiner@usu.edu
* Correspondence: rafal.wysokinski@pwr.edu.pl (R.W.); wiktor.zierkiewicz@pwr.edu.pl (W.Z.); mariusz.michalczyk@pwr.edu.pl (M.M.)

Abstract: [PdCl$_4$]$^{2-}$ dianions are oriented within a crystal in such a way that a Cl of one unit approaches the Pd of another from directly above. Quantum calculations find this interaction to be highly repulsive with a large positive interaction energy. The placement of neutral ligands in their vicinity reduces the repulsion, but the interaction remains highly endothermic. When the ligands acquire a unit positive charge, the electrostatic component and the full interaction energy become quite negative, signalling an exothermic association. Raising the charge on these counterions to +2 has little further stabilizing effect, and in fact reduces the electrostatic attraction. The ability of the counterions to promote the interaction is attributed in part to the H-bonds which they form with both dianions, acting as a sort of glue.

Keywords: π-hole; counterions; molecular electrostatic potential; AIM; energy decomposition

1. Introduction

More than a century of study of the hydrogen bond (HB) [1,2] has yielded a myriad of facts, ideas, and principles concerning this crucial linker in the microscopic world. It has been learned [3–25] that Coulombic forces are a critical factor, wherein the polarization of the R-H covalent bond induces a partial positive charge on the H which attracts an approaching nucleophile. This basic attraction is supplemented by a charge transfer from the lone electron pair of the nucleophile to the antibonding σ*(R-H) orbital of the acid. Other contributing factors arise from mutual polarization of the two subunits and London dispersion. The HBs that result from this confluence of phenomena are both ubiquitous and of enormous importance [7,12,14,26–30], essential for life, occurring within such biomolecules as proteins or nucleic acids, enzymatic reaction pathways, catalytic intermediates, and of course in water.

Of more recent interest are a number of other closely related noncovalent interactions, where the bridging H of the HB is replaced by any of a long list of atoms that lie on the right of the periodic table. These bonds are typically classified by the family of the bridging atom, e.g., halogen, chalcogen, tetrel, pnicogen, and triel bonds. However, they share with the HB many of the same contributing factors [31–33]. The bridging atom acquires a positive region, differing from the H only in that this region is more localized, which can similarly attract a nucleophile. Moreover, like the HB, these other noncovalent bonds are likewise stabilized by charge transfer, polarization, and dispersion [34–42].

As a ubiquitous and powerful force, the HB contributes heavily to assembling and preserving the architecture of supramolecular synthons [15,43–55]. Of the sorts of assemblies to which H-bonds contribute, among the most intriguing are those that contain "like–like charge" interactions where ions of like charge lie adjacent to one another. These anion···anion [56–72] and cation···cation [72–77] interactions are counterintuitive and have generated recent and extensive scrutiny [5,77–81], being called among other names an "anti-electrostatic" hydrogen bond (AEHB) [78,82]. In one picture, the Coulombic repulsion is overcome by resonance-type covalency represented by n→π*/n→σ* charge transfer [82]. Another view claims that the nominal point charge-point charge repulsion is oversimplified [83], and the full electrostatic term is more complicated, arising from the charge distribution over the entire subunit, as well as charge penetration effects. Another factor helping to overcome the implicit repulsion is the cooperativity of hydrogen bonding not only in simple dimers, but also within larger clusters [73], with a supporting role played by dispersion. Such cooperativity has been confirmed from both spectroscopic (IR, NMR) and computational perspectives [73].

Our own group has extended the understanding of this question to anion–anion interactions that involve π-holes [59–61,66–71]. These systems were stabilized by an assortment of noncovalent bonds, including pnicogen, triel, spodium, noble gas, and alkali earth bonds. The calculations showed that these complexes were metastable in the gas phase, wherein the dissociation was impeded by an energy barrier, but were fully stable in solution, despite their like charges. The innate attractive forces in systems such as tetrachloridopalladate(II) or trichloridomercurate(II) units [66,68] were demonstrated by AIM, NBO, and NCI analyses, supported by experimental data. The results showed how the presence of counterions could stabilize these anion–anion interactions, in large part through the attenuation of the charges residing on the interacting anions.

One very recent study in particular [66] explored the interaction between a pair of [PdCl$_4$]$^{2-}$ dianions. The double charge on each makes for a particularly repulsive naked Coulombic repulsion. Indeed, in the absence of any surrounding environment, these two dianions strongly repel one another. However, the inclusion of a few of the surrounding counterions, along with the H-bonds which they form to these dianions, enables the entire system to be held together as seen in the crystal. In this case, the two dianions are held together in part by a charge transfer from the Cl lone pair of one unit to the vacant Pd π-orbitals lying above the plane of the other.

The present work is designed to explore the precise mechanism whereby this nominally highly dianion–dianion repulsion can be overcome by such external species. What is the relative importance of the charges on these surrounding molecules as compared to the H-bonds which they form with the anions? Is this a purely electrostatic phenomenon, or are there strong elements of polarization and charge dispersal which are important? Are there any specific stabilizing interactions between the pair of [PdCl$_4$]$^{2-}$ units which can act to hold them together if the overall Coulombic repulsion can be overcome, and how might these noncovalent bonds be affected by the surrounding molecules?

The analysis is designed to focus on a specific system whose crystal structure has been determined as an example. Figure 1 displays the relevant portion of the NETMOO [84] system, which shows some of the most important interactions. One can see the contact between the Cl of the upper unit and the Pd of that below. Quantum calculations attributed this arrangement to a π-hole bond wherein Cl lone pairs of one unit transfer charge to vacant orbitals above the Pd center of its neighbor [66]. It is also apparent that the NH groups of the counterion can engage in NH···Cl H-bonds with either of the dianions. As a starting point, the two [PdCl$_4$]$^{2-}$ anions are placed in the positions which they occupy in the crystal. Then, various models, of various size and complexity, of the counterions are added to the system in stages, monitoring the strength and nature of the interactions. The size of the counterion is examined by the comparison of the full $^+$NH$_3$CH$_2$CH$_2$CH$_2$CH$_2$NH$_3$$^+$ species which occurs in the crystal with shorter versions such as Ca^{2+}. Not only is the latter much smaller, but it is unable to engage in H-bonds. Other model ligands were considered of charge +1 and 0 so as to monitor the effect of the overall ligand charge. For example, removing a proton from $^+$NH$_3$CH$_2$CH$_2$CH$_2$CH$_2$NH$_3$$^+$ yields the very similar but monocationic $^+$NH$_3$CH$_2$CH$_2$CH$_2$CH$_2$NH$_2$ whose effects can likewise be compared with the much smaller NH$_4$$^+$ and with K$^+$ as a non-H-bonding cation. The models can be extended to those with no charge at all, such as $NH_2CH_2CH_2CH_2CH_2NH_2$, NH_3, and Ar. Lastly, one can isolate the effects of a purely electrostatic treatment by replacing any of these species with a series of point charges, incapable of accepting any charge from any of the participating units, or engaging in any noncovalent bonding of any sort.

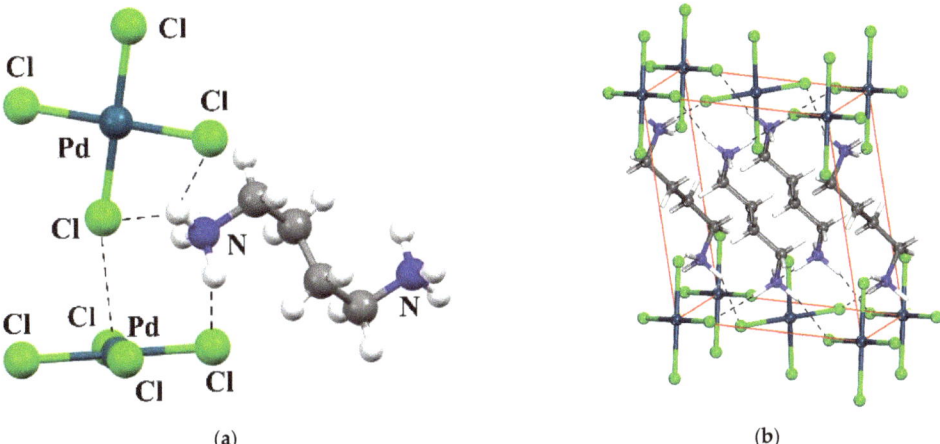

(a) (b)

Figure 1. View of the crystal structure of the studied system (CSD REFCODE: NETMOO [84]). (a): View of the $PdCl_4^{2-}$ anions (b) View of one unit-cell content showing the layered character of the structure. In both views, the shortest hydrogen bonds are shown as dashed lines.

2. Computational Methods

Quantum calculations were carried out with the aid of the Gaussian 16, Rev. C.01 set of codes [85]. DFT computations employed the PBE0-D3 functional with the explicit inclusion of dispersion corrections, along with the def2TZVP [86–88] basis set. The extrema of the molecular electrostatic potentials (MEP) were measured on the 0.001 au isodensity surface using the MultiWFN program [89,90]. NBO analysis (NBO 7.0 [91]) probed the details of charge transfer and supplied natural atomic charges. Bader's AIM methodology [92] elucidated bond paths and quantitative measures of their strength via the AIMAll suite of programs [93]. The decomposition of interaction energies was carried out at the PBE0-D3/ZORA/TZ2P level of theory through the ADF-EDA procedure according to the Morokuma–Ziegler scheme embedded in ADF software [94–96]. The solid-state geometries were accessed through the Cambridge Structural Database (CSD, ver. 5.42 with updates) and supporting CCDC software, Mercury and ConQuest [97,98]. Theoretical computations were based on the NETMOO [84] crystal structure. Heavy atoms were fixed in their crystal coordinates, and the H atom positions optimized. The basis set superposition error (BSSE) was corrected via the standard counterpoise procedure [99].

3. Results

The geometry of the model system, taken directly from the X-ray coordinates, is exhibited in Figure 2a, which surrounds the $PdCl_4^{2-}$ dimer by four counterion ligands. The reader should be aware that a finite excerpt from a full crystal structure is considered here. There are a multitude of H-bonds connecting these counterions to the dianions. To avoid overcomplication of the figure, only very short ones, with R(H···Cl) less than 2.4 Å, are shown explicitly by the broken blue lines. Figure 2b focuses on the dianion dimer itself, showing clearly that the Cl of the top unit approaches the Pd of the lower to within 3.217 Å.

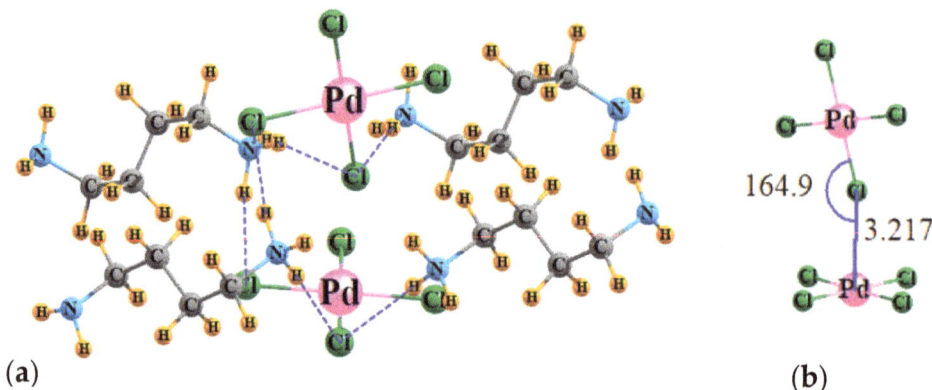

(a) (b)

Figure 2. (a) Geometry of the PdCl$_4{}^{2-}$ dimer, surrounded by four counterions, with coordinates taken directly from X-ray structures; H atom positions optimized. (b) Detailed structure of [PdCl$_4{}^{2-}$]$_2$. Distances in Å, angles in degs.

3.1. Direct Interactions between PdCl$_4$ Units

The initial calculation focuses on the isolated PdCl$_4{}^{2-}$ dimer, in the absence of any counterions. The first row of Table 1 documents the strong repulsion between the two naked dianions. The interaction energy of the pair within their X-ray structure is +212 kcal/mol. Most of that can be attributed to a highly repulsive electrostatic component of +218 kcal/mol. Indeed, it would be difficult to generate any degree of attraction for the negatively charged Cl atom when the maximum of the MEP above the Pd atom is −371 kcal/mol, especially when coupled with the $V_{S,min}$ on the Cl of −387 kcal/mol.

Table 1. Interaction energy and its electrostatic component for interactions between subunits, and the maximum and minimum of the MEP of the uncomplexed subunits (kcal/mol), and total charge (Q, e) assigned to PdCl$_4$ segment within complexes.

	E_{int}	E_{ES}	$V_{s,max}$	$V_{S,min}$	Q [c]
(PdCl$_4$)$^{2-}{}_2$	+212	+218	−371	−387	−2.00
neutral					
+4 Ar	+209	+206	−184	−202	−2.00
+4 NH$_3$	+182	+173	−172	−194	−1.96
+4 L^{0,a}	+157	+159	−162	−185	−1.92
+4 (PC)^{0,b}	+182				−2.00
+1					
+4 K$^+$	−97	−94	−58	−75	−1.91
+4 NH$_4{}^+$	−111	−99	−56	−72	−1.80
+4 L$^+$	−101	−94	−53	−70	−1.82
+4 (PC)$^+$	−98				−2.00
+2					
+2 Ca^{2+}	−121	−44	−27	−63	−1.73
+2 L^{2+}	−124	−64	−64	−84	−1.85
+2 (PC)$^{2+}$	−108				−2.00

[a] L refers to the butyl ligands with amino groups on both ends. L^0 has NH$_2$ on both ends, L$^+$ has NH$_3{}^+$ on one end near the PdCl$_4$, L^{2+} has NH$_3{}^+$ on both ends for total charge of +2. [b] PC refers to the constellation of point charges that approximate L^0, L$^+$, or L^{2+}, respectively. [c] total charge on each PdCl$_4$ unit (average of two).

The next rows of Table 1 indicate the effects of adding four neutral ligands around this dianion pair. For the purposes of examining the interactions of the two principal dianions, two of these ligands were assigned to each PdCl$_4$ unit to compose a [PdCl$_4$]$^{2-}$L$_2$ subunit. The MEP was computed for each [PdCl$_4$]$^{2-}$L$_2$ subunit, and the interaction energy between them was computed as the energy of the dimerization reaction (1)

$$2\,[PdCl_4]^{2-}L_2 \rightarrow [PdCl_4]^{2-}{}_2 L_4 \qquad (1)$$

The Ar atoms were placed at the positions of the proximate N atoms of the NH$_3$-(CH$_2$)$_4$-NH$_3{}^{2+}$ counterions within the X-ray structure, as were the N atoms of the NH$_3$ units. The H atoms of the latter were optimized and thus engaged in NH···Cl H-bonds with the anions. The Ar atoms have essentially no effect on the repulsive energy between the two anions, diminishing it by only 3 kcal/mol. The Ar atoms do reduce the negative value of V$_{S,max}$, lowering its magnitude from −371 to −184 kcal/mol. They also strongly reduce the negative potential on Cl, lowering V$_{S,min}$ from −387 to −202 kcal/mol. However, the electrostatic component of the interaction is little changed, dropping from +218 to +206 kcal/mol. Nor does Ar absorb any of the negative charge of these anions, leaving their total charge at −2.00.

The H-bonds connected with NH$_3$ have a larger impact, albeit still fairly small. V$_{S,max}$ drops a bit more, down to −172 kcal/mol, and V$_{S,min}$ is reduced as well, causing a drop in E$_{ES}$ to +173 kcal/mol. The NH$_3$ units absorb a small amount of density, leaving the charges on the PdCl$_4$ dianions at −1.96. Nevertheless, the interaction energy remains high at +182 kcal/mol. Extending the NH$_3$ units to the full NH$_2$(CH$_2$)$_4$NH$_2$ ligands, likewise capable of engaging in NH···Cl H-bonds, has a further stabilizing effect. These longer species absorb a bit more of the anion's charge, reducing it to −1.92, and raises V$_{S,max}$ a small amount, up to −162 kcal/mol and also reducing the magnitude of V$_{S,min}$. The electrostatic component and interaction energy are accordingly reduced as well, both down below +160 kcal/mol.

In order to distinguish the effects of this longer ligand arising from purely electrostatic considerations, from H-bonding, polarization, dispersion, and so on, these four NH$_2$(CH$_2$)$_4$NH$_2$ ligands were each replaced by a series of point charges. There was a one-to-one replacement of each atom of the ligand by such a charge, which was superimposed on the atomic position, and was assigned the natural charge equal to that of the ligand within the complex. The next row of Table 1 shows that this constellation of point charges has a small stabilizing effect on the dianion repulsion, less than that of the true ligand, and only roughly equivalent to the much smaller NH$_3$ molecule. Of course, as simply a collection of charges, these pseudoligands cannot absorb any charge, so that of each ligand remains at −2.00. So, it is clear that the H-bonds connected with the full ligands, as well as any charge which they can accept from the anions, have a significant effect on stabilizing the anion pair, albeit far too small to make this interaction attractive.

A second iteration of this analysis would involve placing a positive charge on each ligand. The simplest such counterion, and one incapable of engaging in a H-bond, would be a monatomic cation such as K$^+$. As exhibited in the next row of Table 1, the inclusion of four such cations makes the interaction exothermic with negative values of E$_{int}$. The presence of these cations also strongly reduces the negative values of both V$_{S,max}$ and V$_{S,min}$, both smaller in magnitude than −80 kcal/mol. These changes are partly responsible for the negative, attractive electrostatic component at −94 kcal/mol.

The E$_{int}$ of −97 kcal/mol is enhanced to −111 kcal/mol if the K$^+$ is morphed into the NH$_4{}^+$ ion of roughly the same size, but with the added capability of forming NH···Cl H-bonds. This mutation has a small reducing effect on the MEP extrema and drops the formal charge on the PdCl$_4$ unit to −1.80, adding to a slightly more negative E$_{ES}$, which contributes to the more negative interaction energy. Enlarging the ammoniums to the much longer NH$_2$(CH$_2$)$_4$NH$_3{}^+$ was done in such a way that it is the positively charged NH$_3{}^+$ unit that is placed close to the PdCl$_4$ species. This counterion enlargement has a slightly deleterious effect, increasing E$_{int}$ from −111 to −101 kcal/mol, as well as dropping the electrostatic attraction energy by 5 kcal/mol. This rise in E_{int} may be due to the lesser concentration of the positive charge in the larger cation. Each of these counterions, whether NH_4^+ or $NH_2(CH_2)_4NH_3^+$, involves itself in two NH··Cl H-bonds for a total of four such bonds with each $PdCl_4$ unit. The replacement of the ligand atoms by their corresponding point charges is just slightly less effective, with E_{int} becoming less exothermic by 3 kcal/mol.

It may be noted further from Table 1 that the electrostatic components are all quite attractive when any of these monocations are added, nearly -100 kcal/mol. On the other hand, even with these counterions present, $V_{S,max}$ on Pd remains negative by between 53 and 58 kcal/mol. This contradiction argues against taking a positive $V_{S,max}$ as a condition for an exothermic association. In sum, adding a +1 charge to the surrounding ligands enables them to absorb a bit more of the $(PdCl_4)^{2-}$ dianion's negative charge. Most effective in this regard is the set of four ammonium cations which drop the dianion's charge down to -1.80.

The last few rows of Table 1 refer to the addition of two dications instead of the four monocations. (The smaller number of the former is necessary in order to maintain overall electroneutrality. Equation (1) must be modified to describe each subunit as $[PdCl_4]^{2-}L^{2+}$). Despite the reduction in their number, the dications prove more effective at promoting a more exothermic association. Whether the small monatomic Ca^{2+} or the much larger $NH_3(CH_2)_4NH_3^{2+}$ dications, E_{int} drops below -120 kcal/mol. Even the collection of point charges designed to mimic the long ligand dication is effective in this regard, with E_{int} of -108 kcal/mol. The comparison of the Ca^{2+} and the L^{2+} systems enables some assessment of H-bonds, which are only possible for the latter. It is intriguing that although the H-bonding ligands leave both $V_{S,max}$ and $V_{S,min}$ more negative, the total electrostatic term is nonetheless more attractive when compared to Ca^{2+}.

In fact, upon moving the analysis to the dications, the electrostatic component diverges substantially from the full E_{int}. These two quantities differ by some 60–80 kcal/mol. More specifically, while the transition from monatomic K^+ monocation to Ca^{2+} dication makes E_{int} more negative, and the same sort of change can be seen from L^+ to L^{2+}, it has the opposite effect on E_{ES}, which becomes substantially less negative. This change to a less attractive electrostatic term contrasts with the less negative $V_{S,max}$ quantities associated with the monatomic ions. Despite their smaller number, the Ca^{2+} dications are much better at dispersing the negative charge on the central dianions than K^+, dropping this charge down to -1.73. There is much less distinction between the longer ligands, where the dications are slightly poorer at absorbing this charge than the monocations.

3.2. Secondary Interactions

It must be understood that the interaction energy between the two subunits is not wholly due to the Pd···Cl bond. The electrostatic term arises from interactions between the entire charge distributions of each subunit, which includes not only the central $PdCl_4$ but also any ligands appended to it. There are also polarization and dispersion energies that involve the entire electron clouds. Added to that are a number of specific noncovalent contacts as well. For example, the NH groups on the small NH_3 and NH_4^+ entities on one subunit can engage in H-bonds with the Cl atoms of the $PdCl_4$ of the other, but also NH··N H-bonds with one another. The same is true of the larger ligands comprised of NH and CH proton donors. Even the monatomic counterions, such as K^+ and Ca^{2+}, are capable of forming specific bonding contacts with the Cl atoms of the opposite subunit.

One can elucidate such interactions via the examination of AIM bond paths. In order to convey some sense of the number of these bonds, the AIM diagram of the system containing four full monocationic ligands is provided in Figure 3. There is a multitude of H-bonds and other noncovalent interactions between these ligands and both $PdCl_4$ units, and even between one another. The inset to Figure 3 focuses on the dianion pair and one of these ligands for greater clarity. Figure 4 places clearly in evidence the various H-bonds that arise when these ligands are replaced by the smaller NH_4^+, NH_3, and K^+ species. The chief markers of the strengths of these various interactions are contained in Table 2. The first column displays the density of the Pd···Cl bond path between the two subunits, which seems to be relatively constant at 0.013 au. This nearly fixed amount is not surprising in view of the fact that the intermolecular Pd···Cl distance was held constant at its X-ray value regardless of the addition of any ligands, and ρ_{BCP} has been shown repeatedly in the literature [12,100,101] to be very sensitive to this interatomic distance. Prior works in the literature [102] have found and used a relationship that ties the energy of a noncovalent bond to $\frac{1}{2}$ V, where V represents the potential energy density at the bond critical point. The use of this relationship leads to an estimate of the Pd···Cl bond energy as roughly 3 kcal/mol, as listed in the penultimate column of Table 2.

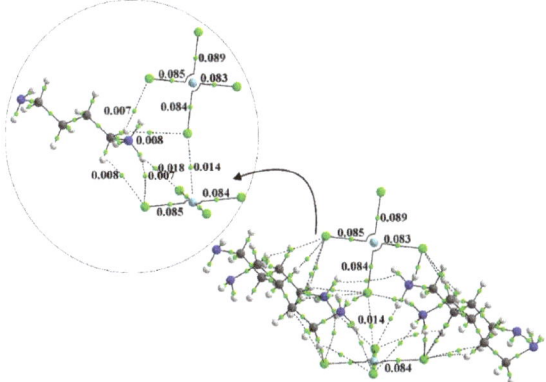

Figure 3. The AIM diagram of [NH$_3$-(CH$_2$)$_4$-NH$_2$]$_4$[PdCl$_4$]$_2$. Numbers refer to bond path critical point in au.

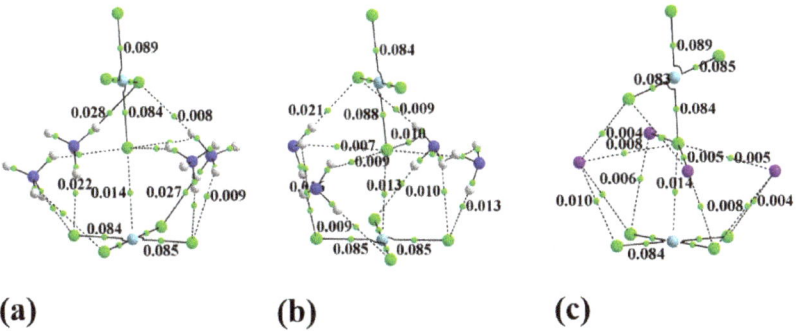

Figure 4. The AIM diagram of dianion surrounded by (a) 4 NH$_4^+$, (b) 4 NH$_3$, and (c) 4 K$^+$.

The other two columns of Table 2 report the bond critical point density and potential energy density as a sum of all bond paths that stretch between the two subunits, exclusive of Pd···Cl. These values suggest that the Pd···Cl bond is only part of the story, and that in a number of cases, AIM would suggest that the sum of the other noncovalent interactions exceeds this primary component. This energy sum from the last column of Table 2 shows a clear progression from monatomic species, such as Ar and K$^+$, to the small H-bonding NH$_3$ and NH$_4^+$, up to the largest ligands containing the connecting butyl chain. Notice also that this auxiliary sum drops off for the dicationic species.

Table 2. AIM properties of bond critical points between subunits in complexes.

	ρ, au		$-\frac{1}{2}$ V, kcal/mol	
	Pd···Cl	Σothers [a]	Pd···Cl	Σothers [a]
$(PdCl_4)^{2-}{}_2$	0.014	-	2.67	-
neutral				
+4 Ar	0.013	0.027	2.73	4.81
+4 NH_3	0.013	0.053	2.77	8.93
+4 $L^{0,a}$	0.013	0.058	2.80	10.03
+4 $(PC)^{0,b}$	0.013	-	2.74	-
+1				
+4 K^+	0.014	0.028	3.00	4.54
+4 NH_4^+	0.014	0.066	3.01	10.79
+4 L^+	0.014	0.071	3.01	11.60
+4 $(PC)^+$	0.014	-	2.97	-
+2				
+2 Ca^{2+}	0.014	0.012	2.89	1.69
+2 L^{2+}	0.014	0.033	2.84	5.90
+2 $(PC)^{2+}$	0.013	-	2.81	-

[a] between $[PdCl_4]^{2-}$-L_n units, $n = 2$ for neutral and monocationic ligands, 1 for dications. [b] PC refers to the constellation of point charges that approximate L^0, L^+, or L^{2+}, respectively.

Given the magnitude of the collective auxiliary noncovalent bonds within these complexes, it is perhaps not surprising that neither the total interaction energy nor even the full electrostatic term in Table 1 can always be closely related to the magnitudes of the MEP extrema on the Pd and Cl atoms, which concern only one of several noncovalent bonds. The ability of the counterions within the crystal structure to stabilize the entire lattice is mainly due to their effects on the $PdCl_4$ units. Moreover, these ligands also act as a glue between $PdCl_4$ units by forming H-bonds with both. This glue is further augmented by H-bonds between the counterions themselves.

As the electrostatic is the leading force stabilizing the current crystal, it is worth comparing the bonding between adjacent, oppositely charged atoms here with that which occurs within a common salt such as NaCl. The atoms of the $[PdCl_4]^{2-}$ dianions that come closest to one another are the Cl of one unit and the Pd of the other. This R(Pd···Cl) distance is 3.217 Å in the system described above, which is considerably shorter than 3.97 Å which corresponds to the sum of Pd and Cl vdW radii [103]. In NaCl, the R(Na···Cl) distance is only 2.8 Å in the crystal, also smaller than their vdW radii sum of 4.3 Å, so in this sense NaCl offers local attractive behavior that is parallel to that in the system under investigation here. If one now extracts a structure similar to the crystallographic arrangement of two $[PdCl_4]^{2-}$ from the NaCl lattice, i.e., a $[NaCl_4]^{3-}$···$[NaCl_4]^{3-}$ arrangement on lattice sites, it will become repulsive between the two units, just like in the Pd case. Adding neutral solvents will not cure this, but adding counterions will. So, qualitatively, a simple salt crystal will behave similarly to the Pd dianion system upon increasing the fragments investigated. Of course, the covalency of the PdCl bonds in the dianions along with the presence of the hydrogen bonds make a difference, but only on a quantitative level, which is explored in this work.

4. Conclusions

The interaction between the two naked $PdCl_4^{2-}$ dianions is clearly highly repulsive. The introduction of neutral ligands reduces the magnitudes of the negative MEP maxima and minima, which helps lower the electrostatic repulsion energy, but the interaction energy remains quite positive nonetheless, roughly equal to its total electrostatic component. Adding a positive charge to the ligands further reduces the magnitudes of the MEP extrema, although they remain negative. Nevertheless, these cations reverse the sign of the electrostatic and interaction energies, turning the latter exothermic by roughly 100 kcal/mol. Changing from four monocationic ligands to two dications reduces the total electrostatic attractive component but makes the total interaction energy a bit more exothermic. The stabilizing effect of the counterions is only partly due to the dispersal of the negative charges on the $PdCl_4^{2-}$ units or the reduction of the negative value of the π-hole on Pd. In a more global sense, the addition of the cationic ligands changes the formal charge on the entire subunit from -2 for naked $PdCl_4^{2-}$ unit to 0 after their introduction. The ability of these counterions to engage in H-bonds with both $PdCl_4$ units further acts as a glue holding them together. A partial contribution to this structure cohesion is achieved via NH···Cl hydrogen bonds.

Author Contributions: Conceptualization, M.M, W.Z. and R.W.; data curation, R.W. and M.M.; supervision, S.S.; visualization, R.W., S.S., T.M. and M.M.; writing—original draft, R.W. and M.M.; writing—review & editing, S.S., R.W., W.Z., T.M. and M.M. All authors have read and agreed to the published version of the manuscript.

Funding: This research was funded by the Polish Ministry of Science and Higher Education for the Faculty of Chemistry of Wroclaw University of Science and Technology under Grant No. 8211104160/K19W03D10 and by the US National Science Foundation under Grant No. 1954310.

Institutional Review Board Statement: Not applicable.

Informed Consent Statement: Not applicable.

Data Availability Statement: The data presented in this study are available on request from the corresponding authors.

Acknowledgments: A generous allotment of computer time from the Wroclaw Supercomputer and Networking Center is acknowledged.

Conflicts of Interest: There are no conflict to declare.

Sample Availability: Samples of the compounds are not available from the authors.

References

1. Scheiner, S. The Hydrogen Bond: A Hundred Years and Counting. *J. Indian Inst. Sci.* **2020**, *100*, 61–76. [CrossRef]
2. Latimer, W.M.; Rodebush, W.H. Polarity and ionization from the standpoint of the Lewis theory of valence. *J. Am. Chem. Soc.* **1920**, *42*, 1419–1433. [CrossRef]
3. Scheiner, S. Forty years of progress in the study of the hydrogen bond. *Struct. Chem.* **2019**, *30*, 1119–1128. [CrossRef]
4. Grabowski, S.J. Hydrogen bonds and other interactions as a response to protect doublet/octet electron structure. *J. Mol. Model.* **2018**, *24*, 38. [CrossRef] [PubMed]
5. Hobza, P.; Havlas, Z. Improper, blue-shifting hydrogen bond. *Theor. Chem. Acc.* **2002**, *108*, 325–334. [CrossRef]
6. Arunan, E.; Desiraju, G.R.; Klein, R.A.; Sadlej, J.; Scheiner, S.; Alkorta, I.; Clary, D.C.; Crabtree, R.H.; Dannenberg, J.J.; Hobza, P.; et al. Definition of the hydrogen bond (IUPAC Recommendations 2011). *Pure Appl. Chem.* **2011**, *83*, 1637–1641. [CrossRef]
7. Pihko, P.M. *Hydrogen Bonding in Organic Synthesis*; John Wiley & Sons: Hoboken, NJ, USA, 2009; pp. 1–383.
8. Gilli, G.; Gilli, P. *The Nature of the Hydrogen Bond: Outline of a Comprehensive Hydrogen Bond Theory*; Oxford University Press: Oxford, UK, 2009; pp. 1–336.
9. Espinosa, E.; Mata, I.; Alkorta, I.; Molins, E. Molecular structure and properties: Looking at hydrogen bonds. *Acta Crystallogr. A-Found. Adv.* **2007**, *63*, S37. [CrossRef]
10. Jablonski, M.; Kaczmarek, A.; Sadlej, A.J. Estimates of the energy of intramolecular hydrogen bonds. *J. Phys. Chem. A* **2006**, *110*, 10890–10898. [CrossRef]
11. Grabowski, S.J. *Hydrogen Bonding—New Insights*; Springer: Berlin/Heidelberg, Germany, 2006; Volume 3.
12. Desiraju, G.R.; Steiner, T. *The Weak Hydrogen Bond: In Structural Chemistry and Biology*; International Union of Crystallography: Chester, UK, 2006.
13. Zierkiewicz, W.; Jurecka, P.; Hobza, P. On differences between hydrogen bonding and improper blue-shifting hydrogen bonding. *Chemphyschem* **2005**, *6*, 609–617. [CrossRef]
14. Metrangolo, P.; Neukirch, H.; Pilati, T.; Resnati, G. Halogen Bonding Based Recognition Processes: A World Parallel to Hydrogen Bonding. *Acc. Chem. Res.* **2005**, *38*, 386–395. [CrossRef]

15. Steiner, T. The hydrogen bond in the solid state. *Angew. Chem. Int. Ed.* **2002**, *41*, 48–76. [CrossRef]
16. Hobza, P.; Havlas, Z. Blue-Shifting Hydrogen Bonds. *Chem. Rev.* **2000**, *100*, 4253–4264. [CrossRef] [PubMed]
17. Ghanty, T.K.; Staroverov, V.N.; Koren, P.R.; Davidson, E.R. Is the Hydrogen Bond in Water Dimer and Ice Covalent? *J. Am. Chem. Soc.* **2000**, *122*, 1210–1214. [CrossRef]
18. Isaacs, E.D.; Shukla, A.; Platzman, P.M.; Hamann, D.R.; Barbiellini, B.; Tulk, C.A. Covalency of the Hydrogen Bond in Ice: A Direct X-ray Measurement. *Phys. Rev. Lett.* **1999**, *82*, 600–603. [CrossRef]
19. Espinosa, E.; Molins, E.; Lecomte, C. Hydrogen bond strengths revealed by topological analyses of experimentally observed electron densities. *Chem. Phys. Lett.* **1998**, *285*, 170–173. [CrossRef]
20. Scheiner, S. *Hydrogen Bonding: A Theoretical Perspective*; Oxford University Press: New York, NY, USA, 1997.
21. Etter, M.C. Encoding and Decoding Hydrogen-Bond Patterns of Organic Compounds. *Acc. Chem. Res.* **1990**, *23*, 120–126. [CrossRef]
22. Zundel, G.; Sandorfy, C.; Schuster, P. *The Hydrogen Bond: Recent Developments in Theory and Experiments*; North-Holland: Amsterdam, The Netherland; Oxford, UK, 1976.
23. Riley, K.E.; Hobza, P. Noncovalent interactions in biochemistry. *Wires Comput. Mol. Sci.* **2011**, *1*, 3–17. [CrossRef]
24. Hobza, P.; Muller-Dethlefs, K. *Non-Covalent Interactions: Theory and Experiment*; Royal Society of Chemistry: London, UK, 2009.
25. Juanes, M.; Saragi, R.T.; Caminati, W.; Lesarri, A. The Hydrogen Bond and Beyond: Perspectives for Rotational Investigations of Non-Covalent Interactions. *Chem. A Eur. J.* **2019**, *25*, 11402–11411. [CrossRef]
26. Strekowski, L.; Wilson, B. Noncovalent interactions with DNA: An overview. *Mutat. Res. Fundam. Mol. Mech. Mutagenesis* **2007**, *623*, 3–13. [CrossRef]
27. Grabowski, S.J. Hydrogen Bond and Other Lewis Acid-Lewis Base Interactions as Preliminary Stages of Chemical Reactions. *Molecules* **2020**, *25*, 4668. [CrossRef]
28. Mahmudov, K.T.; Gurbanov, A.V.; Guseinov, F.I.; da Silva, M.F.C.G. Noncovalent interactions in metal complex catalysis. *Coordin. Chem. Rev.* **2019**, *387*, 32–46. [CrossRef]
29. Bhattacharyya, M.K.; Saha, U.; Dutta, D.; Frontera, A.; Verma, A.K.; Sharma, P.; Das, A. Unconventional DNA-relevant π-stacked hydrogen bonded arrays involving supramolecular guest benzoate dimers and cooperative anion–π/π–π/π–anion contacts in coordination compounds of Co(ii) and Zn(ii) phenanthroline: Experimental and theoretical studies. *New J. Chem.* **2020**, *44*, 4504–4518. [CrossRef]
30. Newberry, R.W.; Raines, R.T. Secondary Forces in Protein Folding. *ACS Chem. Biol.* **2019**, *14*, 1677–1686. [CrossRef] [PubMed]
31. Hennemann, M.; Murray, J.S.; Politzer, P.; Riley, K.E.; Clark, T. Polarization-induced sigma-holes and hydrogen bonding. *J. Mol. Model.* **2012**, *18*, 2461–2469. [CrossRef]
32. Politzer, P.; Murray, J.S.; Lane, P. sigma-hole bonding and hydrogen bonding: Competitive interactions. *Int. J. Quantum Chem.* **2007**, *107*, 3046–3052. [CrossRef]
33. Alkorta, I.; Elguero, J.; Oliva-Enrich, J.M. Hydrogen vs. Halogen Bonds in 1-Halo-Closo-Carboranes. *Materials* **2020**, *13*, 2163. [CrossRef]
34. Clark, T.; Hennemann, M.; Murray, J.S.; Politzer, P. Halogen bonding: The sigma-hole. *J. Mol. Model.* **2007**, *13*, 291–296. [CrossRef]
35. Alkorta, I.; Elguero, J.; Frontera, A. Not Only Hydrogen Bonds: Other Noncovalent Interactions. *Crystals* **2020**, *10*, 180. [CrossRef]
36. Murray, J.S.; Politzer, P. sigma-Holes and Si···N intramolecular interactions. *J. Mol. Model.* **2019**, *25*, 101. [CrossRef]
37. Politzer, P.; Murray, J.S.; Clark, T. The π-hole revisited. *Phys. Chem. Chem. Phys.* **2021**, *23*, 16458–16468. [CrossRef]
38. Politzer, P.; Murray, J.S.; Clark, T.; Resnati, G. The sigma-hole revisited. *Phys. Chem. Chem. Phys.* **2017**, *19*, 32166–32178. [CrossRef] [PubMed]
39. Politzer, P.; Murray, J.S. sigma-holes and π-holes: Similarities and differences. *J. Comput. Chem.* **2018**, *39*, 464–471. [CrossRef] [PubMed]
40. Politzer, P.; Murray, J.S.; Clark, T. sigma-Hole Bonding: A Physical Interpretation. *Top Curr. Chem.* **2015**, *358*, 19–42. [CrossRef] [PubMed]
41. Politzer, P.; Murray, J.S.; Clark, T. Halogen bonding and other sigma-hole interactions: A perspective. *Phys. Chem. Chem. Phys.* **2013**, *15*, 11178–11189. [CrossRef] [PubMed]
42. Murray, J.S.; Lane, P.; Clark, T.; Riley, K.E.; Politzer, P. sigma-Holes, pi-holes and electrostatically-driven interactions. *J. Mol. Model.* **2012**, *18*, 541–548. [CrossRef]
43. Grabowski, S.J. Interactions Steering Arrangement of Molecules in Crystals. *Crystals* **2020**, *10*, 130. [CrossRef]
44. Varadwaj, P.R.; Varadwaj, A.; Marques, H.M.; Yamashita, K. Significance of hydrogen bonding and other noncovalent interactions in determining octahedral tilting in the CH3NH3PbI3 hybrid organic-inorganic halide perovskite solar cell semiconductor. *Sci. Rep.* **2019**, *9*, 50. [CrossRef]
45. Riley, K.E.; Pitonak, M.; Cerny, J.; Hobza, P. On the Structure and Geometry of Biomolecular Binding Motifs (Hydrogen-Bonding, Stacking, X-H center dot center dot center dot pi): WFT and DFT Calculations. *J. Chem. Theory Comput.* **2010**, *6*, 66–80. [CrossRef]
46. Takahashi, O.; Kohno, Y.; Nishio, M. Relevance of Weak Hydrogen Bonds in the Conformation of Organic Compounds and Bioconjugates: Evidence from Recent Experimental Data and High-Level ab Initio MO Calculations. *Chem. Rev.* **2010**, *110*, 6049–6076. [CrossRef]
47. Rissanen, K. Weak Intermolecular Interactions in the Solid State. *Croat. Chem. Acta* **2010**, *83*, 341–347.

48. Matczak-Jon, E.; Slepokura, K.; Zierkiewicz, W.; Kafarski, P.; Dabrowska, E. The role of hydrogen bonding in conformational stabilization of 3,5,6-and 3,5-substituted (pyridin-2-yl)aminomethane-1,1-diphosphonic acids and related (pyrimidin-2-yl) derivative. *J. Mol. Struct.* **2010**, *980*, 182–192. [CrossRef]
49. Braga, D.; Grepioni, F.; Tedesco, E. X–H—π (X = O, N, C) Hydrogen bonds in organometallic crystals. *Organometallics* **1998**, *17*, 2669–2672. [CrossRef]
50. Braga, D.; Grepioni, F. C-H center dot center dot center dot O hydrogen bonds in organometallic crystals. *Intermol. Interact.* **1998**, 83–96.
51. Resnati, G.; Scilabra, P.; Terraneo, G. Chalcogen Bonding in Crystal Engineering. *Acta Cryst.* **2019**, *75*, e488. [CrossRef]
52. Maiti, M.; Thakurta, S.; Pilet, G.; Bauza, A.; Frontera, A. Two new hydrogen-bonded supramolecular dioxo-molybdenum(VI) complexes based on acetyl-hydrazone ligands: Synthesis, crystal structure and DFT studies. *J. Mol. Struct.* **2021**, *1226*, 129346. [CrossRef]
53. Politzer, P.; Murray, J.S. Analysis of Halogen and Other sigma-Hole Bonds in Crystals. *Crystals* **2018**, *8*, 42. [CrossRef]
54. Frontera, A.; Bauzá, A. On the importance of σ–hole interactions in crystal structures. *Crystals* **2021**, *11*, 1205. [CrossRef]
55. Frontera, A.; Bauza, A. On the importance of pnictogen and chalcogen bonding interactions in supramolecular catalysis. *Int. J. Mol. Sci.* **2021**, *22*, 12550. [CrossRef]
56. Priola, E.; Mahmoudi, G.; Andreo, J.; Frontera, A. Unprecedented [d9]Cu$\cdots$[d10]Au coinage bonding interactions in {Cu(NH3)4[Au(CN)2]}+[Au(CN)2]−salt. *Chem. Commun.* **2021**, *57*, 7268–7271. [CrossRef]
57. Gomila, R.M.; Bauza, A.; Mooibroek, T.J.; Frontera, A. π-Hole spodium bonding in tri-coordinated Hg(ii) complexes. *Dalton Trans.* **2021**, *50*, 7545–7553. [CrossRef]
58. Daolio, A.; Pizzi, A.; Terraneo, G.; Frontera, A.; Resnati, G. Anion$\cdots$Anion Interactions Involving σ-Holes of Perrhenate, Pertechnetate and Permanganate Anions. *ChemPhysChem* **2021**, *22*, 2281–2285. [CrossRef] [PubMed]
59. Zierkiewicz, W.; Wysokinski, R.; Michalczyk, M.; Scheiner, S. On the Stability of Interactions between Pairs of Anions–Complexes of MCl3$^-$ (M=Be, Mg, Ca, Sr, Ba) with Pyridine and CN$^-$. *Chemphyschem* **2020**, *21*, 870–877. [CrossRef] [PubMed]
60. Wysokinski, R.; Zierkiewicz, W.; Michalczyk, M.; Scheiner, S. How Many Pnicogen Bonds can be Formed to a Central Atom Simultaneously? *J. Phys. Chem. A* **2020**, *124*, 2046–2056. [CrossRef] [PubMed]
61. Wysokinski, R.; Zierkiewicz, W.; Michalczyk, M.; Scheiner, S. Anion$\cdots$Anion Attraction in Complexes of MCl3$^-$ (M=Zn, Cd, Hg) with CN$^-$. *Chemphyschem* **2020**, *21*, 1119–1125. [CrossRef]
62. Mo, O.; Montero-Campillo, M.M.; Yanez, M.; Alkorta, I.; Elguero, J. Are Anions of Cyclobutane Beryllium Derivatives Stabilized through Four-Center One-Electron Bonds? *J. Phys. Chem. A* **2020**, *124*, 1515–1521. [CrossRef]
63. Miranda, M.O.; Duarte, D.J.R.; Alkorta, I. Anion-Anion Complexes Established between Aspartate Dimers. *Chemphyschem* **2020**, *21*, 1052–1059. [CrossRef]
64. Azofra, L.M.; Elguero, J.; Alkorta, I. A Conceptual DFT Study of Phosphonate Dimers: Dianions Supported by H-Bonds. *J. Phys. Chem. A* **2020**, *124*, 2207–2214. [CrossRef]
65. Azofra, L.M.; Elguero, J.; Alkorta, I. Stabilisation of dianion dimers trapped inside cyanostar macrocycles. *Phys. Chem. Chem. Phys.* **2020**, *22*, 11348–11353. [CrossRef]
66. Zierkiewicz, W.; Michalczyk, M.; Maris, T.; Wysokiński, R.; Scheiner, S. Experimental and theoretical evidence of attractive interactions between dianions: [PdCl4]$^{2-}\cdots$[PdCl4]$^{2-}$. *Chem. Commun.* **2021**, *57*, 13305–13308. [CrossRef]
67. Wysokiński, R.; Zierkiewicz, W.; Michalczyk, M.; Scheiner, S. Anion (MX3−)2 dimers (M = Zn, Cd, Hg; X = Cl, Br, I) in different environments. *Phys. Chem. Chem. Phys.* **2021**, *23*, 13853–13861. [CrossRef]
68. Wysokinski, R.; Zierkiewicz, W.; Michalczyk, M.; Scheiner, S. Crystallographic and Theoretical Evidences of Anion$\cdots$Anion Interaction. *Chemphyschem* **2021**, *22*, 818–821. [CrossRef] [PubMed]
69. Wysokinski, R.; Michalczyk, M.; Zierkiewicz, W.; Scheiner, S. Anion–anion and anion–neutral triel bonds. *Phys. Chem. Chem. Phys.* **2021**, *23*, 4818–4828. [CrossRef] [PubMed]
70. Michalczyk, M.; Zierkiewicz, W.; Wysokinski, R.; Scheiner, S. Triel Bonds within Anion$\cdots$Anion Complexes. *Phys. Chem. Chem. Phys.* **2021**, *23*, 25097–25106. [CrossRef] [PubMed]
71. Grabarz, A.; Michalczyk, M.; Zierkiewicz, W.; Scheiner, S. Anion-Anion Interactions in Aerogen-Bonded Complexes. Influence of Solvent Environment. *Molecules* **2021**, *26*, 2116. [CrossRef]
72. Quiñonero, D.; Alkorta, I.; Elguero, J. Metastable dianions and dications. *ChemPhysChem* **2020**, *21*, 1597–1607. [CrossRef]
73. Niemann, T.; Stange, P.; Strate, A.; Ludwig, R. Like-likes-Like: Cooperative Hydrogen Bonding Overcomes Coulomb Repulsion in Cationic Clusters with Net Charges up to Q=+6e. *Chemphyschem* **2018**, *19*, 1691–1695. [CrossRef]
74. Niemann, T.; Strate, A.; Ludwig, R.; Zeng, H.J.; Menges, F.S.; Johnson, M.A. Spectroscopic Evidence for an Attractive Cation-Cation Interaction in Hydroxy-Functionalized Ionic Liquids: A Hydrogen-Bonded Chainlike Trimer. *Angew. Chem. Int. Edit.* **2018**, *57*, 15364–15368. [CrossRef]
75. Efimenko, I.A.; Churakov, A.V.; Ivanova, N.A.; Erofeeva, O.S.; Demina, L.I. Cationic–anionic palladium complexes: Effect of hydrogen bond character on their stability and biological activity. *Russ. J. Inorg. Chem.* **2017**, *62*, 1469–1478. [CrossRef]
76. Chalanchi, S.M.; Alkorta, I.; Elguero, J.; Quinonero, D. Hydrogen Bond versus Halogen Bond in Cation-Cation Complexes: Effect of the Solvent. *Chemphyschem* **2017**, *18*, 3462–3468. [CrossRef]
77. Wang, C.W.; Fu, Y.Z.; Zhang, L.N.; Danovich, D.; Shaik, S.; Mo, Y.R. Hydrogen- and Halogen-Bonds between Ions of like Charges: Are They Anti-Electrostatic in Nature? *J. Comput. Chem.* **2018**, *39*, 481–487. [CrossRef]

78. Weinhold, F. Anti-Electrostatic Pi-Hole Bonding: How Covalency Conquers Coulombics. *Molecules* **2022**, *27*, 377. [CrossRef] [PubMed]
79. Holthoff, J.M.; Weiss, R.; Rosokha, S.V.; Huber, S.M. "Anti-electrostatic" Halogen Bonding between Ions of Like Charge. *Chem. A Eur. J.* **2021**, *27*, 16530–16542. [CrossRef] [PubMed]
80. Zapata, F.; Gonzalez, L.; Bastida, A.; Bautista, D.; Caballero, A. Formation of self-assembled supramolecular polymers by anti-electrostatic anion-anion and halogen bonding interactions. *Chem. Commun.* **2020**, *56*, 7084–7087. [CrossRef] [PubMed]
81. Holthoff, J.M.; Engelage, E.; Weiss, R.; Huber, S.M. "Anti-Electrostatic" Halogen Bonding. *Angew. Chem. Int. Ed.* **2020**, *59*, 11150–11157. [CrossRef]
82. Weinhold, F.; Klein, R.A. Anti-Electrostatic Hydrogen Bonds. *Angew. Chem. Int. Ed.* **2014**, *53*, 11214–11217. [CrossRef]
83. Murray, J.S.; Politzer, P. Can Counter-Intuitive Halogen Bonding Be Coulombic? *Chemphyschem* **2021**, *22*, 1201–1207. [CrossRef]
84. Maris, T.; Bravic, G.; Chanh, N.B.; Leger, J.M.; Bissey, J.C.; Villesuzanne, A.; Zouari, R.; Daoud, A. Structures and thermal behavior in the series of two-dimensional molecular composites NH3-(CH2)(4)-NH3 MCl(4) related to the nature of the metal M. Part 1: Crystal structures and phase transitions in the case M=Cu and Pd. *J. Phys. Chem. Solids* **1996**, *57*, 1963–1975. [CrossRef]
85. Frisch, M.J.; Trucks, G.W.; Schlegel, H.B.; Scuseria, G.E.; Robb, M.A.; Cheeseman, J.R.; Scalmani, G.; Barone, V.; Petersson, G.A.; Nakatsuji, H.; et al. *Gaussian 16 Revision C.01*; Gaussian, Inc.: Wallingford, CT, USA, 2016.
86. Adamo, C.; Barone, V. Toward reliable density functional methods without adjustable parameters: The PBE0 model. *J. Chem. Phys.* **1999**, *110*, 6158–6170. [CrossRef]
87. Weigend, F. Accurate Coulomb-fitting basis sets for H to Rn. *Phys. Chem. Chem. Phys.* **2006**, *8*, 1057–1065. [CrossRef]
88. Weigend, F.; Ahlrichs, R. Balanced basis sets of split valence, triple zeta valence and quadruple zeta valence quality for H to Rn: Design and assessment of accuracy. *Phys. Chem. Chem. Phys.* **2005**, *7*, 3297–3305. [CrossRef]
89. Lu, T.; Chen, F. Quantitative analysis of molecular surface based on improved Marching Tetrahedra algorithm. *J. Mol. Graph. Model.* **2012**, *38*, 314–323. [CrossRef] [PubMed]
90. Lu, T.; Chen, F. Multiwfn: A multifunctional wavefunction analyzer. *J. Comput. Chem.* **2012**, *33*, 580–592. [CrossRef] [PubMed]
91. Glendening, E.D.; Badenhoop, J.K.; Reed, A.E.; Carpenter, J.E.; Bohmann, J.A.; Morales, C.M.; Karafiloglou, P.; Landis, C.R.; Weinhold, F. *NBO 7.0*; Theoretical Chemistry Institute, University of Wisconsin: Madison, WI, USA, 2018.
92. Bader, R. *Atoms in Molecules. A Quantum Theory*; Clarendon Press: Oxford, UK, 1990.
93. Keith, A.T. *AIMAll (Version 14.11.23)*; TK Gristmill Software: Overland Park, KS, USA, 2014.
94. *ADF2014, SCM, Theoretical Chemistry*; Vrije Universiteit: Amsterdam, The Netherlands, 2014.
95. Kitaura, K.; Morokuma, K. A new energy decomposition scheme for molecular interactions within the Hartree-Fock approximation. *Int. J. Quantum Chem.* **1976**, *10*, 325–340. [CrossRef]
96. Te Velde, G.; Bickelhaupt, F.M.; Baerends, E.J.; Guerra, C.F.; Van Gisbergen, S.J.A.; Snijders, J.G.; Ziegler, T. Chemistry with ADF. *J. Comput. Chem.* **2001**, *22*, 931–967. [CrossRef]
97. Groom, C.R.; Bruno, I.J.; Lightfoot, M.P.; Ward, S.C. The Cambridge Structural Database. *Acta Crystallogr. Sect. B Struct. Sci. Cryst. Eng. Mater.* **2016**, *72*, 171–179. [CrossRef]
98. Macrae, C.F.; Bruno, I.J.; Chisholm, J.A.; Edgington, P.R.; McCabe, P.; Pidcock, E.; Rodriguez-Monge, L.; Taylor, R.; van de Streek, J.; Wood, P.A. Mercury CSD 2.0—New features for the visualization and investigation of crystal structures. *J. Appl. Crystallogr.* **2008**, *41*, 466–470. [CrossRef]
99. Boys, S.F.; Bernardi, F. Calculation of Small Molecular Interactions by Differences of Separate Total Energies—Some Procedures with Reduced Errors. *Mol. Phys.* **1970**, *19*, 553–566. [CrossRef]
100. Boonseng, S.; Roffe, G.W.; Spencer, J.; Cox, H. The nature of the bonding in symmetrical pincer palladacycles. *Dalton Trans.* **2015**, *44*, 7570–7577. [CrossRef]
101. Boonseng, S.; Roffe, G.W.; Jones, R.N.; Tizzard, G.J.; Coles, S.J.; Spencer, J.; Cox, H. The Trans Influence in Unsymmetrical Pincer Palladacycles: An Experimental and Computational Study. *Inorganics* **2016**, *4*, 25. [CrossRef]
102. Espinosa, E.; Alkorta, I.; Elguero, J.; Molins, E. From weak to strong interactions: A comprehensive analysis of the topological and energetic properties of the electron density distribution involving X-H center dot center dot center dot F-Y systems. *J. Chem. Phys.* **2002**, *117*, 5529–5542. [CrossRef]
103. Alvarez, S. A cartography of the van der Waals territories. *Dalton Trans.* **2013**, *42*, 8617–8636. [CrossRef] [PubMed]

Article

Intramolecular Interactions in Derivatives of Uracil Tautomers

Paweł A. Wieczorkiewicz [1,*], Tadeusz M. Krygowski [2] and Halina Szatylowicz [1,*]

1 Faculty of Chemistry, Warsaw University of Technology, Noakowskiego 3, 00-664 Warsaw, Poland
2 Department of Chemistry, University of Warsaw, Pasteura 1, 02-093 Warsaw, Poland
* Correspondence: pawel.wieczorkiewicz.dokt@pw.edu.pl (P.A.W.); halina.szatylowicz@pw.edu.pl (H.S.)

Abstract: The influence of solvents on intramolecular interactions in 5- or 6-substituted nitro and amino derivatives of six tautomeric forms of uracil was investigated. For this purpose, the density functional theory (B97-D3/aug-cc-pVDZ) calculations were performed in ten environments $(1 > \varepsilon > 109)$ using the polarizable continuum model (PCM) of solvation. The substituents were characterized by electronic (charge of the substituent active region, cSAR) and geometric parameters. Intramolecular interactions between non-covalently bonded atoms were investigated using the theory of atoms in molecules (AIM) and the non-covalent interaction index (NCI) method, which allowed discussion of possible interactions between the substituents and N/NH endocyclic as well as =O/−OH exocyclic groups. The nitro group was more electron-withdrawing in the 5 than in the 6 position, while the opposite effect was observed in the case of electron donation of the amino group. These properties of both groups were enhanced in polar solvents; the enhancement depended on the *ortho* interactions. Substitution or solvation did not change tautomeric preferences of uracil significantly. However, the formation of a strong NO···HO intramolecular hydrogen bond in the 5-NO$_2$ derivative stabilized the dienol tautomer from +17.9 (unsubstituted) to +5.4 kcal/mol (substituted, energy relative to the most stable diketo tautomer).

Keywords: substituent effect; solvent effect; hydrogen bond; tautomers; nitro group; amino group

1. Introduction

Uracil is a common and naturally occurring pyrimidine derivative. The best known occurrences of uracil are probably nucleic acids, as it is one of the five bases of the nucleic acid. In RNA, uracil forms a complementary base pair with adenine, while its 5-methylated derivative, called thymine, is an equivalent base in DNA [1]. Uracil and its derivatives have also found applications in other branches of biochemistry. For example, 5-fluorouracil is used in treatment of several cancer types by chemotherapy [2,3], while 5-bromo and iodo uracil derivatives are studied as radiosensitizers for radiotherapy [4–7]. In 2013, a computational study of various 5-substituted uracil derivatives (X = CN, SCN, NCS, NCO, OCN, SH, N$_3$, NO$_2$) was performed in order to identify the most suitable radiosensitizers for experimental studies [8]. The most promising derivatives with high electron affinities, 5-(N-Trifluoromethylcarboxy)aminouracil [9], 5-thiocyanatouracil [10] and 5-selenocyanatouracil [11], were synthesized. Among them, 5-thiocyanatouracil has already been tested against prostate cancer cells with promising results [12]. Some uracil derivatives show antifungal and antimicrobial properties, whereas others act as inhibitors of specific enzymes [13]. On the other hand, some of them are mutagenic, for example, 5-hydroxyuracil [14]. An interesting novel class of compounds that are derived from nucleic acid base molecules, including uracil, are ferrocene-like complexes in which the nitrogen base molecule is attached to one of the cyclopentadienyl ligands [15]. It is a relatively new class of compounds that may find applications in pharmacy, biology and electrochemistry.

An important issue regarding nucleic acid bases is tautomerism. Each of the bases can exist in several forms that differ in the position of the labile hydrogen atom. In general, one of these forms is more stable than the others, and most of the molecules exist in

that form [16–18]. For this reason, RNA and DNA base pairs are built only from N9H tautomer of purine bases and N1H of pyrimidine bases [1]. However, relative stability of the tautomers can significantly change upon oxidation, reduction [17], substitution of the nucleobase [19,20], polarity of the environment [17] and even interaction with a metal cation [21,22]. Tautomerism of nucleobases is of interest in knowledge of biochemical processes. Importantly, it has been proposed that the existence of rare tautomeric forms can cause mutations of genetic code recorded in the DNA or alter functions performed by different variants of RNA [23–27]. Therefore, much effort has been put into studying the properties of uracil and its tautomers, including both theoretical and experimental studies ([16,28] and references therein). As mentioned above, various uracil derivatives are used or currently being studied for medical applications, where they are introduced into the human body. For this reason, investigating which factors can affect the tautomeric equilibria of uracil (and how) is a relevant research topic.

Uracil consists of a pyrimidine ring and two attached −OH groups at the 2 and 4 positions. However, the most stable tautomeric form has both hydrogen atoms of the −OH groups attached to the nitrogen atoms in the pyrimidine ring. The four most stable uracil tautomers (**u1–u4**) and their two rotamers (**u5, u6**), along with their relative stabilities, are shown in Figure 1. Based on calorimetric experiments [29], it was found that the dienol form is 20 ± 10 kcal/mol less stable than **u1**, while **u3** by 19 ± 6 kcal/mol. In addition, both diketo (**u1**) and keto-enol tautomers (**u2, u3**) were identified using the dispersed fluorescence spectra, although the precise structure of the latter was not determined [30]. The most stable keto-enol tautomer was estimated to have about 9.6 kcal/mol higher energy than the diketo form (**u1**).

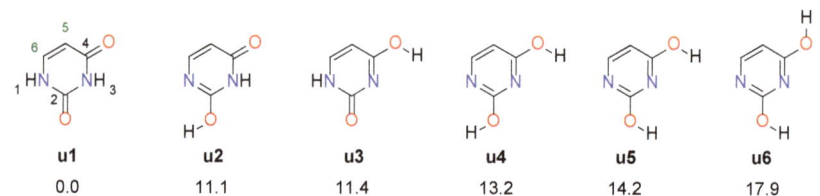

Figure 1. Four most stable tautomers of uracil (**u1–u4**) and two rotamers of **u4** (**u5, u6**). The numbers given below are their relative energies in kcal/mol.

The aim of the research is to investigate both the intramolecular interactions in uracil derivatives and their sensitivity to solvent change, as well as their ability to change tautomeric preferences. Similar studies on adenine and purine derivatives were recently carried out [31,32]; our computational results were in agreement with the experimental NMR data of 8-halopurines obtained by other groups [19,20].

For this study, we selected the 5- and 6-substituted nitro and amino derivatives of the six tautomeric forms of uracil (Figure 1). The nitro and amino groups represent model electron-withdrawing and electron-donating substituents, respectively. In addition, the nitro group rotated by 90 degrees from the plane of the ring was taken into account. This group interacts with the substituted system only inductively, as opposed to the planar NO_2 group, which acts through induction and resonance.

Two substitution positions, 5 and 6, differ in through-space *ortho* interactions and through-bond interactions with endocyclic N atoms/NH groups as well as −OH/=O groups. In position 5, depending on the tautomeric form, the substituent can interact through-space with the C4=O or C4−OH group. In turn, the substituent in position 6 can interact through-space with the N or NH group in the 1 position. Regarding the through-space interactions, in some cases, formation of an intramolecular hydrogen bond is possible. Thus, the question arises whether it can alter tautomeric preferences.

Regarding the through-bond interactions, the 5 position is *meta*-related towards two endocyclic N/NH groups and *ortho*- and *para*-related towards two exocyclic −OH/=O groups.

Conversely, the 6 position is *meta*-related to the −OH/=O and *ortho*- and *para*-related towards N/NH. Here, it is important to mention that in pyrimidines, the position of the substituent in relation to the endocyclic N atoms has a profound influence on the substituent–substituted system interaction, which affects the electron-withdrawing/donating strength of substituents. This topic is discussed in our recent paper [33].

It should be emphasized that the −OH and =O groups have opposite electronic properties: the −OH group is an electron-donating substituent, whereas =O is an electron-withdrawing substituent. Therefore, the tautomeric form should be important for the intramolecular interactions in uracil derivatives.

2. Methodology

Quantum chemical DFT calculations [34,35] were performed in the Gaussian 16 program [36]. We used the B97-D3/aug-cc-pVDZ method, in accordance with our recent research regarding purine and adenine derivatives [31,32,37]. The optimized geometries correspond to the minima on the potential energy surface since no imaginary vibrational frequencies were found. In the constrained optimization cases, i.e., systems with the NO_2 group rotated by 90 degrees, one imaginary frequency corresponding to the rotation along the C-N bond was found.

Electronic properties of substituents were evaluated using the charge of the substituent active region (cSAR) parameter [38,39]. Its definition is presented in Figure 2. Positive cSAR values indicate the deficit of electrons in the substituent active region, i.e., the substituent is electron-donating. Negative values represent an excess of electrons in the active region of the substituent, indicating its electron-withdrawing properties. To allow comparison with our other results, the atomic charges used to calculate cSAR were obtained by the Hirshfeld method [40].

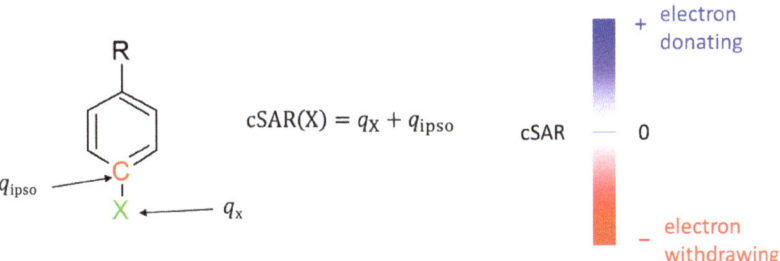

Figure 2. Definition of cSAR and interpretation of its value. q_X is the sum of atomic charges of all atoms forming a substituent X, while q_{ipso} is the atomic charge at the *ipso* atom.

In order to study solvent effects, the IEF-PCM implicit model of solvation was used [41–43]. Calculations were performed in ten media, listed in Table 1 along with their dielectric constants. It should be mentioned that the PCM has been used many times in computational studies of nucleic acid bases [6,17,19]. In the AT and GC base pairs, the molecular geometries obtained with the PCM were in good agreement with the experimental data and the calculations using the H_2O microsolvation model [44].

Table 1. Media in which calculations were performed, and their dielectric constants, ε.

Formamide	Water	DMSO	Ethanol	Pyridine	THF	o-Cresol	Chloroform	Toluene	Gas Phase
108.94	78.36	46.83	24.85	12.98	7.43	6.76	4.71	2.37	1.00

Analysis of electron density using the atoms in molecules (AIM) theory [45] was performed in the AIMAII program [46]. The main goal of this analysis was the search for possible bond critical points (BCPs) of non-covalent intramolecular interactions. When such a BCP was present, we estimated the interaction energy according to the formula of Afonin et al. (Equation (1)) [47], derived from the Espinosa equation [48].

$$E_{HB} = 0.277 \cdot V_{BCP} + 0.450 \quad (1)$$

where V_{BCP} is the potential energy density (in kcal · mol^{-1} · bohr^{-3}) at the BCP. This equation can be applied to OH···O, OH···N, OH···halogen, NH···O, NH···N, CH···O, CH···N and CH···halogen interactions.

Intramolecular interactions between non-covalently bonded atoms were also investigated using the non-covalent interaction index (NCI) method [49]. The nature of a given interaction was assigned and color-coded according to the value of sgn(λ_2) · ρ(r), where λ_2 is the second eigenvalue of the Hessian matrix of electron density (ρ(r)). Points on reduced density gradient isosurfaces with a value of sgn(λ_2)·ρ(r) > 0 indicate non-bonding (steric) contacts (in red), with sgn(λ_2)·ρ(r)~0 indicating weakly attractive interactions (e.g., van der Waals, in green) and sgn(λ_2)·ρ(r) < 0 indicating strongly attractive interactions (e.g., hydrogen and halogen bonding, in blue). For more information on the NCI analysis, see Johnson et al. [49]. In our case, NCI calculations were performed in Multiwfn 3.8 software [50] and the visualization in the VMD program [51].

3. Results and Discussion

3.1. Electronic Properties of Substituents

The raw data generated in this study and used in statistical analyses are available in the Supplementary Materials. Table 2 presents the cSAR values of the substituents in all studied systems. In the case of amino derivatives, the NH$_2$ substituent in position 6 has more than twice, in the cSAR scale, stronger electron-donating properties than in position 5. In nitro derivatives, the NO$_2$ group in position 5 is more electron-withdrawing than in position 6. Therefore, the substitution position, i.e., the position in relation to the nitrogen atoms in the ring, has a decisive influence on the properties of the substituent. In contrast, the effect of the tautomeric form of uracil is clearly less significant. It is also worth noting that in polar solvents, the characteristic properties of both NO$_2$ and NH$_2$ groups are enhanced, as shown by the difference between cSAR(X) values in the water and gas phase (Δ).

Table 2. Values of cSAR(X) (in elementary charge units, e) for X = NH$_2$, NO$_2$ groups in the gas phase. Δ indicates a difference between the cSAR(X) values in the aqueous solution (PCM) and the gas phase.

Taut.	5-NH$_2$	Δ	5-NO$_2$	Δ	5-NO$_2$ (90°)	Δ	6-NH$_2$	Δ	6-NO$_2$	Δ	6-NO$_2$ (90°)	Δ
u1	0.067	0.011	−0.163	−0.070	−0.121	−0.045	0.215	0.102	−0.001	0.002	0.001	0.003
u2	0.078	0.003	−0.166	−0.082	−0.127	−0.053	0.201	0.052	−0.052	−0.035	−0.030	−0.034
u3	0.057	0.032	−0.180	−0.047	−0.136	−0.026	0.234	0.103	0.004	0.003	0.013	0.005
u4	0.068	0.025	−0.173	−0.055	−0.138	−0.037	0.206	0.048	−0.042	−0.036	−0.021	−0.034
u5	0.068	0.025	−0.172	−0.055	−0.139	−0.037	0.214	0.038	−0.035	−0.044	−0.015	−0.041
u6	0.042	0.055	−0.136	−0.018	−0.157	−0.009	0.208	0.045	−0.036	−0.040	−0.018	−0.037
range	0.037	0.052	0.045	0.064	0.036	0.044	0.032	0.065	0.056	0.047	0.043	0.046

In 5-NH$_2$ derivatives, electron-donating strength of the amino group decreases in the sequence: u2 > u5~u4~u1 > u3 > u6. The clearly lower cSAR(X) for u6 is a consequence of the rotation of the NH$_2$ group by 90 degrees and the formation of the hydrogen bond, H$_2$N···HO, which is discussed in more detail later in the paper. In this case, the large influence of the solvent on the cSAR(NH$_2$) value is due to the rotation of the NH$_2$ group to more planar conformation with respect to the ring in polar solvents. This strengthens the resonance effect.

In 6-NH$_2$ derivatives, electron-donating strength of the amino group decreases in the sequence: u3 > u1~u5 > u6~u4 > u2. Two systems containing the NH endocyclic group at the ortho position, u3 and u1, have the greatest electron-donating properties. An interesting difference is present between the **u5** form and its rotamers: **u4** and **u6**. Among them, the highest cSAR(NH$_2$) value and the lowest Δ occur in **u5**, where the two OH groups are facing in the same direction. When they are in opposite directions, as in **u4** and **u6**, the value of cSAR(X) is lower, while Δ is higher. This may be due to the differences in the dipole moments in these two cases, as the conformation of the OH groups has a significant impact on the value and direction of molecular dipole moment (Table S1). By far the strongest solvent effect on cSAR(X) among the 6-NH$_2$ derivatives occurs in **u1** and **u3** (highest Δ). These systems also have the highest values of the dipole moment (Table S1). All cSAR(NH$_2$) values in 5-NH$_2$ derivatives are lower than in aniline (0.094), while in 6-NH$_2$ they are higher.

Generally, in all 5-NO$_2$ tautomers, the NO$_2$ group is withdrawing electrons more strongly than in nitrobenzene, where the cSAR(X) is higher, −0.140. Its rotation by 90 degree increases cSAR(NO$_2$) by about 0.4 units. The only exception is **u6**, where a decrease in cSAR is observed; however, this is caused by the hydrogen bonding between the NO$_2$ and *ortho* OH groups. In 5-NO$_2$ systems, electron-withdrawing strength of the nitro group decreases in the sequence: **u3** > **u4**~**u5** > **u2** > **u1** > **u6**. The systems with the strongest electron-withdrawing NO$_2$ groups (**u3**, **u4** and **u5**) have an electron-donating OH group in the *ortho* position, but its hydrogen atom is directed to the endocyclic N atom, so that NO⋯OH interaction can be expected. When NO⋯HO interaction is present (5-NO$_2$ **u6**), the electron-withdrawing ability of the NO$_2$ group is the weakest. Again, the greatest variability of cSAR(X) due to solvation occurs in the derivatives with the highest values of the dipole moments (**u1** and **u2**).

In the 6-NO$_2$ derivatives, the cSAR(NO$_2$) values are high, indicating weak electron-withdrawing properties. This is caused by the disturbance of the resonance interactions by ring nitrogen atoms in *ortho* and *para* positions. Weak resonance is also evidenced by a smaller increase in cSAR due to the rotation of NO$_2$ by 90° as compared to the 5-NO$_2$ derivatives. This increase is by about 0.2 units, with the exception of **u1** and **u3** where cSAR(NO$_2$) is positive and its change due to rotation is smaller. Electron-withdrawing strength decreases in the sequence: **u2** > **u4** > **u6**~**u5** > **u1** > **u3**. The loss of electron-withdrawing properties (cSAR close to 0.0) of the 6-NO$_2$ group occurs in the **u1** and **u3** derivatives, where the NH group is in the *ortho* position. Thus, apart from the relative position of the endo N atoms and the substituent, the NO⋯HN through-space interaction has an effect as well. The summary of the cSAR analysis in the form of a bar chart is shown in Figure 3.

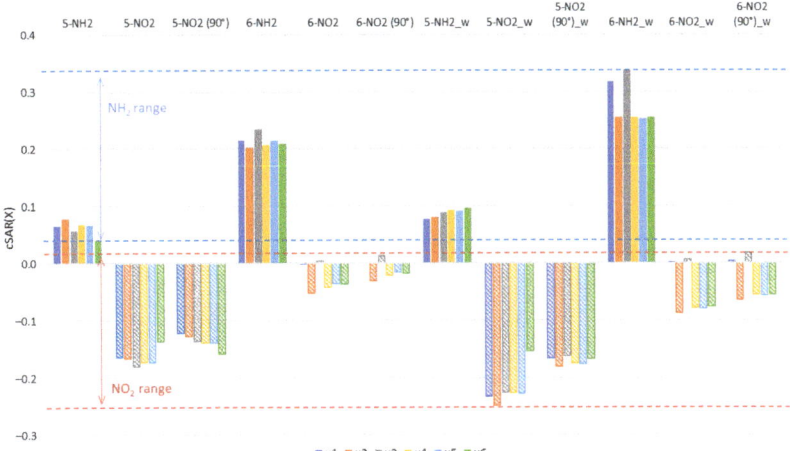

Figure 3. Values of cSAR(X) (in *e*) for X = NH$_2$, NO$_2$ groups in the gas phase and in the aqueous solution (_w). The 90° in parentheses indicates nitro derivatives where 90° rotation around the CN bond was forced.

In most cases, the dependences of cSAR(X) on $1/\varepsilon$ are well approximated by a linear function. The parameters of resulting cSAR $(X) = a \cdot (1/\varepsilon) + b$ functions are summarized in Table 3. The slope value, a, informs about the sensitivity of the electronic properties of the substituent in a given derivative to the solvent effect. In general, except for **u6** 5-NH$_2$, large absolute values of the coefficient occur in systems with a large dipole moment, and small ones in systems with a small dipole moment (Table 3 and Table S1). In 6-substituted systems (6-NO$_2$ and 6-NH$_2$), the values of a in **u1** and **u3** (*ortho* NH) clearly differ from other tautomers (*ortho* N). This can be attributed to the influence of *ortho* interactions with endocylic N/NH groups. It can be concluded that the repulsive *ortho* interaction, NH···HN for 6-NH$_2$ and NO···N for 6-NO$_2$, causes high sensitivity of the substituent properties to the solvent effect, whereas the attractive interaction causes low sensitivity. A similar effect was observed in adenine and purine derivatives [29,30].

Table 3. Parameters of cSAR(X) = $a \cdot (1/\varepsilon) + b$ linear regressions: slopes, a, and determination coefficients, R^2; unit cSAR(X) is e.

Tautomer	5-NH$_2$		5-NO$_2$		5-NO$_2$ (90°)		6-NO$_2$		6-NO$_2$ (90°)		6-NH$_2$	
	a	R^2	a	R^2	a	R^2	a	R^2	a	R^2	a	R^2
u1	−0.011	0.993	0.071	0.957	0.044	0.996	−0.002	0.772	−0.002	0.683	−0.103	0.952
u2	−0.003	0.831	0.084	0.961	0.052	0.996	0.036	0.981	0.034	0.998	−0.053	0.962
u3	−0.032	0.974	0.047	0.969	0.026	1.000	−0.003	0.654	−0.004	0.838	−0.104	0.950
u4	−0.025	0.976	0.055	0.967	0.037	0.998	0.036	0.981	0.034	0.999	−0.049	0.965
u5	−0.025	0.974	0.056	0.968	0.037	0.998	0.045	0.972	0.041	0.997	−0.039	0.979
u6	−0.056	0.968	0.018	0.983	0.010	0.989	0.040	0.974	0.037	0.998	−0.046	0.963

Properties of the =O/−OH groups of all studied forms of uracil, quantified by cSAR, are shown in Figure 4. Negative values correspond to the electron-withdrawing =O group, whereas positive values to the electron-donating −OH. Both the interactions with the substituent and the type of tautomer can affect the electron-donating (−OH) or -withdrawing (=O) properties of these groups. The electron-withdrawing properties of the =O groups are greater in the amino derivatives than in the nitro derivatives, which is shown by the more negative cSAR(=O) values in the 5-NH$_2$ and 6-NH$_2$ derivatives. In turn, the electron-donating properties of the −OH groups are greater in the nitro than in the amino derivatives. This is due to charge transfer between groups with opposite electronic properties. Global ranges of variation of cSAR are 0.143 for the =O group and 0.107 for the −OH group. The ranges for the =O group in C4 and C2 positions are 0.096 and 0.078, respectively, while the average values are −0.134 for C4 and −0.115 for C2. In the case of the −OH group, the ranges are 0.099 for C4 and 0.103 for C2 positions; the average values are 0.159 for C4 and 0.199 for C2. Thus, the characteristic electronic properties of the −OH group are on average stronger in the C2 position, while those of the =O group are stronger in the C4 position. Stronger electronic properties are accompanied by higher ranges of their variability.

The C2 position of the uracil ring is double *ortho* with respect to the two *endo* N/NH atoms/groups, while the C4 position is *ortho* and *para*. So, two electronegative atoms in the *ortho* position of the −OH group might enhance its electron-donating properties, while diminishing the electron-withdrawing by the =O group. A similar effect of *ortho* N atoms on the substituent properties was observed in our recent studies on nitro and amino derivatives of pyridine, pyrimidine, pyrazine and triazine [33].

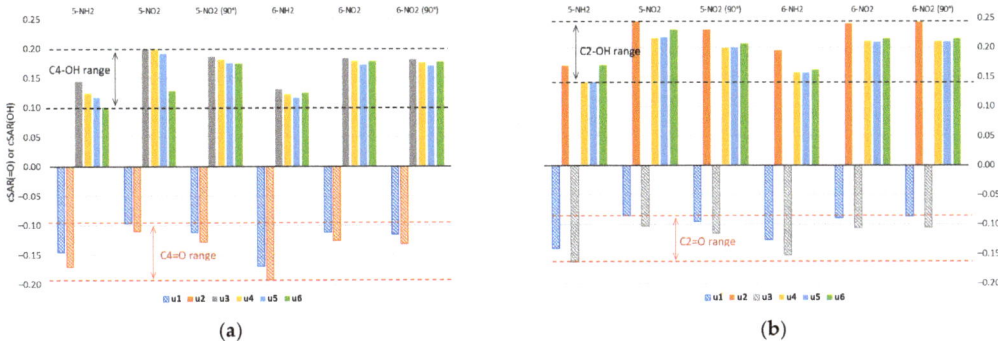

Figure 4. Values of cSAR(=O) or cSAR(OH) (in *e*) for two groups in (**a**) C4 and (**b**) C2 position of uracil molecule. Data for NH_2- and NO_2-substituted uracil derivatives in the gas phase.

3.2. Geometry

Analysis of geometry will be focused on the lengths of CN bonds connecting the NO_2 and NH_2 substituents and the substituted system. As shown in Figure 5a, they vary depending on the substitution position and the tautomeric form. In the case of 5-NH_2 derivatives, the shortest CN bond occurs in the **u2** tautomer and the longest in **u6**. The **u2** tautomer is also characterized by the highest electron-donating strength of the NH_2 group among 5-NH_2 derivatives (Table 2 and Figure 3). In the case of the **u6** tautomer in the gas phase, the NH_2 group is rotated by 90 degrees in order to form a $H_2N\cdots HO$ hydrogen bond with the OH group in the *ortho* position. This is accompanied by a significant extension of the CN bond, which reaches the length observed for the 5-NO_2 group in **u6**. In 6-NH_2 derivatives, CN bonds are shorter than in 5-NH_2, which is connected with the strong electron-donating 6-NH_2 group. A slightly longer bond relative to other tautomers occurs in **u1** and **u3**. This may be due to the presence of the NH group in the *ortho* position resulting in $NH\cdots HN$ steric interaction.

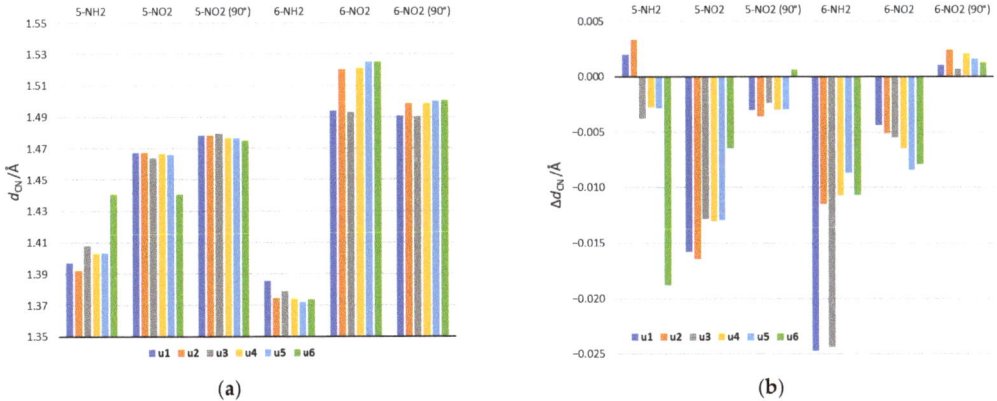

Figure 5. (**a**) The lengths of the CN bonds, connecting the substituent and substituted system in the gas phase and (**b**) differences between their lengths in formamide (the most polar solvent) and in the gas phase. Positive values of Δd_{CN} indicate longer bond in formamide than in the gas phase, while negative values indicate shorter.

In NO_2 derivatives, shorter CN bonds are found in 5-NO_2 than in 6-NO_2 systems. This is in line with the electron-withdrawing strength of the 5-NO_2 and 6-NO_2 groups.

In position 5, the shortest bond occurs in **u6**, where a strong NO···HO hydrogen bond is formed, while the second shortest is in **u3**, in which the NO_2 group has the strongest electron-accepting properties among all systems. For 6-NO_2 tautomers, clearly the shortest bonds occur in **u1** and **u3**, where the NH group is in the *ortho* position. This results from the attractive NO···HN interaction.

The rotation of the NO_2 group in 5-NO_2 derivatives causes the elongation of CN bonds, which is related to the disturbance of the resonance effect of the NO_2 group. The largest elongation occurs in the **u6** 5-NO_2 derivative. It is caused by breaking of the NO···HO hydrogen bond as a result of NO_2 rotation. In the 6-NO_2 systems, in four tautomers: **u2, u4, u5** and **u6** (*ortho* N), the NO_2 rotation clearly shortens the CN bond. This is caused by the weakening of through-space repulsive interactions with the *ortho* endocyclic N atom. Thus, the main factor determining the CN bond lengths in the 5-NO_2 derivatives is the resonance between the NO_2 group and the substituted system, while in the 6-NO_2 derivatives it is the *ortho* interaction.

The solvation effect is also reflected in the CN bond lengths. Figure 5b shows the difference in CN bond lengths between the values obtained in the aqueous solution and the gas phase. In NH_2 derivatives, a stronger solvent effect occurs in 6-NH_2 systems, while in the case of NO_2 derivatives, in 5-NO_2 systems. This is connected with the greater variability of the substituent's electronic properties in these systems (see, for example, Table 3). Thus, the bond shortening is related to an increase in the characteristic electronic properties of a given substituent, due to the increase in the solvent polarity.

3.3. Intramolecular Interactions between Non-Covalently Bonded Atoms

An important aspect of the interaction between the substituent and the substituted system are through-space *ortho* interactions, which in some cases could already be seen by the cSAR(X) values and CN bond lengths. In order to identify these interactions, the lengths of two NH/NO bonds of the NH_2/NO_2 groups were plotted against each other (Figure 6). Deviations from the equal length of these two bonds may indicate the existence of an asymmetric through-space interaction. Such plots also provide information about the attractive/repulsive nature of these interactions, based on the location of a point above or below the y = x line.

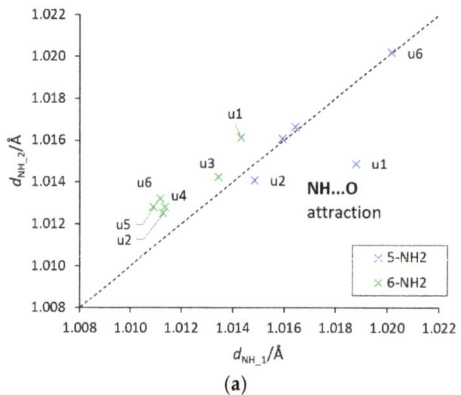
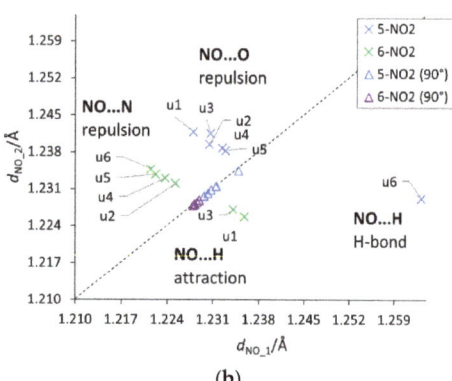

Figure 6. Plots between the lengths of the two (**a**) NH and (**b**) NO bonds of the NH_2/NO_2 groups. The dashed y = x line indicates a symmetry between the bonds. A system where asymmetry is present and the H-bond is detected (**u6** 5-NO_2) has been marked appropriately. NH_1 and NO_1 are the bonds facing towards the lower atom numbers in the ring (for example, 4 in 5 substitution), while NH_2 and NO_2 towards higher numbers (see the atom numbering in Figure 1).

First of all, it should be noticed that the asymmetry in the bond lengths of the NO_2 group is about four times greater than that of the NH_2 group. Moreover, for the nitro group, the obtained results indicate greater variability of interactions, but as expected in systems with rotated groups, the lengths of both NO bonds are similar. Both repulsive and attractive interactions as well as hydrogen bonds are observed. In the latter case, the systems in which the interaction meets the Koch–Popelier criteria for hydrogen bonding [52] are depicted as H-bonds in Figure 6. Only one system (in the gas phase), visible in the plot, **u6** 5-NO_2, fulfills the criteria. Additionally, an increase in the polarity of the solvent weakens the through-space interactions—an increase in the O$\cdots$H distance and a decrease in O$\cdots$HO angle, as shown in Figure 7. An interesting system in which, despite the symmetry between NH bonds, there is a strong H-bond is **u6** 5-NH_2. In this case, the NH_2 group rotates by 90°, and forms a $H_2N\cdots HO$ hydrogen bond. Moreover, the NH_2 group in the formamide solution rotates slightly towards the coplanar conformation (76.7° dihedral angle) and the H-bond is weakened. This rotation is an interesting example of competition of attractive through-space interactions and the resonance between the group and the substituted system. In the gas phase, the H-bond has a greater influence on the structure, but in the polar solvent, due to the weakening of the H-bond, stabilization by resonance forces the group to be coplanar. The structures of **u6** 5-NO_2 and **u6** 5-NH_2 are shown in Figure 7.

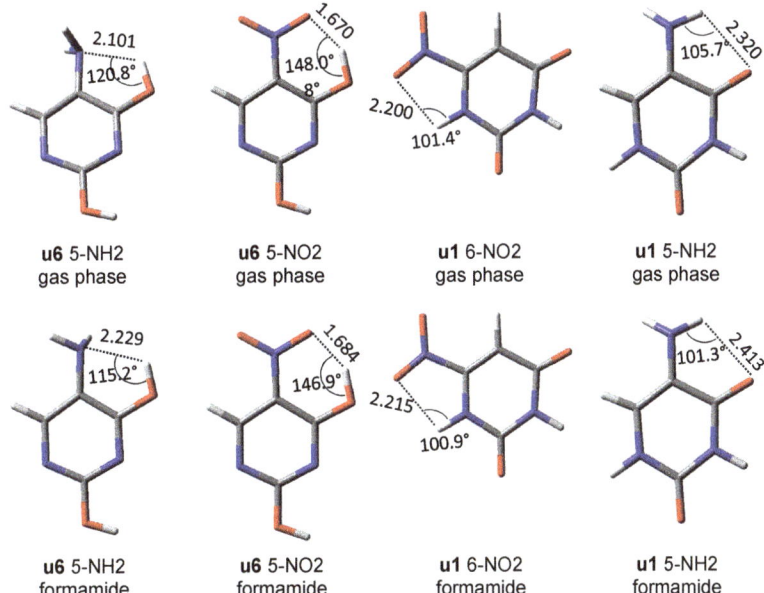

Figure 7. Structures of several systems, in which an interesting intramolecular interaction is present, and geometric data for this interaction (distances in Å).

Based on the potential energy density at the critical point of each hydrogen bond, their energy was calculated from the Afonin equation (Equation (1)). For comparison, the hydrogen bond energy was also calculated using the rotational method [53], i.e., the difference between **u6** and **u5** rotamers. Both methods give similar results (Table 4), especially in the case of stronger hydrogen bonding in **u6** 5-NO_2.

Table 4. Energies (in kcal/mol) of intramolecular hydrogen bonds in the gas phase calculated by means of rotational method (energy of **u6** minus **u5**) and from the Afonin equation (Equation (1)).

	Rotational	Afonin
u6 5-NO$_2$	−7.49	−7.55
u6 5-NH$_2$	−3.36 *	−2.61

* Calculated by rotating the OH group in the 4 position by 180° with NH$_2$ group frozen in **u6** 5-NH$_2$ conformation (perpendicular relative to the plane of the ring).

Figure 8 shows the energy scan along the dihedral angle between the amino group and the uracil ring plane. The global minimum corresponds to the conformer shown in Figure 7, the minimum near scan coordinate 300 corresponds to the form rotated by 180° from the global minimum, so that NH$_2 \cdots$HO bifurcated contact is present. Two maxima correspond to forms with close NH$\cdots$HO contacts (1.956 Å). Rotational barrier height is 5.08 kcal/mol, while the difference in energy between the two minima is 4.16 kcal/mol.

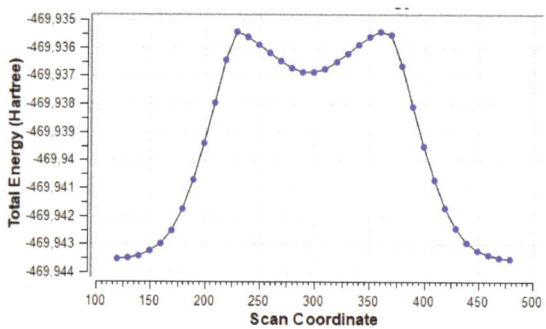

Figure 8. Energy scan for rotation of the NH$_2$ group about the CN bond in the **u6** 5-NH$_2$ system, shown in Figure 7.

The NCI analysis, shown in Figure 9, was performed to visualize all non-covalent interactions. In most cases, only weak interactions (green-shaded isosurfaces) are present. However, in systems where the asymmetry of two NH/NO bonds (Figure 6) was high, a blue color can be noticed on the isosurfaces between the interacting atoms. This indicates a stronger attractive character of these interaction. The **u1** 5-NH$_2$ system, which has the highest bond length asymmetry (Figure 6) among the amino derivatives, has very slight blue features on the isosurface between NH and =O, which indicated stronger attractive interaction than in **u2**–**u5** 5-NH$_2$ systems. The intramolecular H-bond in **u6** 5-NH$_2$, discussed earlier, appears as a mostly blue isosurface between H$_2$N and HO. The H-bond in the **u6** 5-NO$_2$ system is so strong that the NCI analysis treats it as a partially covalent interaction, as the hole is pierced through the isosurface along the H$\cdots$O line. In **u1** and **u3** 6-NO$_2$ systems, some blue accents are noticeable on the isosurface corresponding to the NO$\cdots$HN contact. Bond critical points of non-covalent interactions were found only in **u6** 5-NH$_2$ and **u6** 5-NO$_2$.

Interestingly, in several nitropurines, NO$\cdots$HN interactions have a bond critical point [30]. It is possible that this interaction is on the edge of being classified as H-bonding. The reasons are probably low values of O$\cdots$HN angles (105.6° in 1H 6-nitropurine vs. 101.4° in **u1** 6-NO$_2$ uracil), which are close to the limit of 110° proposed by Desiraju [54], and rather high O$\cdots$H distances (2.107 Å in 1H 6-nitropurine vs. 2.200 Å in **u1** 6-NO$_2$ uracil).

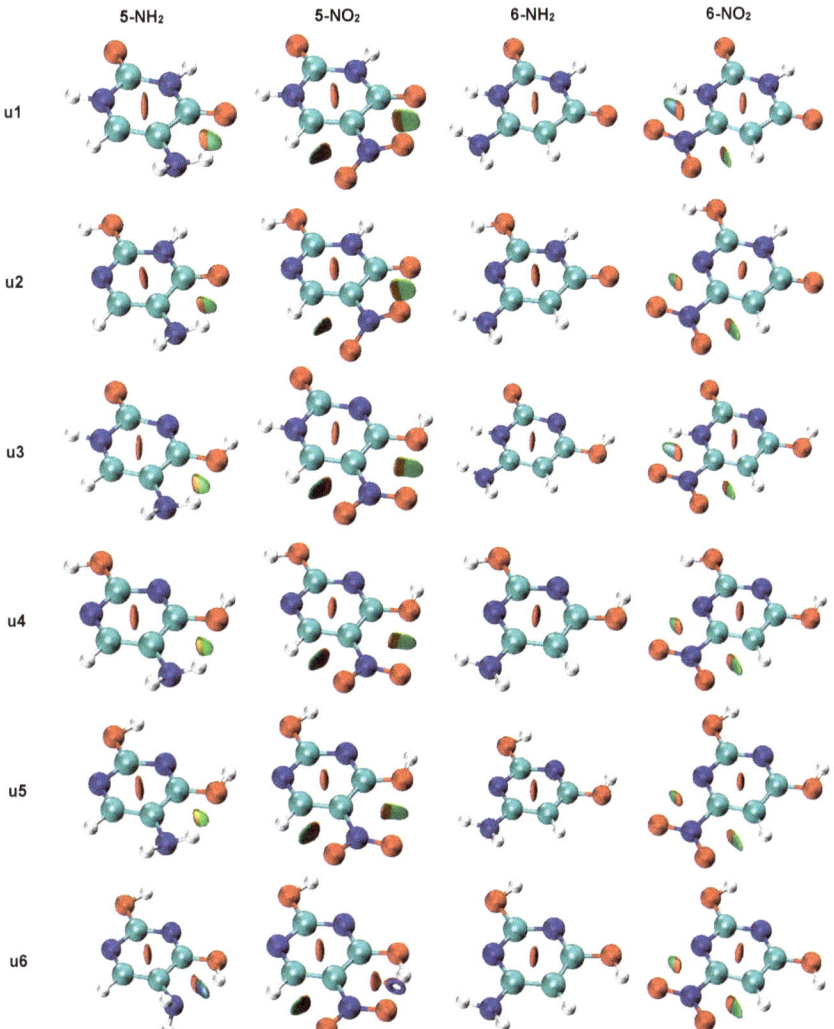

Figure 9. NCI plots for all studied systems (gas phase geometry). Isosurfaces correspond to the value of reduced density gradient function of 0.5. Red shading indicates non-bonding (steric) contacts, green weakly attractive interactions (e.g., van der Waals) and blue strongly attractive interactions (e.g., hydrogen bonding).

3.4. Tautomer Stability

The last section is devoted to the effects of substitution and solvation on the stability of uracil tautomers. Table 5 presents electronic energies of each system relative to the **u1** tautomer. In all cases, this tautomer remains the most stable, irrespective of substitution and solvation. Considering the 5-NO$_2$ substitution, the **u6** 5-NO$_2$ derivative is a particularly interesting case. Formation of a strong NO···HO H-bond results in a large stabilization relative to the unsubstituted **u6** tautomer (by 12.5 kcal/mol). Consequently, among the 5-NO$_2$ tautomers, **u6** becomes the second most stable tautomer after **u1**, despite the fact that **u6** is the least stable tautomer for unsubstituted uracil. Rotating the 5-NO$_2$ group by 90 degrees and breaking the hydrogen bond increases the relative energy of **u6** by 10.4 kcal/mol and in 5-NO$_2$ (90°), **u6** is again the least stable tautomer.

Table 5. Energies (in kcal/mol) relative to the **u1** tautomer. Δ indicates a difference in relative energies between the aqueous phase and the gas phase, Δ = E_{rel}(aq) − E_{rel}(gas).

Taut.	H	5-NH$_2$	Δ	5-NO$_2$	Δ	5-NO$_2$ (90°)	Δ	6-NH$_2$	Δ	6-NO$_2$	Δ	6-NO$_2$ (90°)
u1	0.0	0.0	0.0	0.0	0.0	0.0	0.0	0.0	0.0	0.0	0.0	0.0
u2	11.1	9.5	2.7	10.1	2.1	9.1	2.5	6.6	3.8	9.9	1.1	5.4
u3	11.4	14.6	−2.2	10.2	1.1	11.8	0.2	10.3	−0.6	11.4	0.4	11.1
u4	13.2	14.4	3.6	11.8	5.9	11.6	5.6	8.1	6.5	11.7	1.6	7.2
u5	14.2	15.5	2.9	12.9	5.0	12.6	4.8	8.5	6.4	12.9	0.5	8.1
u6	17.9	18.1	2.1	5.4	7.3	15.8	4.8	12.4	3.9	16.2	1.3	11.7

In the case of 5-NH$_2$ substitution, the energy difference between the **u1** and **u2** tautomers decreases compared to the unsubstituted systems, while between **u1** and others it increases. In 6-NH$_2$, the relative energies are smaller than for unsubstituted systems. A noteworthy increase in stability relative to **u1** is observed for **u2**, **u4**, **u5** and **u6** tautomers (between 5 and 6 kcal/mol), while much less for **u3** (1.1 kcal/mol). In the case 6-NO$_2$ tautomers, apart from **u3**, the relative energies decrease slightly, but not as much as in 6-NH$_2$. In all cases, the relative energies between the **u1** tautomer and the second most stable one are above 5.4 kcal/mol; therefore, it is unlikely that substitution with NH$_2$ or NO$_2$ groups can significantly affect the tautomeric equilibrium. Solvation. in most cases, further increases the difference between **u1** and the other forms, as evidenced by the positive values of Δ (apart of two cases) in Table 5. The only cases where Δ is negative are the two NH$_2$ derivatives of the **u3** tautomer: **u3** 5-NH$_2$ (Δ = −2.2 kcal/mol) and **u3** 6-NH$_2$ (Δ = −0.6 kcal/mol).

Similarly to the cSAR (X), electronic energy can be plotted against 1/ε and relations approximated with straight lines can be obtained (Table 6). In this case, the slopes (*a*) inform about the sensitivity of the energy of a given system to the solvent effect. In most cases, the **u1** and **u3** tautomers are the most sensitive, these two tautomers have an endo NH group in the 1 position of the uracil ring. The only exception is the 6-NO$_2$ substitution, where the **u2** and **u6** tautomers are most sensitive to the solvent effect. The **u4** and **u5** tautomers are in all but one case (H-bond forming **u6** 5-NO$_2$) the least sensitive. In amino derivatives, the sensitivity to the solvent effect seems to be correlated with the dipole moments of the molecules, i.e., a large dipole moment is associated with a large value of *a*. However, no such relation can be observed in the case of nitro derivatives.

Table 6. Slopes, *a*, of $E_{rel} = a \cdot (1/\varepsilon) + b$ linear regressions (in all cases $R^2 > 0.97$) and molecular dipole moments in the gas phase, μ (E_{rel} in kcal/mol, μ in Debye).

Tautomer	5-NH$_2$		5-NO$_2$		5-NO$_2$ (90°)		6-NO$_2$		6-NO$_2$ (90°)		6-NH$_2$	
	a	μ	a	μ	a	μ	a	μ	a	μ	a	μ
u1	0.0180	4.5	0.0234	4.9	0.0207	4.7	0.0152	0.5	0.0164	1.1	0.0231	6.2
u2	0.0135	2.3	0.0200	7.0	0.0168	6.3	0.0170	4.5	0.0155	4.2	0.0168	4.8
u3	0.0216	5.9	0.0217	2.2	0.0203	2.3	0.0159	2.8	0.0167	2.6	0.0242	6.6
u4	0.0120	2.6	0.0137	3.8	0.0120	2.8	0.0126	3.3	0.0112	2.8	0.0125	3.2
u5	0.0133	3.3	0.0152	4.7	0.0132	3.7	0.0144	5.8	0.0125	5.4	0.0127	1.9
u6	0.0145	3.8	0.0115	1.8	0.0130	1.3	0.0173	4.7	0.0157	4.5	0.0168	4.2

Plotting the relative energy, E_{rel}, against the cSAR(X) for all systems in all solvents (Figure 10) reveals linearly correlated groups of points for each tautomer. The linearity comes from the fact that both E_{rel} and cSAR change linearly with 1/ε (see Tables 3 and 6). The ranges on the y and x axes for particular tautomers are a visual representation of the strength of the solvent effect on E_{rel} and cSAR, respectively. It is clearly visible that, in general, the greatest changes in both parameters occur for the 5-NO$_2$ and 6-NH$_2$ derivatives.

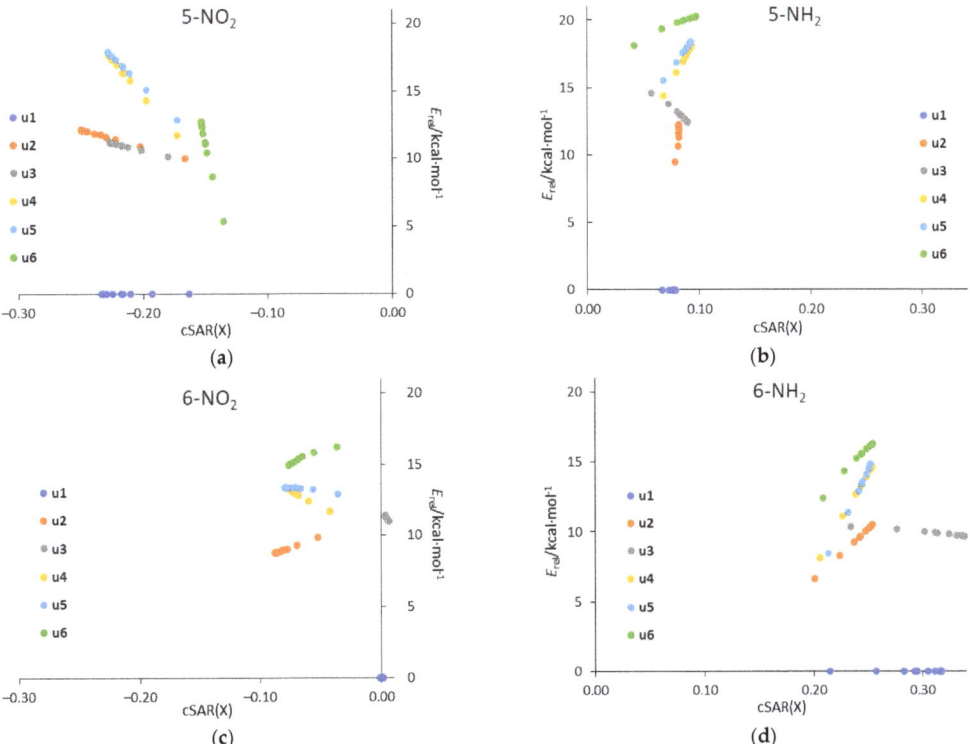

Figure 10. Plots of relative energy of tautomer against the cSAR(X) (in *e*) for (**a**) 5-NO$_2$, (**b**) 5-NH$_2$, (**c**) 6-NO$_2$, (**d**) 6-NH$_2$ systems in all considered solvents.

4. Conclusions

This work is devoted to the influence of the substituent and solvent on the tautomeric preferences and intramolecular interactions of uracil. For this purpose, the four most stable uracil tautomers and two rotamers of the dienol form, substituted by nitro and amino groups at C5 and C6 positions in ten environments, were studied. In addition, changes in the properties of the substituents were also realized by rotating the NO$_2$ group 90 degrees about the CN bond. The research was carried out using the DFT-D method and the polarizable continuum solvent model (PCM).

In uracil derivatives, the properties of the substituents depend primarily on their position with respect to endocyclic N atoms and less on the tautomeric form. Changing the =O to −OH group in the C2 or/and C4 position has less effect on the electronic properties of the substituent (and geometry), despite their opposite electronic properties. Therefore, the relationships between the relative position of endocyclic N atoms and the substituent on its electronic properties and geometry observed in simple monosubstituted N-heterocycles (pyridine, pyrimidine, pyrazine, etc.) [33] can be applied to more complex systems, such as uracil. Thus, the NH$_2$ substituent at the 6 position of uracil has more than twice (in the cSAR scale) stronger electron-donating properties than at the C5 position. In contrast, the NO$_2$ group has more electron-withdrawing power in position C5 than in position C6. The characteristic properties of both NO$_2$ and NH$_2$ groups are enhanced in polar solvents. The strength of the solvation effects on the substituent properties depends on through-space *ortho* interactions. This has also been previously observed in purine and adenine derivatives [29,30].

Regarding the intramolecular interactions between non-covalently bonded atoms, both repulsive and attractive interactions, including hydrogen bonds, are observed. This is evidenced by the results of the NCI and AIM analyses and geometric parameters. Interesting hydrogen bonding interactions, NO···HO and H_2N···HO (with NH_2 rotated by 90°), were found in **u6** 5-NO_2 and 5-NH_2 derivatives, respectively. The NO···HO interaction is strong and it highly stabilizes the **u6** tautomeric form of 5-NO_2 derivative, with respect to other tautomers. The attractive interactions between the 6-NO_2 group and endocyclic NH group (NO···HN) are visible on the NCI plots and in the geometry data, but do not have the bond critical point. Interaction between the NH_2 group and endocyclic N atom (NH···N) is not detectable by any method.

The substitution of the uracil molecule, as well as the solvation effects, does not significantly alter its tautomeric preferences. This differs from what has been reported for purine and adenine derivatives, where substitution and solvation significantly affected the tautomeric equilibrium [17,19,29,30]. However, the observed decrease in the relative energy of **u6** and **u2** uracil tautomers due to the 5-NO_2 and 6-NH_2 substitution, respectively, may cause an increase in the amount of enol tautomers in the equilibrium mixture.

Supplementary Materials: The following supporting information can be downloaded at: https://www.mdpi.com/article/10.3390/molecules27217240/s1, Table S1: Dipole moments; Tables S2–S21: Relative and solvation electronic energies and Gibbs energies of all systems; Tables S22 and S23: Statistical data on G and G_{solv} vs. $1/\varepsilon$ correlations; Tables S24–S29: cSAR(X) values for nitro and amino substituents in all systems; Table S30: Geometry data in the gas phase and formamide; Table S31: Differences between CN bond length in formamide and the gas phase; Table S32: CO bond lengths of C=O/C-OH groups in 2 and 4 position of the uracil molecule; Table S33: cSAR(X) values of C=O/C-OH groups; Table S34: Ranges and average values of cSAR(X) of C=O/C-OH groups in 2 and 4 position.

Author Contributions: Conceptualization, H.S. and T.M.K.; methodology, H.S. and P.A.W.; validation, H.S. and P.A.W.; formal analysis, H.S. and P.A.W.; investigation, P.A.W.; data curation, P.A.W.; writing—original draft preparation, H.S and P.A.W.; writing—review and editing, T.M.K. and P.A.W.; visualization, P.A.W.; supervision, H.S.; funding acquisition, T.M.K. and H.S. All authors have read and agreed to the published version of the manuscript.

Funding: H.S. and P.A.W. thank the Warsaw University of Technology for financial support. The APC was funded by MDPI.

Institutional Review Board Statement: Not applicable.

Informed Consent Statement: Not applicable.

Data Availability Statement: The data presented in this study are available in the article and in the associated Supplementary Materials.

Acknowledgments: The authors would like to thank the Wrocław Center for Networking and Supercomputing and the Interdisciplinary Center for Mathematical and Computational Modeling (Warsaw, Poland) for providing computer time and facilities.

Conflicts of Interest: The authors declare no conflict of interest.

References

1. Neidle, S.; Sanderson, M. *Principles of Nucleic Acid Structure*, 2nd ed.; Elsevier: San Diego, CA, USA, 2021.
2. Chu, C.K.; Baker, D.C. (Eds.) *Nucleosides and Nucleotides as Antitumor and Antiviral Agents*; Springer US: Boston, MA, USA, 1993.
3. Pozharskiĭ, A.F.; Katritzky, A.R.; Soldatenkov, A.T. *Heterocycles in Life and Society: An Introduction to Heterocyclic Chemistry, Biochemistry, and Applications*, 2nd ed.; Wiley: Chichester, UK, 2011.
4. Gong, L.; Zhang, Y.; Liu, C.; Zhang, M.; Han, S. Application of Radiosensitizers in Cancer Radiotherapy. *Int. J. Nanomed.* **2021**, *16*, 1083–1102. [CrossRef] [PubMed]
5. Wang, S.; Zhao, P.; Zhang, C.; Bu, Y. Mechanisms Responsible for High Energy Radiation Induced Damage to Single-Stranded DNA Modified by Radiosensitizing 5-Halogenated Deoxyuridines. *J. Phys. Chem. B* **2016**, *120*, 2649–2657. [CrossRef] [PubMed]

6. Rak, J.; Chomicz, L.; Wiczk, J.; Westphal, K.; Zdrowowicz, M.; Wityk, P.; Żyndul, M.; Makurat, S.; Golon, Ł. Mechanisms of Damage to DNA Labeled with Electrophilic Nucleobases Induced by Ionizing or UV Radiation. *J. Phys. Chem. B* **2015**, *119*, 8227–8238. [CrossRef] [PubMed]
7. Poštulka, J.; Slavíček, P.; Fedor, J.; Fárník, M.; Kočišek, J. Energy Transfer in Microhydrated Uracil, 5-Fluorouracil, and 5-Bromouracil. *J. Phys. Chem. B* **2017**, *121*, 8965–8974. [CrossRef] [PubMed]
8. Chomicz, L.; Zdrowowicz, M.; Kasprzykowski, F.; Rak, J.; Buonaugurio, A.; Wang, Y.; Bowen, K.H. How to Find Out Whether a 5-Substituted Uracil Could Be a Potential DNA Radiosensitizer. *J. Phys. Chem. Lett.* **2013**, *4*, 2853–2857. [CrossRef]
9. Spisz, P.; Kozak, W.; Chomicz-Mańka, L.; Makurat, S.; Falkiewicz, K.; Sikorski, A.; Czaja, A.; Rak, J.; Zdrowowicz, M. 5-(N-Trifluoromethylcarboxy)Aminouracil as a Potential DNA Radiosensitizer and Its Radiochemical Conversion into N-Uracil-5-Yloxamic Acid. *Int. J. Mol. Sci.* **2020**, *21*, 6352. [CrossRef]
10. Zdrowowicz, M.; Chomicz, L.; Żyndul, M.; Wityk, P.; Rak, J.; Wiegand, T.J.; Hanson, C.G.; Adhikary, A.; Sevilla, M.D. 5-Thiocyanato-2′-Deoxyuridine as a Possible Radiosensitizer: Electron-Induced Formation of Uracil-C5-Thiyl Radical and Its Dimerization. *Phys. Chem. Chem. Phys.* **2015**, *17*, 16907–16916. [CrossRef]
11. Sosnowska, M.; Makurat, S.; Zdrowowicz, M.; Rak, J. 5-Selenocyanatouracil: A Potential Hypoxic Radiosensitizer. Electron Attachment Induced Formation of Selenium Centered Radical. *J. Phys. Chem. B* **2017**, *121*, 6139–6147. [CrossRef]
12. Zdrowowicz, M.; Datta, M.; Rychłowski, M.; Rak, J. Radiosensitization of PC3 Prostate Cancer Cells by 5-Thiocyanato-2′-Deoxyuridine. *Cancers* **2022**, *14*, 2035. [CrossRef]
13. Pałasz, A.; Cież, D. In Search of Uracil Derivatives as Bioactive Agents. Uracils and Fused Uracils: Synthesis, Biological Activity and Applications. *Eur. J. Med. Chem.* **2015**, *97*, 582–611. [CrossRef]
14. Tsuda, H.; Ohshima, Y.; Nomoto, H.; Fujita, K.; Matsuda, E.; Iigo, M.; Takasuka, N.; Moore, M.A. Cancer Prevention by Natural Compounds. *Drug Metab. Pharmacokinet.* **2004**, *19*, 245–263. [CrossRef] [PubMed]
15. Kowalski, K. Ferrocenyl-Nucleobase Complexes: Synthesis, Chemistry and Applications. *Coord. Chem. Rev.* **2016**, *317*, 132–156. [CrossRef]
16. Raczyńska, E.D. Quantum-Chemical Studies on the Favored and Rare Isomers of Isocytosine. *Comput. Theor. Chem.* **2017**, *1121*, 58–67. [CrossRef]
17. Raczyńska, E.D.; Kamińska, B. Variations of the Tautomeric Preferences and π-Electron Delocalization for the Neutral and Redox Forms of Purine When Proceeding from the Gas Phase (DFT) to Water (PCM). *J. Mol. Model.* **2013**, *19*, 3947–3960. [CrossRef] [PubMed]
18. Raczyńska, E.D.; Makowski, M.; Hallmann, M.; Kamińska, B. Geometric and Energetic Consequences of Prototropy for Adenine and Its Structural Models—A Review. *RSC Adv.* **2015**, *5*, 36587–36604. [CrossRef]
19. Chen, Y.-L.; Wu, D.-Y.; Tian, Z.-Q. Theoretical Investigation on the Substituent Effect of Halogen Atoms at the C_8 Position of Adenine: Relative Stability, Vibrational Frequencies, and Raman Spectra of Tautomers. *J. Phys. Chem. A* **2016**, *120*, 4049–4058. [CrossRef]
20. Laxer, A.; Major, D.T.; Gottlieb, H.E.; Fischer, B. ($^{15}N_5$)-Labeled Adenine Derivatives: Synthesis and Studies of Tautomerism by ^{15}N NMR Spectroscopy and Theoretical Calculations. *J. Org. Chem.* **2001**, *66*, 5463–5481. [CrossRef]
21. Lippert, B.; Gupta, D. Promotion of Rare Nucleobase Tautomers by Metal Binding. *Dalton Trans.* **2009**, *24*, 4619–4634. [CrossRef]
22. Raczyńska, E.D.; Gal, J.-F.; Maria, P.-C.; Kamińska, B.; Igielska, M.; Kurpiewski, J.; Juras, W. Purine Tautomeric Preferences and Bond-Length Alternation in Relation with Protonation-Deprotonation and Alkali Metal Cationization. *J. Mol. Model.* **2020**, *26*, 93. [CrossRef]
23. Singh, V.; Fedeles, B.I.; Essigmann, J.M. Role of Tautomerism in RNA Biochemistry. *RNA* **2015**, *21*, 1–13. [CrossRef]
24. Khuu, P.; Ho, P.S. A Rare Nucleotide Base Tautomer in the Structure of an Asymmetric DNA Junction. *Biochemistry* **2009**, *48*, 7824–7832. [CrossRef] [PubMed]
25. Brovarets', O.O.; Hovorun, D.M. Renaissance of the Tautomeric Hypothesis of the Spontaneous Point Mutations in DNA: New Ideas and Computational Approaches. In *Mitochondrial DNA-New Insights*; Seligmann, H., Ed.; InTech: London, UK, 2018.
26. Srivastava, R. The Role of Proton Transfer on Mutations. *Front. Chem.* **2019**, *7*, 536. [CrossRef] [PubMed]
27. Slocombe, L.; Al-Khalili, J.S.; Sacchi, M. Quantum and Classical Effects in DNA Point Mutations: Watson–Crick Tautomerism in AT and GC Base Pairs. *Phys. Chem. Chem. Phys.* **2021**, *23*, 4141–4150. [CrossRef] [PubMed]
28. Jalbout, A.F.; Trzaskowski, B.; Xia, Y.; Li, Y.; Hu, X.; Li, H.; El-Nahas, A.; Adamowicz, L. Structures, Stabilities and Tautomerizations of Uracil and Diphosphouracil Tautomers. *Chem. Phys.* **2007**, *332*, 152–161. [CrossRef]
29. Beak, P.; White, J.M. Relative Enthalpies of 1,3-Dimethyl-2,4-Pyrimidinedione, 2,4-Dimethoxypyrimidine, and 4-Methoxy-1-Methyl-1-2-Pyrimidinone: Estimation of the Relative Stabilities of Two Protomers of Uracil. *J. Am. Chem. Soc.* **1982**, *104*, 7073–7077. [CrossRef]
30. Tsuchiya, Y.; Tamura, T.; Fujii, M.; Ito, M. Keto-Enol Tautomer of Uracil and Thymine. *J. Phys. Chem.* **1988**, *92*, 1760–1765. [CrossRef]
31. Jezuita, A.; Wieczorkiewicz, P.A.; Szatylowicz, H.; Krygowski, T.M. Effect of the Solvent and Substituent on Tautomeric Preferences of Amine-Adenine Tautomers. *ACS Omega* **2021**, *6*, 18890–18903. [CrossRef]
32. Jezuita, A.; Wieczorkiewicz, P.A.; Szatylowicz, H.; Krygowski, T.M. Solvent Effect on the Stability and Reverse Substituent Effect in Nitropurine Tautomers. *Symmetry* **2021**, *13*, 1223. [CrossRef]

33. Wieczorkiewicz, P.A.; Szatylowicz, H.; Krygowski, T.M. Energetic and Geometric Characteristics of the Substituents: Part 2: The Case of NO_2, Cl, and NH_2 Groups in Their Mono-Substituted Derivatives of Simple Nitrogen Heterocycles. *Molecules* **2021**, *26*, 6543. [CrossRef]
34. Becke, A.D. Perspective: Fifty Years of Density-Functional Theory in Chemical Physics. *J. Chem. Phys.* **2014**, *140*, 18A301. [CrossRef]
35. Jones, R.O. Density functional theory: Its origins, rise to prominence, and future. *Rev. Mod. Phys.* **2015**, *87*, 897–923. [CrossRef]
36. Frisch, M.J.; Trucks, G.W.; Schlegel, H.B.; Scuseria, G.E.; Robb, M.A.; Cheeseman, J.R.; Scalmani, G.; Barone, V.; Petersson, G.A.; Nakatsuji, H.; et al. *Gaussian 16, Revision C.01*; J. Gaussian Inc.: Wallingford, CT, USA, 2016.
37. Marek, P.H.; Szatylowicz, H.; Krygowski, T.M. Stacking of Nucleic Acid Bases: Optimization of the Computational Approach—The Case of Adenine Dimers. *Struct. Chem.* **2019**, *30*, 351–359. [CrossRef]
38. Sadlej-Sosnowska, N. On the Way to Physical Interpretation of Hammett Constants: How Substituent Active Space Impacts on Acidity and Electron Distribution in p-Substituted Benzoic Acid Molecules. *Polish J. Chem.* **2007**, *81*, 1123–1134.
39. Sadlej-Sosnowska, N. Substituent Active Region—A Gate for Communication of Substituent Charge with the Rest of a Molecule: Monosubstituted Benzenes. *Chem. Phys. Lett.* **2007**, *447*, 192–196. [CrossRef]
40. Hirshfeld, F.L. Bonded-Atom Fragments for Describing Molecular Charge Densities. *Theoret. Chim. Acta* **1977**, *44*, 129–138. [CrossRef]
41. Tomasi, J.; Mennucci, B.; Cancès, E. The IEF version of the PCM solvation method: An overview of a new method addressed to study molecular solutes at the QM ab initio level. *J. Mol. Struct. Theochem* **1999**, *464*, 211–226. [CrossRef]
42. Cancès, E.; Mennucci, B.; Tomasi, J. A New Integral Equation Formalism for the Polarizable Continuum Model: Theoretical Background and Applications to Isotropic and Anisotropic Dielectrics. *J. Chem. Phys.* **1997**, *107*, 3032–3041. [CrossRef]
43. Tomasi, J.; Mennucci, B.; Cammi, R. Quantum Mechanical Continuum Solvation Models. *Chem. Rev.* **2005**, *105*, 2999–3094. [CrossRef]
44. Romero, E.E.; Hernandez, F.E. Solvent Effect on the Intermolecular Proton Transfer of the Watson and Crick Guanine–Cytosine and Adenine–Thymine Base Pairs: A Polarizable Continuum Model Study. *Phys. Chem. Chem. Phys.* **2018**, *20*, 1198–1209. [CrossRef]
45. Bader, R.F.W. *Atoms in Molecules: A Quantum Theory*; Clarendon Press: Oxford, UK, 1994.
46. Todd, A.; Keith, T.K. Gristmill Software, Overland Park KS, AIMAll (Version 19.10.12), USA. 2019. Available online: Aim.tkgristmill.com (accessed on 23 September 2022).
47. Afonin, A.V.; Vashchenko, A.V.; Sigalov, M.V. Estimating the Energy of Intramolecular Hydrogen Bonds from ^{1}H NMR and QTAIM Calculations. *Org. Biomol. Chem.* **2016**, *14*, 11199–11211. [CrossRef]
48. Espinosa, E.; Molins, E.; Lecomte, C. Hydrogen Bond Strengths Revealed by Topological Analyses of Experimentally Observed Electron Densities. *Chem. Phys. Lett.* **1998**, *285*, 170–173. [CrossRef]
49. Johnson, E.R.; Keinan, S.; Mori-Sánchez, P.; Contreras-García, J.; Cohen, A.J.; Yang, W. Revealing Noncovalent Interactions. *J. Am. Chem. Soc.* **2010**, *132*, 6498–6506. [CrossRef] [PubMed]
50. Lu, T.; Chen, F. Multiwfn: A Multifunctional Wavefunction Analyzer. *J. Comput. Chem.* **2012**, *33*, 580–592. [CrossRef] [PubMed]
51. Humphrey, W.; Dalke, A.; Schulten, K. VMD: Visual Molecular Dynamics. *J. Mol. Graph.* **1996**, *14*, 33–38. [CrossRef]
52. Koch, U.; Popelier, P.L.A. Characterization of C-H-O Hydrogen Bonds on the Basis of the Charge Density. *J. Phys. Chem.* **1995**, *99*, 9747–9754. [CrossRef]
53. Jabłoński, M. A Critical Overview of Current Theoretical Methods of Estimating the Energy of Intramolecular Interactions. *Molecules* **2020**, *25*, 5512. [CrossRef]
54. Desiraju, G.R. A Bond by Any Other Name. *Angew. Chem. Int. Ed.* **2011**, *50*, 52–59. [CrossRef]

Article

Theoretical Studies on the Structure and Intramolecular Interactions of Fagopyrins—Natural Photosensitizers of *Fagopyrum*

Sebastian Szymański * and Irena Majerz

Faculty of Pharmacy, Wroclaw Medical University, Borowska 211a, 50-556 Wroclaw, Poland; irena.majerz@umw.edu.pl
* Correspondence: sebastian.szymanski@umw.edu.pl; Tel.: +48-71-784-0305; Fax: +48-71-784-0307

Simple Summary: The study determines the spatial structure and intramolecular interactions of fagopyrins—natural photosensitizers of *Fagopyrum* species. In silico calculations show many fagopyrin conformers characterized by the formation of strong intramolecular interactions.

Abstract: Compounds characterized by a double-anthrone moiety are found in many plant species. One of them are fagopyrins—naturally occurring photosensitizers of *Fagopyrum*. The photosensitizing properties of fagopyrins are related to the selective absorption of light, which is a direct result of their spatial and electronic structure and many intramolecular interactions. The nature of the interactions varies in different parts of the molecule. The aim of this study is to determine the structure and intramolecular interactions of fagopyrin molecules. For this purpose, in silico calculations were used to perform geometry optimization in the gas phase. QTAIM and NCI analysis suggest the formation of the possible conformers in the fagopyrin molecules. The presence of a strong OHO hydrogen bond was shown in the anthrone moiety of fagopyrin. The minimum energy difference for selected conformers of fagopyrins was 1.1 kcal·mol^{-1}, which suggested that the fagopyrin structure may exist in a different conformation in plant material. Similar interactions were observed in previously studied structures of hypericin and sennidin; however, only fagopyrin showed the possibility of brake the strong OHO hydrogen bond in favor of forming a new OHN hydrogen bond.

Keywords: fagopyrins; conformation; hydrogen bond; QTAIM; NCI

1. Introduction

Fagopyrins are a group of compounds of natural origin found in plants of the genus Fagopyrum. Parts of these plants are commonly consumed by humans and animals throughout the world [1]. There are many species of Fagopyrum; the most consumed and studied are *Fagopyrum esculentum*, *F. tataricum*, and *F. cymosum* [2,3]. Parts of these plants provide a low-calorie, gluten-free food and a source of many elements and organic compounds of biological interest, such as rutin, quercetin, and fagopyrins [4–8].

Fagopyrins are anthraquinone derivatives characterized by a polycyclic system, which is interesting from the chemical point of view. The structure and intramolecular interactions of polycyclic compounds affect the physical and chemical properties and can influence potential applications of the compounds in pharmacy and medicine [9–12]. As shown in earlier work on hypericin [13] and sennidines [14], a highly substituted polycyclic system can exhibit a non-planar structure due to a variety of intramolecular interactions. At the time of writing this paper, data unambiguously define the spatial and electronic structures of fagopyrins. An explanation of this may be found in difficult and multi-stage processes needed to obtain the pure substance from plant material [15]. Additionally, the existence of unstable protofagopyrins [16,17] and the possibility of the existence of many derivatives [17,18] can be a problem.

The interest in fagopyrins is mainly due to their spectroscopic properties. They exhibit absorption of electromagnetic radiation in the wavelength range of the light around λ_{max} 550 and 590 nm [15,19,20]. Upon excitation, they are able to transfer energy to the oxygen molecule, thereby producing reactive oxygen species (ROS) [21]. Singlet oxygen and other ROS are responsible for cell damage. Easy light activation of fagopyrin molecules shows the potential to be used in photodynamic therapy [20]. On the other hand, high consumption of the *Fagopyrum* plants can lead to the light sensitivity in animals, called fagopyrism [22]. Additionally, there are reports of the possible hepatotoxic effect of consumption of food rich in *Fagopyrum* plants on dogs [23]. Leaving aside the potential dangers of the photo-sensitizing properties of fagopyrins, their potential for pharmacological use appears to be high. Easy excitation with energy from the visible range, confirmed antifungal [24] and antimicrobial [25] properties, and natural origin are promising for use in targeted photodynamic therapy.

Although general studies on the structure of fagopyrins were undertaken [17], no crystal structures are available, given the current state of knowledge. Additionally, such a strongly substituted double anthrone moiety has many possibilities for intramolecular interactions. The introduction of piperidine and pyrrolidine substituents in the hypericin molecule that, in fact, forms fagopyrin allows for the appearance of new intramolecular interactions.

Considering the intramolecular interactions occurring in fagopyrins, the OHO hydrogen bond system formed by hydroxyl groups bound to carbonyl oxygen is the most characteristic. As the strength of the hydrogen bond is determined primarily by the electronegativity of the atoms with which the proton is bound [26], the hydrogen bond of the OHO type is the strongest. The strength of the hydrogen bond is expressed in changes in geometry, consisting in shortening the distance between the donor and the acceptor of the proton and the location of the proton close to the center of the distance between the donor and the acceptor [27]. The OHO hydrogen bond system in fagopyrin is additionally strengthened by the participation of the hydrogen bonds in closed bond cycles in which double and aromatic bonds are present [28]. However, in the case of fagopyrins containing the piperidine and pyrrolidine substituents with the nitrogen atom and the possibility of conformational changes of the molecules, the presence of weaker hydrogen bonds of OHN type should be taken into account.

Therefore, it seems necessary to investigate the possible intramolecular interactions and structure of fagopyrins and compare them with the present state of knowledge using the example of hypericin.

The purpose of this paper is to use in silico methods to determine the molecular structure and interactions of still unknown fagopyrin molecules. Conformational analysis is carried out to determine the probable spatial and electronic structure of fagopyrins and, importantly, the intramolecular interactions of substituents that determine the pharmacological properties of these natural compounds. The results calculated for the gas phase will provide valuable knowledge about these interesting derivatives and will contribute to the future exploration of the pharmacological properties of fagopyrins. The conformational analysis will determine the minimum energy structure and possible formation of the double-anthrone system in plant material.

2. Materials and Methods

Determination of the spatial structure and conformational analysis of fagopyrin molecules were carried out using the Gaussian16 package [29]. Calculations were carried out using DFT/B3LYP/6-311++G(d,p) model with Grimme dispersion [30]. The optimized structures correspond to a minimum on local potential energy surfaces. QTAIM analysis was performed using the AIMALL program [31]. NCI analyses [32] were performed using the Multiwfn program [33]. NCI graphics were printed using the VMD program [34]. UV-VIS spectra and orbital analysis were performed with the ADF program [35].

3. Results and Discussion

Six structures of fagopyrins A–F (Scheme 1) proposed by Benković et al. [17] were optimized in the gas phase. Conformational analysis was performed for the A–F structures by searching for the minimum on local potential energy surfaces.

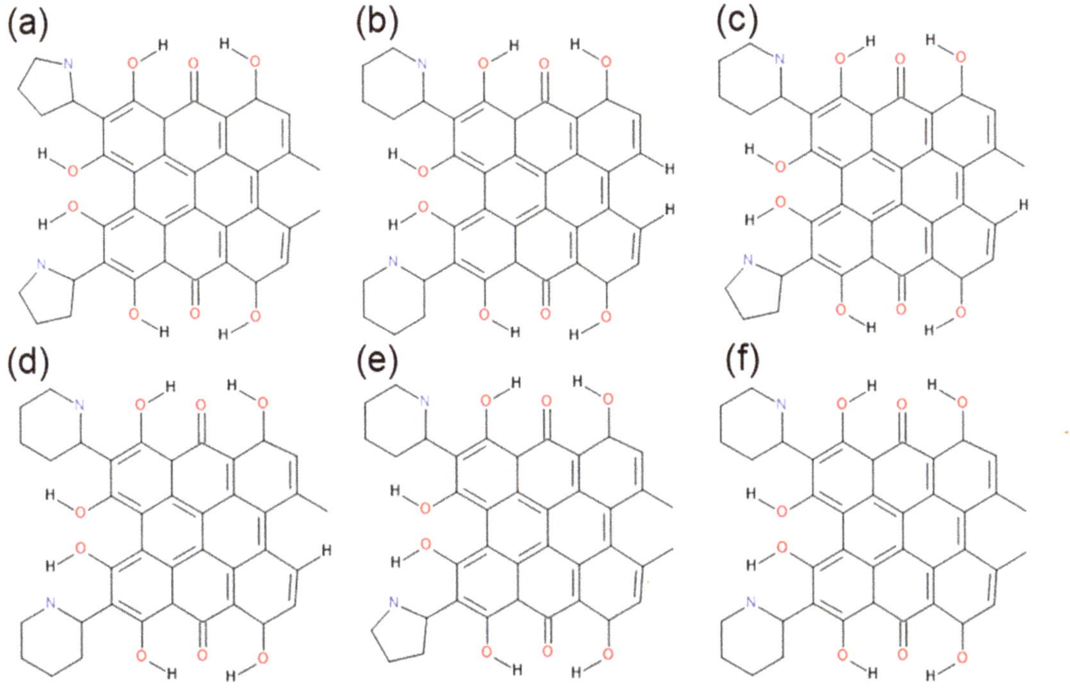

Scheme 1. Analyzed structures of (**a**) fagopyrin A, (**b**) fagopyrin B, (**c**) fagopyrin C, (**d**) fagopyrin D, (**e**) fagopyrin E, and (**f**) fagopyrin F.

Conformers representing different arrangements and interactions of substituent hydroxyl groups, carbonyl oxygen, piperidine, and pyrrolidine rings were compiled based on a structure with minimum energy. The close position of the substituents allowed for the formation of intramolecular interactions at the "peri" (O-H···O···H-O) and "bay" region of the molecules (O-H···O-H). Additionally, the presence of piperidine and pyrrolidine substituents allowed for the formation of new interactions with the nitrogen atom.

3.1. Conformational Analysis of Fagopyrins

The structure of fagopyrin consists of a polycyclic system of eight rings. As shown in Scheme 2, each ring, named A–H, consists of six carbon atoms. Rings A–H are characterized by the presence of at least one substituent. In the A + B + C and F + G + H regions of the molecule, there are two hydroxyl groups and carbonyl oxygen. Such a close position of the substituents allows the formation of a hydrogen bond system in which hydroxyl groups are directed to the centrally located carbonyl oxygen. As shown for the hypericin molecule [13], such formation of strong hydrogen bonds is energetically preferred, and these hydrogen bonds are difficult to break. What is new in the structure of fagopyrin is the close position of piperidine and pyrrolidine substituents. These substituents can occur at positions named R1 and R2 (Scheme 2). The rings containing the substituents with nitrogen atoms give an additional possibility to form of intramolecular OHN hydrogen bond and possible breaking of the strong OHO hydrogen bonds. So far, no studies have been found on the arrangement of these substituents in fagopyrin molecules. Another interesting part of the fagopyrin molecule is the "bay" region consisting of the A + D + F ring system. In hypericin molecule, the preferred arrangement of the substituents in the "bay" region is to form an OHO hydrogen bond between two hydroxyl groups. In the fagopyrin molecule, the addition of piperidine and pyrrolidine rings allows the interactions to be directed to the nitrogen atom forming new OHN interactions. Another part of fagopyrin molecule that may affect the overall structure is the presence of R3 and R4 substituents. Depending on the type of fagopyrin A–F molecule, these parts can be substituted by protons or methyl groups. As shown for hypericin, the close distance of two methyl groups may cause strain in the entire molecule and can strongly affect the planarity of the polycyclic system. Such a variety of substituents and possible strain effects from methyl groups make the structure and intramolecular interactions in fagopyrin molecules worth describing.

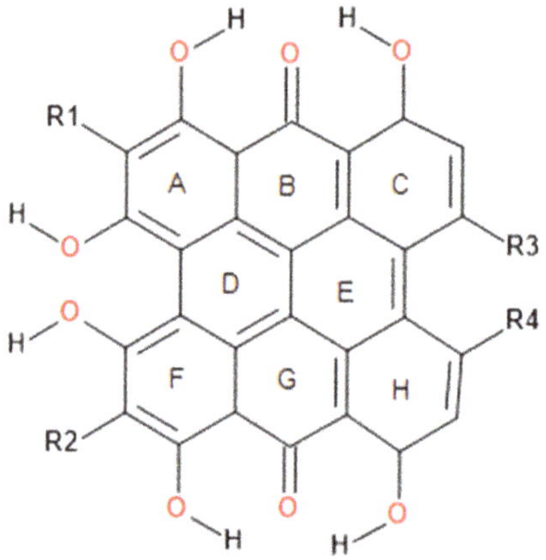

Scheme 2. Double anthrone polycyclic system of fagopyrin A–H.

Six low-energy conformers of fagopyrin A have been obtained, and structure 2 is the minimum energy conformer (Figure 1). The hydrogen bonds in the "peri" region of the molecule show alignment to the carbonyl oxygen. Two OHN hydrogen bonds in the "bay" region are preferred; however, breaking one OHN hydrogen bond in the "bay" region and forming OHO hydrogen bond between hydroxyl groups results in a total energy change of only 3.8 kcal·mol^{-1} (structure 1). A similar change in the total energy of the system ($\Delta E \approx 4.9$ kcal·mol^{-1}) is caused by breaking a strong OHO hydrogen bond in the "peri" region and the formation of an OHN bond with the pyrrolidine substituent (structure 3). Breaking the OHO hydrogen bond in the "peri" region without the formation of another interaction destabilizes the fagopyrin structure and raises its energy (structure 4, 6).

For fagopyrin B, six low-energy conformers (Figure 2) have been obtained. Structure 8 showing the lowest energy is characterized by the OHO hydrogen bond arrangement in the "peri" region typical for anthrones. In the "bay" region, the OHN hydrogen bonds linking the hydroxyl group and the nitrogen atom are formed. The energy differences between the structures 7, 8, and 9 show the energy difference up to 10.0 kcal·mol^{-1}. The energy difference for these conformers is larger than the analogous difference for structures 1, 2, and 3 of fagopyrin A. Formation of OHN hydrogen bonds with the piperidine ring (fagopyrin A) shows larger energy differences than the formation of OHN interactions with the pyrrolidine ring (fagopyrin B). Additionally, it can be seen that the piperidine ring in the fagopyrin B prefers a "chair" conformation; however, interaction with the hydroxyl substituent in the "peri" and "bay" region can disrupt the chair conformation (structure 7–12). The presence of a free hydroxyl group (structure 10, 12) results in a significant increase in the energy of fagopyrin B. In contrast, the lack of the methyl groups brings the double anthrone system closer to planarity.

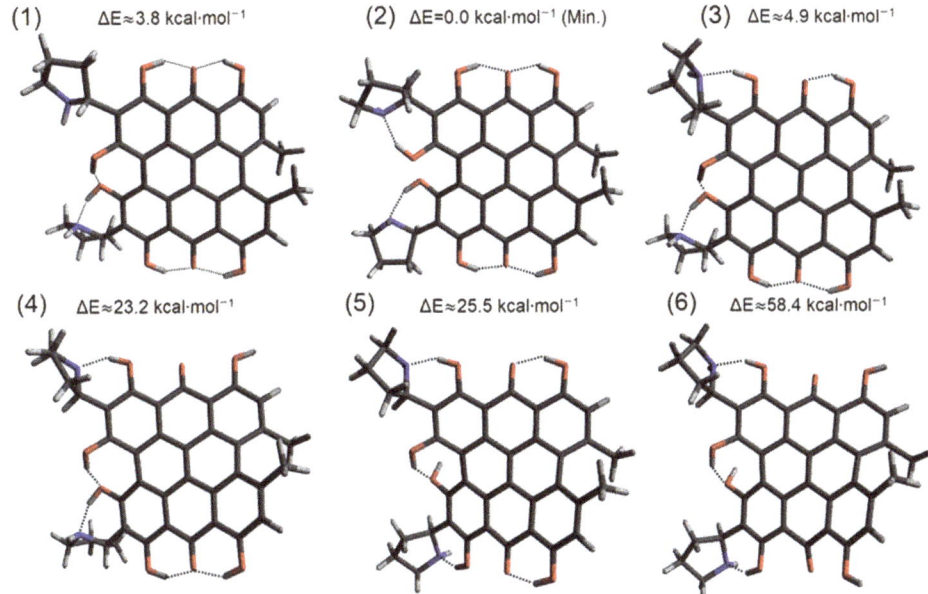

Figure 1. Conformers (**1–6**) of fagopyrin A.

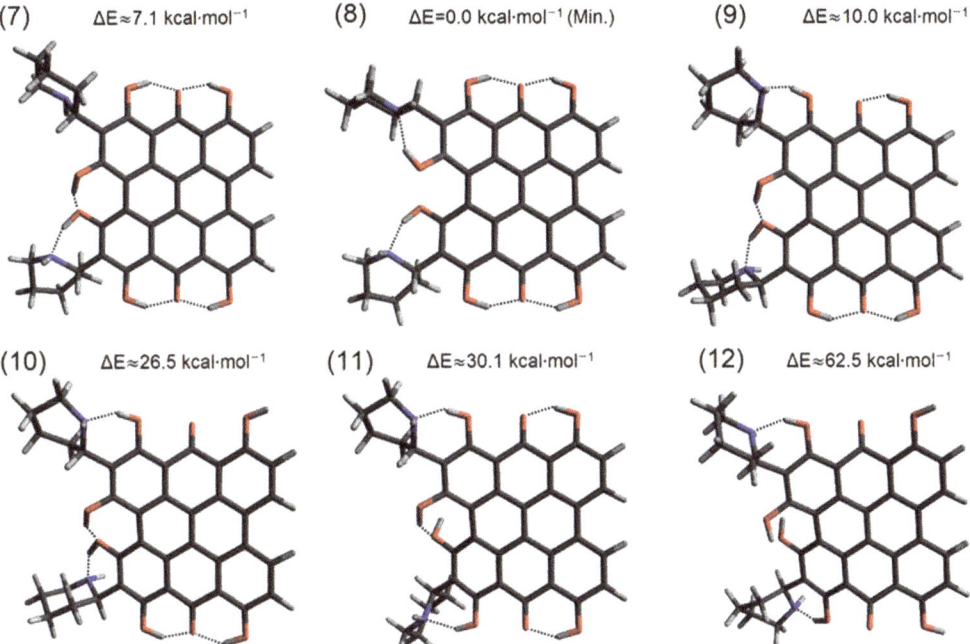

Figure 2. Conformers (**7–12**) of fagopyrin B.

Six conformers that were obtained for fagopyrin C (Figure 3) are characterized by low energy. The structure with the lowest energy (structure 14) favors the formation of

OHN hydrogen bonds and the breaking of the OHO hydrogen bonds in the "bay" region of the molecule. The piperidine ring shows a "chair" conformation for all the obtained structures (structure 13–18). Breaking of the OHN hydrogen bond located in the "bay" region results in leaving the piperidine ring free and increasing the energy of the molecule by 7.0 kcal·mol^{-1} (structure 13). Breaking of the OHO hydrogen bond in the "bay" region together with the formation of the OHN hydrogen bond with a hydroxyl group located in the "peri" region (structure 15) is associated with an energy increase of 9.6 kcal·mol^{-1}. Leaving the "free" hydroxyl group in the "peri" region results in a significant increase in the energy $\Delta E \approx 27.7$ kcal·mol^{-1} (structure 16) and $\Delta E \approx 65.8$ kcal·mol^{-1} (structure 18).

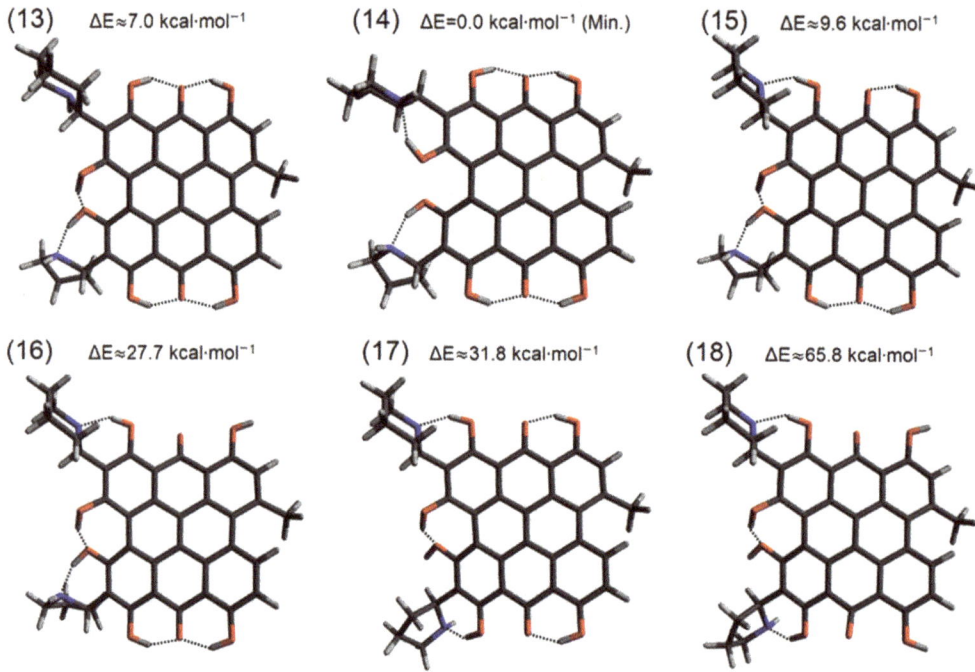

Figure 3. Conformers (13–18) of fagopyrin C.

Six low-energy conformers (Figure 4) were obtained for fagopyrin D. The lowest energy structure (structure 20) is characterized by the formation of an OHN hydrogen bond in the "bay" region. Structure 19 is characterized by a "hypericin-like" arrangement of the hydroxyl groups in the "bay", and the "peri" region differs in energy by 7.0 kcal·mol^{-1} from the lowest energy structure. The "chair" conformation is preferred for both piperidine rings in fagopyrin D. Structure 21 is characterized by the breaking of the strong OHO hydrogen bond in the "peri" region and the formation of an OHN hydrogen bond to the piperidine ring. Such transfer of the hydrogen interaction results in the energy difference of 9.6 kcal·mol^{-1} to the minimum energy structure (structure 20). As in the fagopyrin A–C structure, the "free" hydroxyl group (22, 24) increases the energy of the conformer; however, in such a polycyclic system, this may not be a direct expression of breaking the OHN hydrogen bond but also due to possible structural changes of the multi-ring molecule. The formation of the OHN hydrogen bond in the "peri" region stabilizes the fagopyrin D molecule.

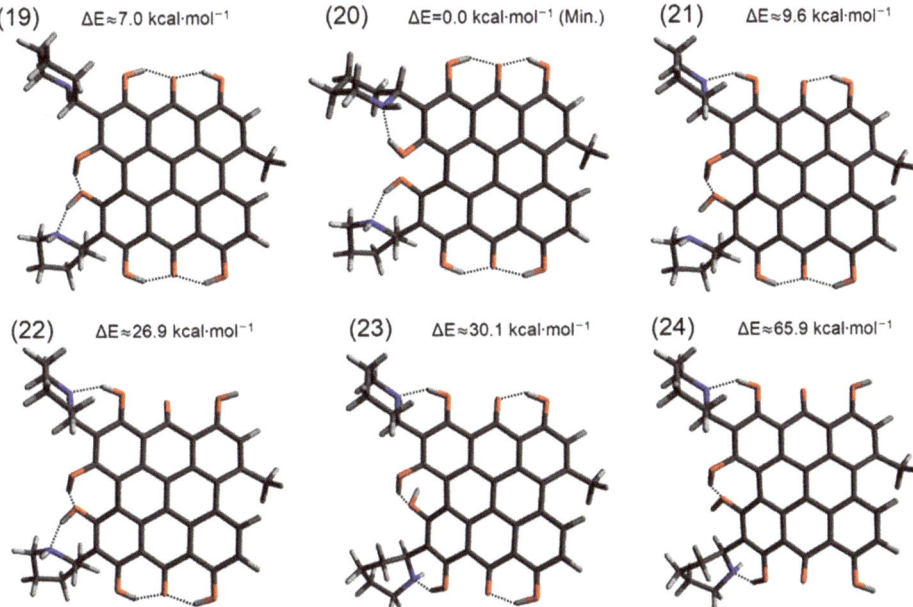

Figure 4. Conformers (**19–24**) of fagopyrin D.

Six conformers (Figure 5) were obtained for fagopyrin E. The lowest-energy conformer (structure 26) shows hydrogen bonding in the "bay" region of the molecule. The OHN hydrogen bonds are formed by the hydroxyl groups to both nitrogen atoms in the piperidine and pyrrolidine substituent. In the minimum-energy conformer, the hydrogen bonds in the "peri" region are directed to the carbonyl oxygen. The "chair" conformation of the piperidine substituent is preferred. Conformer characterized by the "free" piperidine group (structure 25) differs in the energy of 7.5 kcal·mol^{-1}. Additionally, breaking of OHO hydrogen bond in the "peri" region and transferring it to the "free" piperidine substituent (structure 27) raises the energy relative to conformer 25 by 2.1 kcal·mol^{-1}. As in the fagopyrin structures described previously, breaking of a strong OHO hydrogen bond in the "peri" region and leaving the hydroxyl group unbound raises the total energy of the polycyclic system (structure 28 and structure 30).

For fagopyrin F, six low-energy conformers were obtained. The lowest-energy structure again is characterized by the formation of the OHN hydrogen bonds in the "bay" region (structure 32). The chair conformation of the piperidine substituents is preferred. The energetically similar conformers 31 and 33 are characterized by an energy difference of 7.5 and 9.7 kcal·mol^{-1}, relatively to the minimum. As in conformers of fagopyrin E, it is possible to break the OHN hydrogen bond in the "bay" region and form an OHN hydrogen bond in the "peri" region. Breaking of the strong OHO hydrogen bond system in the "peri" region causes the deformation of the polycyclic system and deviates the molecule from planarity (36).

In general, the structure of fagopyrin tends to form OHN hydrogen bonds in the "bay" region. Energetically preferred formation of strong OHO hydrogen bonds to carbonyl oxygen in the "peri" region is evident in most conformers, and breaking of these interactions has the consequence of raising the energy of the system. Nevertheless, it is possible to break the strong OHO hydrogen bonds in the "peri" region in favor of the formation of an OHN hydrogen bond with the piperidine or pyrrolidine substituent. In summary, the introduction of piperidine and pyrrolidine substituents into the hypericin system provides an opportunity to form an OHN hydrogen bond instead of the strongest OHO.

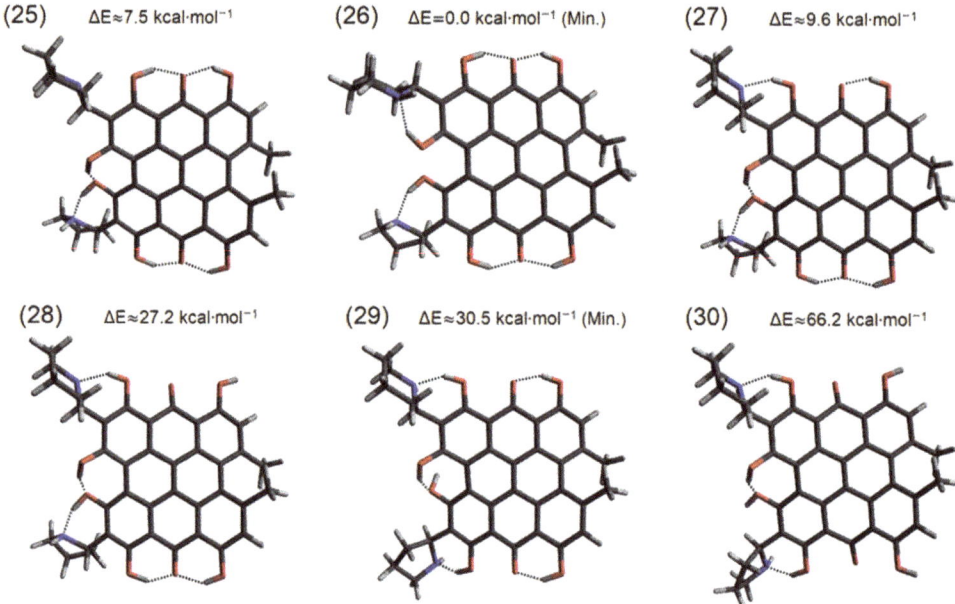

Figure 5. Conformers (25–30) of fagopyrin E.

3.2. Analysis of Geometry of Fagopyrin Structures

A parameter that describes the geometry of the fagopyrin conformers is the angle between the planes formed by the peripheral rings A–C, F–H, A–F, and C–H (Scheme 2). For hypericin, (Table 1) these angles are of degrees: A–C: 13.334, F–H: 12.363, A–F: 23.188, and C–H: 30.095. Selected conformers of fagopyrin A–F show significant similarity to the structure of hypericin. These conformers are 1, 7, 13, 19, 25, and 31. These conformers are characterized by different substitutions at the R1–R4 position but the hydroxyl groups in the "bay" and "peri" regions are oriented as in the hypericin molecule and form the same type OHO hydrogen bonds. The difference in the angle between the ring plane is the greatest for fagopyrin C and D. These fagopyrins have an asymmetric substitution with a methyl group and a proton at the R3 and R4 position.

Transfer of the OHO hydrogen bond in the "peri" region to the piperidine or pyrrolidine results in the formation of an OHN hydrogen bond (structures 3, 9, 15, 21, 27, and 33) and causes little change in the angles between the A–C and A–F planes. Formation of another OHN hydrogen bond in molecules 5, 11, 17, 23, 29, and 35 causes more significant changes in the polycyclic system. The changes are visible in the angle between F–H and C–H planes, so the effect of the OHN hydrogen bonds in the "peri" region on the geometry of the fagopyrin molecule is apparent and may have a real impact on the electron structure. Additionally, breaking the OHO hydrogen bond and leaving the hydroxyl group in the "peri" region as free causes deformation of the polycyclic system (structures 4, 10, 16, 22, 28, and 34). Larger differences can be observed when two free hydroxyl groups in the molecule are present (structures 6, 12, 18, 24, 30, and 36). This arrangement of the hydroxyl groups causes strong deformation of the polycyclic system of fagopyrins, which is reflected in the high energy of these conformers. So far, two similarities of the fagopyrin molecules to the hypericin molecule can be given. These are a strong influence on the geometrical structure of substituents at the R3 and R4 positions (methyl groups) and the preferred formation of the OHO hydrogen bonds in the "peri" region formed by hydroxyl groups and carbonyl oxygen.

Table 1. Angles between the ring planes in fagopyrin and hypericin conformers. σ = 0.001 [deg.].

Compound	Structure	Plane/Angle [Deg.]			
		A–C	F–H	A–F	C–H
Fagopyrin A	1	13.695	12.708	23.717	30.165
	2	10.787	10.001	23.965	29.275
	3	16.005	13.476	25.420	30.763
	4	18.505	15.241	27.456	31.058
	5	16.694	14.646	30.072	30.773
	6	20.601	20.002	28.554	33.411
Fagopyrin B	7	12.227	11.517	23.245	4.320
	8	11.303	11.437	23.632	3.099
	9	17.136	11.257	26.050	5.012
	10	23.343	8.266	26.391	2.675
	11	17.52	15.562	30.230	4.862
	12	20.611	20.796	33.032	5.543
Fagopyrin C	13	10.728	12.357	24.436	19.975
	14	8.323	10.393	24.598	18.545
	15	13.846	12.925	25.928	20.427
	16	17.886	13.166	27.414	18.563
	17	15.129	18.324	30.626	20.972
	18	17.561	20.219	33.223	20.238
Fagopyrin D	19	10.778	12.366	24.422	19.978
	20	8.305	10.341	24.604	18.515
	21	13.847	12.918	25.848	20.431
	22	17.800	13.066	27.297	18.533
	23	14.637	16.649	30.708	20.473
	24	17.558	20.262	33.338	20.209
Fagopyrin E	25	13.469	12.696	24.195	30.168
	26	10.838	10.654	24.508	29.384
	27	15.981	13.491	25.502	30.778
	28	18.502	15.112	27.347	30.987
	29	16.798	14.604	30.088	30.787
	30	21.120	22.148	27.824	33.555
Fagopyrin F	31	13.482	12.721	24.089	30.172
	32	10.805	10.638	24.461	29.384
	33	16.011	13.53	25.430	30.789
	34	18.581	15.220	27.322	30.993
	35	17.361	16.870	30.244	31.195
	36	21.112	21.813	27.898	33.568
Hypericin	-	13.334	12.363	23.188	30.095

The structures corresponding to the energy minima (2, 8, 14, 20, 26, and 32) differ from hypericin in the "bay" region. The hydroxyl groups in the "bay" region are directed to the nitrogen atom in the piperidine and pyrrolidine substituents. In the case of hypericin, the hydroxyl groups prefer the OHO hydrogen bonding. The introduction of piperidine or pyrrolidine rings at the R1 and R2 positions favors the formation of an OHN hydrogen bond and decreasing of the fagopyrin energy to a minimum.

Changes in the angles between the plane rings of the peripheral rings of fagopyrins result from a number of substituents and the intramolecular interactions. As fagopyrin F is the major form in the plant material [18], an analysis of hydrogen bond parameters has been performed for the structures shown in Figure 6. The length and angles of OHO and OHN hydrogen bonds are summarized in Table 2. The parameters of OHN hydrogen bonds directed to the pyrrolidine and piperidine rings are highlighted in bold. The results calculated for the fagopyrin F conformers have been compared with hypericin. The letters in parentheses in Table 2 identify the hydrogen bond location described according to Scheme 2.

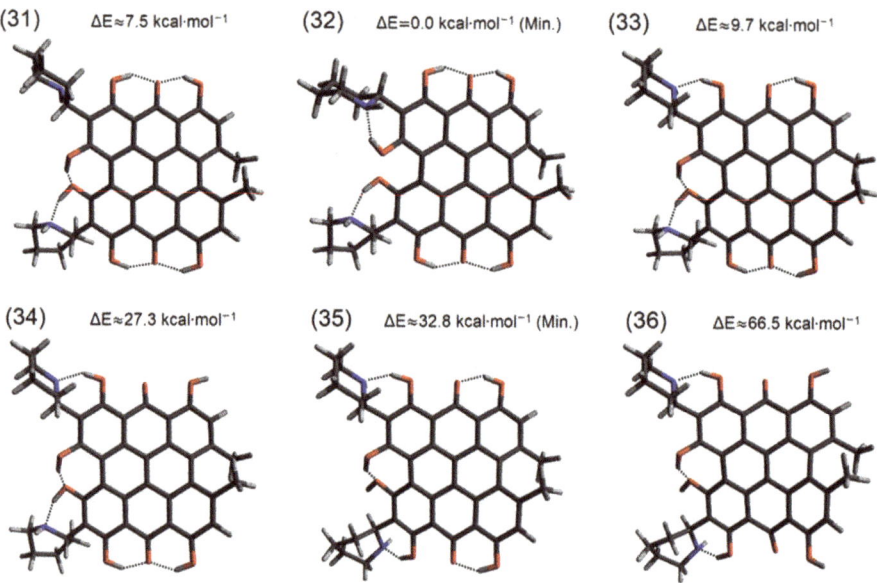

Figure 6. Conformers (**31**–**36**) of fagopyrin F.

Table 2. Hydrogen bonds in possible fagopyrin F conformers. σ = 0.0001 [Å], σ = 0.001 [deg.].

Structure	Hydrogen Bond Location	H···O/H···N [Å]	O···O/O···N [Å]	OHO/OHN [deg.]
31	C(A)-O-H···O=C(B)	1.6047	2.5133	149.530
	C(C)-O-H···O=C(B)	1.6583	2.5485	147.083
	C(F)-O-H···O=C(G)	1.6022	2.5125	149.135
	C(H)-O-H···O=C(G)	1.6721	2.5591	146.198
	C(A)-O-H···O-C(F)	1.5751	2.5155	158.577
	C(F)-O-H···N(R2)	**1.5386**	**2.5278**	**156.683**
32	C(A)-O-H···O=C(B)	1.6220	2.5319	149.412
	C(C)-O-H···O=C(B)	1.6667	2.5580	147.383
	C(F)-O-H···O=C(G)	1.6244	2.5338	149.346
	C(H)-O-H···O=C(G)	1.6677	2.5589	147.388
	C(A)-O-H···N(R1)	**1.7233**	**2.6379**	**150.007**
	C(F)-O-H···N(R2)	**1.6676**	**2.6012**	**152.071**
33	**C(A)-O-H···N(R1)**	**1.7164**	**2.6276**	**149.868**
	C(C)-O-H···O=C(B)	1.5936	2.5098	149.406
	C(F)-O-H···O=C(G)	1.6035	2.5143	149.243
	C(H)-O-H···O=C(G)	1.6720	2.5597	147.032
	C(A)-O-H···O-C(F)	1.5849	2.5207	157.420
	C(F)-O-H···N(R2)	**1.5508**	**2.5354**	**156.626**
34	**C(A)-O-H···N(R1)**	**1.7350**	**2.6378**	**148.993**
	C(F)-O-H···O=C(G)	1.6062	2.5164	149.237
	C(H)-O-H···O=C(G)	1.6750	2.5627	147.055
	C(A)-O-H···O-C(F)	1.5977	2.5319	157.421
	C(F)-O-H···N(R2)	**1.5590**	**2.5402**	**156.438**

Table 2. Cont.

Structure	Hydrogen Bond Location	H···O/H···N [Å]	O···O/O···N [Å]	OHO/OHN [deg.]
35	C(A)-O-H···N(R1)	1.7220	2.6316	149.766
	C(C)-O-H···O=C(B)	1.6017	2.5156	149.256
	C(F)-O-H···N(R2)	1.6940	2.6188	152.744
	C(H)-O-H···O=C(G)	1.6095	2.5203	149.017
	C(A)-O-H···O-C(F)	1.6717	2.5572	149.732
36	C(A)-O-H···N(R1)	1.7438	2.6437	148.744
	C(F)-O-H···N(R2)	1.7087	2.6243	151.524
	C(A)-O-H···O-C(F)	1.6658	2.5536	150.072
Hypericin	C(A)-O-H···O=C(B)	1.6499	2.5422	147.330
	C(C)-O-H···O=C(B)	1.6659	2.5548	146.940
	C(F)-O-H···O=C(G)	1.6400	2.5351	147.385
	C(H)-O-H···O=C(G)	1.6750	2.5606	146.668
	C(A)-O-H···O-C(F)	1.6670	2.5605	151.168

Structure 31 is characterized by an arrangement of substituents similar to hypericin. The length of hydrogen bonds in the "peri" region is similar to the length of analogous bonds in hypericin. In the "bay" region, the OHO bond length is shorter than in the hypericin molecule. The OHN hydrogen bond is characterized by a length of 1.5386 Å and an angle of 156.683°. The lowest energy conformer (structure 32) is characterized by "peri"-OHO hydrogen bond lengths similar to hypericin. In structure 32, two OHN hydrogen bonds are presented in the "bay" region. The bond labeled as C(F)-O-H····N(R2) is characterized by length and angle similar to the OHO "peri" bonds. The bond labeled as C(A)-O-H···N(R1) is elongated up to 1.7233 Å. Additionally, the "peri" OHN hydrogen bond in structure 33 is longer (1.7164 Å) than the "peri" OHO hydrogen bonds.

3.3. Aromaticity of Fagopyrin

The aromaticity of polycyclic compounds is related to their structure and reactivity. There are many indices describing aromaticity; however, the classical HOMA index (Harmonic Oscillator Measure of Aromaticity) is convenient for the description of aromaticity in organic compounds and relates it directly to the structure [36]. In typical aromatic compounds, the values of the HOMA index are in the range from 0 to 1, where 0 corresponds to a non-aromatic ring, and the value of 1 corresponds to a fully delocalized benzene structure, and only in special cases, the HOMA value can exceed the 0–1 range. Figure 7 shows the HOMA values calculated for the rings of the fagopyrin F conformers. The rings are marked according to Scheme 2. For hypericin, the HOMA values for particular rings are: A—0.7186, B—0.3937, C—0.8054, D—0.4712, E—0.5138, F—0.7863, G—0.4010, and H—0.7979. For the fagopyrin F structures, the peripheral rings A, C, F, and H show the highest HOMA value indicating the aromatic character of the ring. The aromaticity of the rings D and E is about 0.5. The HOMA value for rings B and G is the most variable, and for the structures 34 and 36, it is negative. Structure 31 is characterized by a hypericin-like arrangement of the substituents. The HOMA values of rings A and F are lower relatively to the hypericin; thus, the presence of the piperidine ring decreases the aromaticity of the rings. However, the D ring in structure 31 gains aromaticity relative to the hypericin moiety. The formation of the strong OHO hydrogen bonds in the "peri" region stabilizes the polycyclic system and increases the aromaticity of the central B and G rings. In general, structure 31 shows aromaticity of the rings similar to hypericin, with the influence of piperidine substituents on the aromaticity of rings A, D, and F. The lowest energy structure 32 is characterized by the formation of two OHN hydrogen bonds to the piperidine substituent in the "bay" region. This arrangement increases the aromaticity of the A, D, and F ring. Structure 33 is characterized by the breaking of the strong OHO bond in the "peri" region and the formation of an OHN hydrogen bond to the piperidine ring. Such conformation causes an increase in the energy of the system, an increase in the HOMA value of the F ring, and a decrease in the HOMA value of the B ring up to 0.1644. Breaking of another OHO bond in the "peri" region (structure 34) deepens the loss of aromaticity of the B ring. The HOMA parameter below zero indicates a complete loss of aromaticity of the ring. In structures 35 and 36, two OHN bonds in the "peri" region are present and such conformation of the hydroxyl groups causes an increase in the HOMA value in the A and F ring with significant aromaticity decreasing in the B and G ring. These changes cause increasing the total energy of the molecular system (Figure 6).

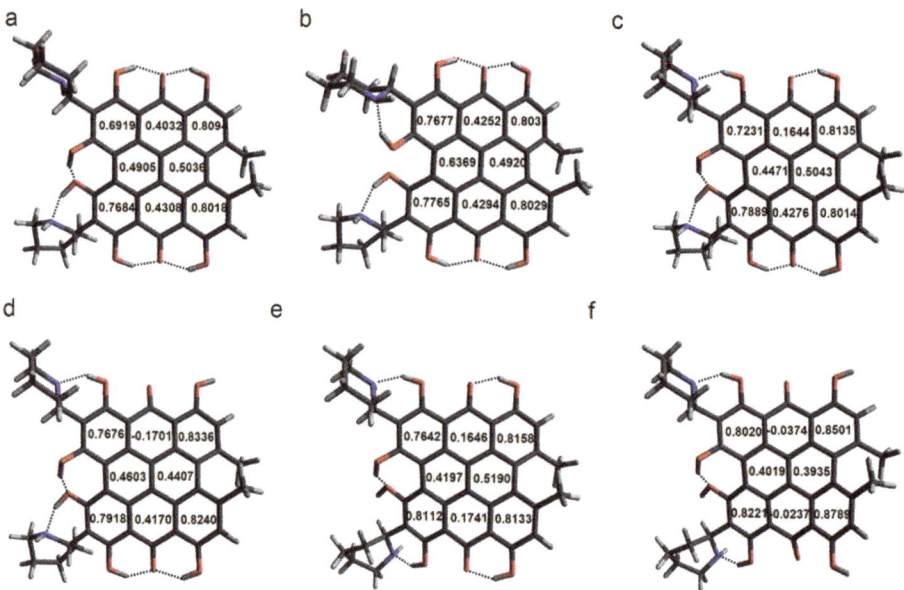

Figure 7. HOMA values calculated for (**a**) **31**, (**b**) **32**, (**c**) **33**, (**d**) **34**, (**e**) **35**, and (**f**) **36** fagopyrin F conformers.

3.4. Analysis of Intramolecular Interactions in Fagopyrin Derivatives

Changes in aromaticity must be related to the changes in electron density of the polycyclic system. To describe the possible intermolecular interactions and arrangement of the electron density, the QTAIM [37] (Quantum Theory of Atoms in Molecules) analysis for the fagopyrin F conformers has been performed. In the frame of the QTAIM theory, a molecule consists of maximum, minimum, and saddle points of the electron density $\rho(r)$. The saddle point indicates bond-critical points (BCPs) or ring-critical points (RCPs). The points representing the maximum electron density correspond to atoms. Figure 8 shows QTAIM graphs of fagopyrin F conformers. The structure of fagopyrin F is characterized by the presence of numerous substituents in a polycyclic system. Such structure allows for the occurrence of numerous intramolecular interactions of diverse nature [38]. QTAIM analysis confirms the presence of the hydrogen bond interactions in the "peri" region of fagopyrin F. Hydroxyl groups directed to carbonyl oxygen form a stable moiety as in the case of hypericin and sennidin [13,14]. QTAIM analysis indicates that OHN interaction can be formed in both the "peri" and "bay" region. The strong OHO hydrogen bonds in the anthrone moiety can be broken and replaced by weaker OHN hydrogen bonds. The electron density values at the bond critical points $\rho(r)$ presented in Figure 8 reflect the strength of the OHO and OHN hydrogen bonds. The structures 33, 34, 35, and 36 show the formation of OHN hydrogen bonds characterized by lower values of ρ relative to the OHO hydrogen bonds in the "peri" region. However, conformer 32 (b) is characterized by the formation of OHN hydrogen bonds in the "bay" region. These bonds are characterized by a similar value of ρ(r) relative to the strong OHO hydrogen bonds in the "peri" region. This conformer shows the lowest energy; thus, the formation of strong OHN hydrogen bonds to the piperidine substituents stabilizes the anthrone system. The proximity of the piperidine ring affects the adjacent hydroxyl groups even if they do not form a direct bond. In addition, the close position of the methyl groups also introduces intermolecular interactions.

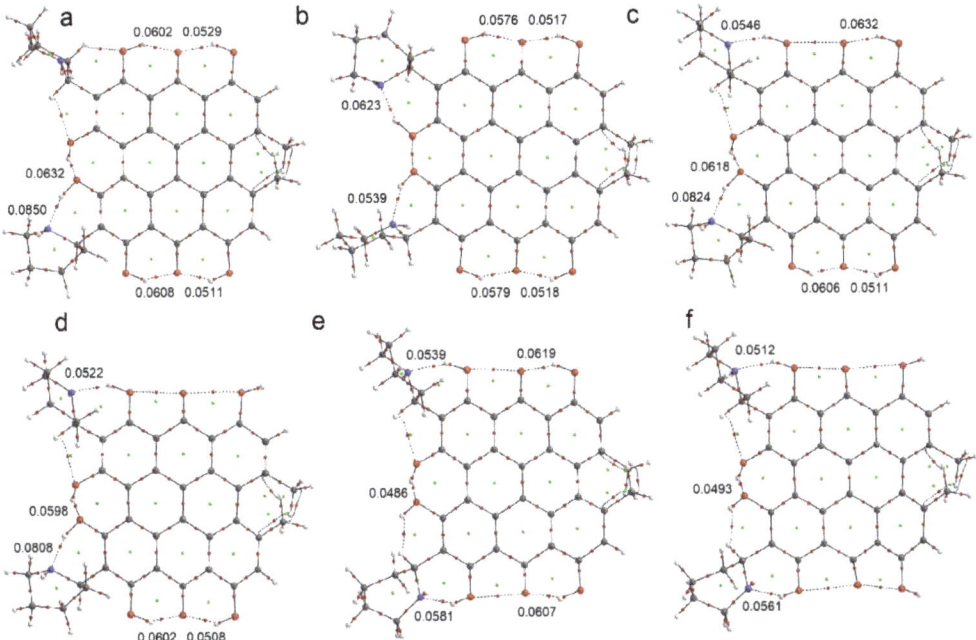

Figure 8. QTAIM plots for (**a**) **31**, (**b**) **32**, (**c**) **33**, (**d**) **34**, (**e**) **35**, and (**f**) **36** fagopyrin F conformers. Green points correspond to RCPs. Red points indicate BCPs.

To confirm the presence of the interactions, the non-covalent interactions [32] (NCI) analysis was performed. Figure 9 shows the conformers of fagopyrin F, showing multiple intramolecular interactions. The blue isosurfaces in the "peri" and "bay" moiety confirm the presence of strong hydrogen bonds in the fagopyrin F molecule. Conformer 31 (a) shows similarity to hypericin in the formation of strong OHO hydrogen bonds. The lowest energy conformer 32 (b) confirms the formation of OHN hydrogen bonds in the "bay" region. Structure 33 (c) confirms the possibility of breaking the strong OHO hydrogen bond in favor of OHN hydrogen bond formation with the nitrogen atom of the piperidine ring. The interactions between the methyl groups can be described as dispersive.

3.5. UV-VIS Spectra of Fagopyrin Conformers

Different conformation of the investigated fagopyrins is reflected in their electron structures. In Figure 10 are presented the HOMO and LUMO orbitals for the conformers of fagopyrin F—the most popular in the plant material. For other fagopyrins, the HOMO and LUMO orbitals are collected in Supplementary Materials. It is characteristic that for all conformers, the HOMO orbital is located mainly on the outer A, C, F, and H rings and on the oxygen atoms of the hydroxyl group. Only for conformers 34 and 36, the HOMO orbital is more concentrated on rings A and F than on C and H. The transfer of electrons to the LUMO orbital is connected with the shifting of electrons to the B and G rings, the oxygen of the carbonyl group, and outer bonds of the A, C, F, and H rings. Since the arrangement of the HOMO and LUMO orbitals is similar for all conformers, the HOMO–LUMO gap energy is also similar. This is persistent for all the analyzed fagopyrins. In Table 3 are collected the HOMO–LUMO gap energies for all the analyzed fagopyrins.

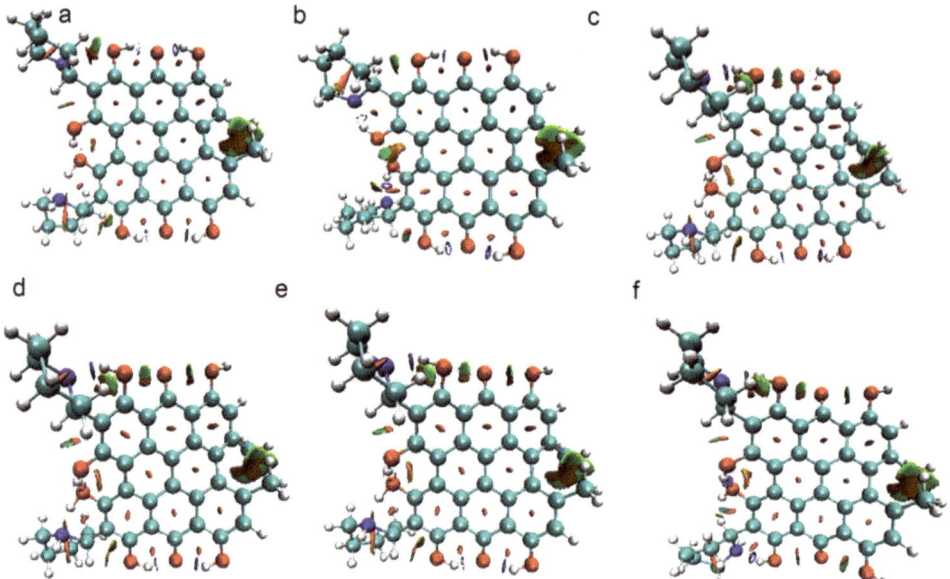

Figure 9. NCI plots for (a) **31**, (b) **32**, (c) **33**, (d) **34**, (e) **35**, and (f) **36** fagopyrin F conformers. Colors used for isosurfaces are: red for repulsive, green for dispersive, and blue for attractive interactions.

Table 3. HOMO–LUMO gap energy for calculated fagopyrin structures.

Structure	HOMO–LUMO Gap [kcal·mol^{-1}]
1	58.4
2	58.4
3	59.2
4	61.0
5	61.0
6	66.5
7	59.7
8	59.9
9	60.2
10	61.7
11	62.7
12	67.9
13	59.1
14	59.3
15	60.0
16	61.6
17	62.2
18	67.3
19	59.0
20	59.3
21	60.0
22	61.6
23	62.1
24	67.1
25	58.4
26	58.5

Table 3. *Cont.*

Structure	HOMO–LUMO Gap [kcal·mol^{-1}]
27	59.5
28	61.1
29	61.1
30	67.1
31	58.4
32	58.5
33	59.4
34	61.1
35	61.2
36	66.9

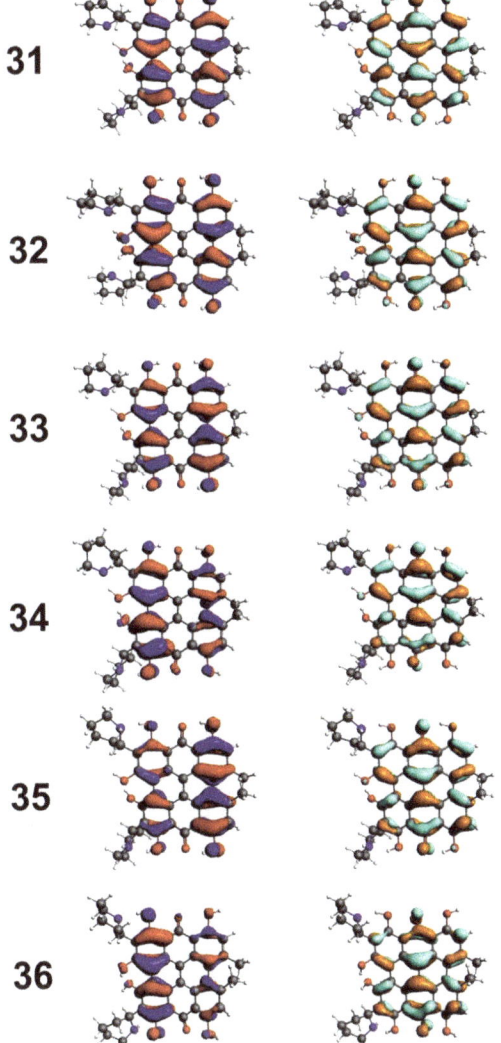

Figure 10. HOMO and LUMO orbitals for the fagopyrin F conformers **31**, **32**, **33**, **34**, **35**, and **36**.

The UV spectra for fagopyrin F shown in Figure 11 are characterized by the presence of two intense bands. For the conformers 31, 32, and 33, the most intensive band shifts from 556 nm to 552 nm. The second intensive band is located at 457, 455, and 459 nm. For conformer 34, except for the most intensive band at 546 nm, two bands with similar intensity at 477 and 427 nm are present. The last band at 427 nm is visible in the UV-VIS spectra of 31, 32, 34, and 36 conformers; however, it is significantly lower compared to other bands. For conformers 31, 32, 33, and 34, the most intensive band is related to HOMO–LUMO transition. For conformer 35, this band is shifted to 539 nm, for conformer 36 to 503 nm, and the intensity of this band is lower than the bands at 437 and 421, respectively. For the conformers 35 and 36, transitions from lower energy orbitals are more intense than for the HOMO–LUMO transition. The electron transition participating in the bands for fagopyrin F conformers are collected in Table 4. The shape of the orbitals involved in the electron transitions in structure 32 (the lowest energy structure of fagopyrin F) is shown in Figure 12. The shape of the orbitals involved in UV-VIS transitions for fagopyrin F conformers except the presented in the text (Figure 12) is shown in Supplementary Materials.

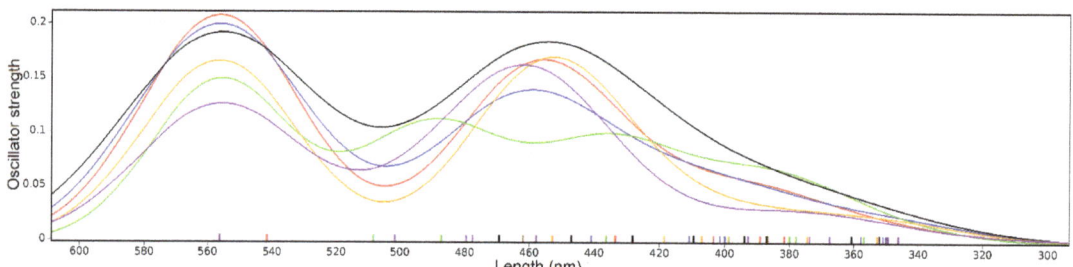

Figure 11. Theoretical excitation spectra for fagopyrin F conformers. Structure **31**—red, **32**—blue, **33**—black, **34**—green, **35**—orange, and **36**—purple.

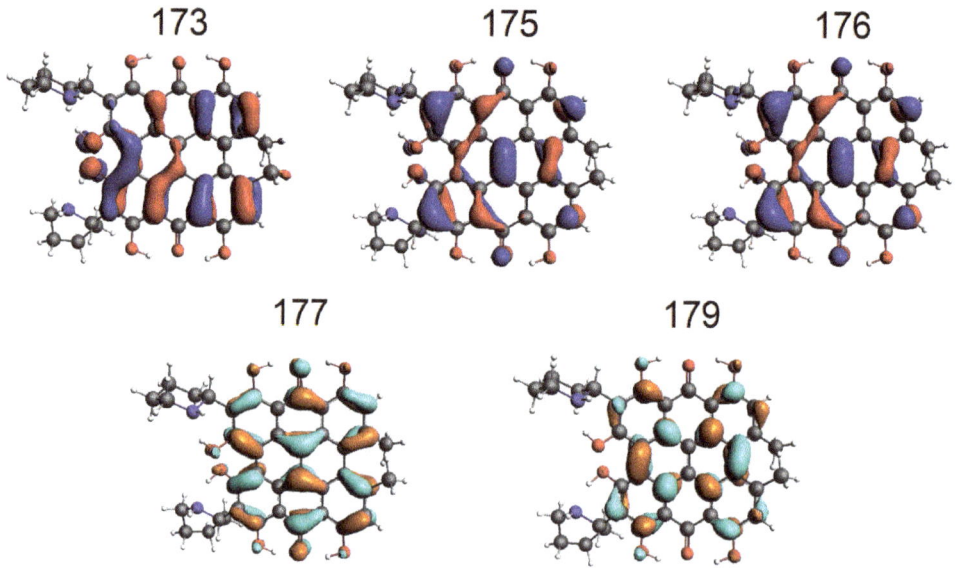

Figure 12. The shape of the orbitals for structure **32** (the lowest energy structure of fagopyrin F). 176—HOMO and 177—LUMO orbitals.

Table 4. The electron transition contribution and orbitals for fagopyrin F conformers.

Structure	Band [nm]	Orbital Transition	Transition Contribution [%]
31	556	176(HOMO) → 177(LUMO)	95.7
		175 → 177	1.3
	457	174 → 177	90.3
		172 → 177	3.0
		176 → 179	2.7
		173 → 177	1.3
32	552	176(HOMO) → 177(LUMO)	96.7
	455	175 → 177	91.2
		173 → 177	3.6
		176 → 179	3.1
33	552	176(HOMO) → 177(LUMO)	97.1
	459	175 → 177	86.7
		174 → 177	7.7
		176 → 179	1.5
		173 → 177	1.4
	436	174 → 177	87.1
		175 → 177	7.0
		176 → 179	2.2
		173 → 177	1.2
34	546	176(HOMO) → 177(LUMO)	96.0
	477	175 → 177	70.7
		174 → 177	17.3
		173 → 177	6.8
		176 → 179	1.1
	427	173 → 177	82.6
		174 → 177	10.2
		176 → 179	1.9
		172 → 177	1.2
		175 → 177	1.1
35	539	176(HOMO) → 177(LUMO)	97.3
	437	175 → 177	74.5
		174 → 177	17.2
		176 → 179	2.7
		172 → 177	1.8
36	503	176(HOMO) → 177(LUMO)	96.8
	421	174 → 177	52.5
		175 → 177	29.4
		172 → 177	9.5
		176 → 179	3.1
		171 → 177	1.1

A comparison of calculated fagopyrin F and experimental spectra [19] for the plant material suggests that in the plant material, many fagopyrin structures may be present. It is not clear which version of fagopyrin in the experimental spectra was registered; however, in the experimental spectra, the most intensive theoretically calculated bands are visible.

4. Conclusions

Theoretical calculations can provide information on the molecular structure when the structure is unknown, which is often the case with plant material. Fagopyrin compounds may exist as conformers characterized by a different energy. The presence of the piperidine and pyrrolidine ring in fagopyrin introduces novel intramolecular interactions compared to the double anthrone molecules. Fagopyrin A–F structures are characterized by the presence of a number of substituents and strong hydrogen bonds in the anthrone moiety. Although the OHO hydrogen bonds in the anthrone moiety are characterized as very strong, both the OHO and OHN hydrogen bonds may exist in the fagopyrin A–F structure. It is possible to break the strong OHO hydrogen bonds in the anthrone moiety in favor of interactions with the nitrogen atom in piperidine or pyrrolidine substituent. Changes in the molecular geometry are related to the changes in the orbital localization, which is reflected in the UV–VIS spectra of fagopyrin conformers.

Supplementary Materials: The following supporting information can be downloaded at: https://www.mdpi.com/article/10.3390/molecules27123689/s1. Figure S1. HOMO (left) and LUMO (right) orbitals for the 1, 2, 3, 4, 5, and 6 Fagopyrin A conformers. Figure S2. HOMO (left) and LUMO (right) orbitals for the 7, 8, 9, 10, 11, and 12 Fagopyrin B conformers. Figure S3. HOMO (left) and LUMO (right) orbitals for the 13, 14, 15, 16, and 17 Fagopyrin C conformers. Figure S4. HOMO (left) and LUMO (right) orbitals for the 19, 20, 21, 22, 23, and 24 Fagopyrin D conformers. Figure S5. HOMO (left) and LUMO (right) orbitals for the 25, 26, 27, 28, 29, and 30 Fagopyrin E conformers. Figure S6. The shape of the orbitals for structure 31 (Fagopyrin F). 176—HOMO and 177—LUMO orbitals. Figure S7. The shape of the orbitals for structure 33 (Fagopyrin F). 176—HOMO and 177—LUMO orbitals. Figure S8. The shape of the orbitals for structure 34 (Fagopyrin F). 176—HOMO and 177—LUMO orbitals. Figure S9. The shape of the orbitals for structure 35 (Fagopyrin F). 176—HOMO and 177—LUMO orbitals. Figure S10. The shape of the orbitals for structure 36 (Fagopyrin F). 176—HOMO and 177—LUMO orbitals.

Author Contributions: S.S., I.M. contributed to the conceptualization, methodology, and writing of the manuscript. All authors have read and agreed to the published version of the manuscript.

Funding: This research was financially supported by a Ministry of Health grant number SUBK.D050.22.003 from the IT Simple system of Wroclaw Medical University.

Institutional Review Board Statement: Not applicable.

Informed Consent Statement: Not applicable.

Data Availability Statement: The data presented in this study are available on request from the corresponding author.

Acknowledgments: The Wroclaw Center for Networking and Supercomputing is acknowledged for generous allocations of computer time.

Conflicts of Interest: The authors declare no conflict of interest. The funders had no role in the design of the study; in the collection, analyses, or interpretation of data; in the writing of the manuscript, or in the decision to publish the results.

Sample Availability: Samples of the compounds are not available from the authors.

References

1. Babu, S.; Yadav, G.S.; Singh, R.; Avasthe, R.K.; Das, A.; Mohapatra, K.P.; Tahashildar, M.; Kumar, K.; Prabha, M.; Thoithoi Devi, M.; et al. Production technology and multifarious uses of buckwheat (*Fagopyrum* spp.): A review. *Indian J. Agron.* **2018**, *63*, 415–427.
2. Sytar, O.; Brestic, M.; Zivcak, M.; Phan Tran, L.-S. The Contribution of Buckwheat Genetic Resources to Health and Dietary Diversity. *Curr. Genom.* **2016**, *17*, 193–206. [CrossRef] [PubMed]
3. Giupponi, L.; Borgonovo, G.; Panseri, S.; Giorgi, A. Multidisciplinary study of a little known landrace of Fagopyrum tataricum Gaertn. of Valtellina (Italian Alps). *Genet. Resour. Crop Evol.* **2019**, *66*, 783–796. [CrossRef]
4. Ožbolt, L.; Kreft, S.; Kreft, I.; Germ, M.; Stibilj, V. Distribution of selenium and phenolics in buckwheat plants grown from seeds soaked in Se solution and under different levels of UV-B radiation. *Food Chem.* **2008**, *110*, 691–696. [CrossRef]
5. Habtemariam, S. Antioxidant and rutin content analysis of leaves of the common buckwheat (*fagopyrum esculentum* moench) grown in the United Kingdom: A case study. *Antioxidants* **2019**, *8*, 160. [CrossRef]
6. Dziedzic, K.; Górecka, D.; Szwengiel, A.; Sulewska, H.; Kreft, I.; Gujska, E.; Walkowiak, J. The Content of Dietary Fibre and Polyphenols in Morphological Parts of Buckwheat (*Fagopyrum tataricum*). *Plant Foods Hum. Nutr.* **2018**, *73*, 82–88. [CrossRef]
7. Kočevar Glavač, N.; Stojilkovski, K.; Kreft, S.; Park, C.H.; Kreft, I. Determination of fagopyrins, rutin, and quercetin in Tartary buckwheat products. *LWT—Food Sci. Technol.* **2017**, *79*, 423–427. [CrossRef]

8. Sytar, O.; Brestic, M.; Rai, M. Possible ways of fagopyrin biosynthesis and production in buckwheat plants. *Fitoterapia* **2013**, *84*, 72–79. [CrossRef]
9. Szymańska, M.; Majerz, I. Geometry and electron density of phenothazines. *J. Mol. Struct.* **2020**, *1200*, 127095. [CrossRef]
10. Szymańska, M.; Majerz, I. Effect of substitution of hydrogen atoms in the molecules of anthrone and anthraquinone. *Molecules* **2021**, *26*, 502. [CrossRef]
11. Edim, M.M.; Enudi, O.C.; Asuquo, B.B.; Louis, H.; Bisong, E.A.; Agwupuye, J.A.; Chioma, A.G.; Odey, J.O.; Joseph, I.; Bassey, F.I. Aromaticity indices, electronic structural properties, and fuzzy atomic space investigations of naphthalene and its aza-derivatives. *Heliyon* **2021**, *7*, e06138. [CrossRef] [PubMed]
12. Galinari, C.B.; Biachi, T.D.P.; Gonçalves, R.S.; Cesar, G.B.; Bergmann, E.V.; Malacarne, L.C.; Kioshima Cotica, É.S.; Bonfim-Mendonça, P.D.S.; Svidzinski, T.I.E. Photoactivity of hypericin: From natural product to antifungal application. *Crit. Rev. Microbiol.* **2022**, *49*, 1–19. [CrossRef] [PubMed]
13. Szymanski, S.; Majerz, I. Aromaticity and Electron Density of Hypericin. *J. Nat. Prod.* **2019**, *82*, 2106–2115. [CrossRef] [PubMed]
14. Szymanski, S.; Majerz, I. In silico studies on sennidines—Natural dianthrones from senna. *Biology* **2021**, *10*, 468. [CrossRef] [PubMed]
15. Eguchi, K.; Anase, T.; Osuga, H. Development of a high-performance liquid chromatography method to determine the fagopyrin content of Tartary buckwheat (*Fagopyrum tartaricum* Gaertn.) and common buckwheat (*F. esculentum* Moench). *Plant Prod. Sci.* **2009**, *12*, 475–480. [CrossRef]
16. Kim, J.; Kim, S.; Hwang, K.T. Determination and photochemical conversion of protofagopyrins and fagopyrins in buckwheat plants. *J. Food Compos. Anal.* **2021**, *100*, 103894. [CrossRef]
17. Benković, E.T.; Žigon, D.; Friedrich, M.; Plavec, J.; Kreft, S. Isolation, analysis and structures of phototoxic fagopyrins from buckwheat. *Food Chem.* **2014**, *143*, 432–439. [CrossRef]
18. Kim, J.; Hwang, K.T. Fagopyrins in different parts of common buckwheat (*Fagopyrum esculentum*) and Tartary buckwheat (*F. tataricum*) during growth. *J. Food Compos. Anal.* **2020**, *86*, 103354. [CrossRef]
19. Kosyan, A.; Sytar, O. Implications of fagopyrin formation in vitro by uv spectroscopic analysis. *Molecules* **2021**, *26*, 2013. [CrossRef]
20. Samel, D.; Donnella-Deana, A.; De Witte, P. The effect of purified extract of Fagopyrum esculentum (buckwheat) on protein kinases involved in signal transduction pathways. *Planta Med.* **1996**, *62*, 106–110. [CrossRef]
21. Kim, J.; Kim, S.; Lee, K.; Kim, R.H.; Hwang, K.T. Antibacterial photodynamic inactivation of fagopyrin f from tartary buckwheat (*Fagopyrum tataricum*) flower against streptococcus mutans and its biofilm. *Int. J. Mol. Sci.* **2021**, *22*, 6205. [CrossRef] [PubMed]
22. Tavčar Benković, E.; Kreft, S. Fagopyrins and Protofagopyrins: Detection, Analysis, and Potential Phototoxicity in Buckwheat. *J. Agric. Food Chem.* **2015**, *63*, 5715–5724. [CrossRef] [PubMed]
23. Lineva, A.; Benković, E.T.; Kreft, S.; Kienzle, E. Remarkable frequency of a history of liver disease in dogs fed homemade diets with buckwheat. *Tierärztliche Prax. Ausg. K Kleintiere-Heimtiere* **2019**, *47*, 242–246. [CrossRef] [PubMed]
24. Sytar, O.; Švedienė, J.; Ložienė, K.; Paškevičius, A.; Kosyan, A.; Taran, N. Antifungal properties of hypericin, hypericin tetrasulphonic acid and fagopyrin on pathogenic fungi and spoilage yeasts. *Pharm. Biol.* **2016**, *54*, 3121–3125. [CrossRef]
25. Zambounis, A.; Sytar, O.; Valasiadis, D.; Hilioti, Z. Effect of photosensitisers on growth and morphology of phytophthora citrophthora coupled with leaf bioassays in pear seedlings. *Plant Prot. Sci.* **2020**, *86*, 74–82. [CrossRef]
26. Pauling, L. *The Nature of the Chemical Bond an Introduction to Modern Structural Chemistry*; Cornell University Press: New York, NY, USA, 1960.
27. Schuster, P.; Zundel, G.; Sandorfy, C. *The Hydrogen Bond, II: Structure and Spectroscopy*; North-Holland Publishing Company: Amsterdam, The Netherlands; New York, NY, USA; Oxford, UK, 1975.
28. Gilli, G.; Gilli, P. *The Nature of the Hydrogen Bond: Outline of a Comprehensive Hydrogen Bond Theory*; Oxford University Press: Oxford, UK, 2009; ISBN 9780191720949.
29. Frisch, G.W.; Schlegel, H.B.; Scuseria, G.E.; Robb, M.A.; Cheeseman, J.R.; Scalmani, G.; Barone, V.; Petersson, G.A.; Nakatsuji, H.; Li, X.; et al. *Gaussian 16, Rev. A.03*; Gaussian, Inc.: Wallingford, CT, USA, 2016.
30. Grimme, S.; Antony, J.; Ehrlich, S.; Krieg, H. A consistent and accurate ab initio parametrization of density functional dispersion correction (DFT-D) for the 94 elements H-Pu. *J. Chem. Phys.* **2010**, *132*, 154104. [CrossRef]
31. Keith, T.A. *AIMALL*, version 19.02.13; TK Gristmill Software: Overland Park, KS, USA, 2019.
32. Johnson, E.R.; Keinan, S.; Mori-Sánchez, P.; Contreras-García, J.; Cohen, A.J.; Yang, W. Revealing noncovalent interactions. *J. Am. Chem. Soc.* **2010**, *132*, 6498–6506. [CrossRef]
33. Lu, T.; Chen, F. Multiwfn: A multifunctional wavefunction analyzer. *J. Comput. Chem.* **2012**, *33*, 580–592. [CrossRef]
34. Humphrey, W.; Dalke, A.; Schulten, K. VMD: Visual molecular dynamics. *J. Mol. Graph.* **1996**, *14*, 33–38. [CrossRef]
35. te Velde, G.; Bickelhaupt, F.M.; Baerends, E.J.; Fonseca Guerra, C.; van Gisbergen, S.J.A.; Snijders, J.G.; Ziegler, T. Chemistry with ADF. *J. Comput. Chem.* **2001**, *22*, 931–967. [CrossRef]
36. Krygowski, T.M. Crystallographic Studies of Inter- and Intramolecular Interactions Reflected in Aromatic Character of π-Electron Systems. *J. Chem. Inf. Comput. Sci.* **1993**, *33*, 70–79. [CrossRef]
37. Richard, F. *Bader: Atoms in Molecules (A Quantum Theory)*; Clarendon Press: Oxford, UK, 1990; ISBN 019-855-1681.
38. Bader, R.F.W. Bond paths are not chemical bonds. *J. Phys. Chem. A* **2009**, *113*, 10391–10396. [CrossRef] [PubMed]

Article

Association Complexes of Calix[6]arenes with Amino Acids Explained by Energy-Partitioning Methods

Emran Masoumifeshani [1], Michał Chojecki [1], Dorota Rutkowska-Zbik [2] and Tatiana Korona [1,*]

[1] Faculty of Chemistry, University of Warsaw, ul. Pasteura 1, 02-093 Warsaw, Poland
[2] Jerzy Haber Institute of Catalysis and Surface Chemistry, Polish Academy of Sciences, ul. Niezapominajek 8, 30-239 Cracow, Poland
* Correspondence: tania@chem.uw.edu.pl

Abstract: Intermolecular complexes with calixarenes are intriguing because of multiple possibilities of noncovalent binding for both polar and nonpolar molecules, including docking in the calixarene cavity. In this contribution calix[6]arenes interacting with amino acids are studied with an additional aim to show that tools such as symmetry-adapted perturbation theory (SAPT), functional-group SAPT (F-SAPT), and systematic molecular fragmentation (SMF) methods may provide explanations for different numbers of noncovalent bonds and of their varying strength for various calixarene conformers and guest molecules. The partitioning of the interaction energy provides an easy way to identify hydrogen bonds, including those with unconventional hydrogen acceptors, as well as other noncovalent bonds, and to find repulsive destabilizing interactions between functional groups. Various other features can be explained by energy partitioning, such as the red shift of an IR stretching frequency for some hydroxy groups, which arises from their attraction to the phenyl ring of calixarene. Pairs of hydrogen bonds and other noncovalent bonds of similar magnitude found by F-SAPT explain an increase in the stability of both inclusion and outer complexes.

Keywords: calixarene; SAPT; molecular fragmentation; intermolecular interaction; hydrogen bond; host–guest complex

1. Introduction

Calixarenes are phenolic-based macrocycles consisting of repeated units of phenol and hydrocarbon groups, which have attracted a lot of attention in recent years owing to their ability to form host–guest complexes for hydrophobic guests, in particular, in the context of their biological and pharmaceutical potential [1–3]. Experimentally known calixarenes usually contain between four and eight phenol units [4]. The simplest calixarene, calix[4]arene, consists of four phenol groups connected through methylene links ($-CH_2-$). A theoretical study of the stability of smaller calix[3]arenes [5] revealed strong structure stress, which disappears only when one more phenol ring is added to the macrocycle. A known feature of calixarenes is their ability to form inclusion complexes in one conformation and release the guest molecule after the conformation change. Such adducts, which are examples of intermolecular complexes, are being studied as possible drug carriers [6–8], models of enzyme active centers [4], the base for supramolecular catalysts [9,10], building elements of nonlinear optics structures [11], structures binding metal ions [5,12] and their clusters [13], and as chemosensors [14–16]. Furthermore, the modification of the calixarenes' basic core and both upper and lower rims extended the range of their applications towards becoming anticancer drugs [17–19], antimicrobial agents [17,20–25], and sensitizers for photodynamic therapy [26].

The smallest stable calixarene is known to have four conformations, which can be unanimously described by just listing relative orientations of the phenol groups with respect to the macroring, i.e., since each phenol ring can assume either upward (u) or downward (*d*) position relative to this ring, one finds four possibilities, denoted as *cone* (*uuuu*), *partial cone* (*uuud*), *1,2-alternate* (*uudd*), and *1,3-alternate* (*udud*)–in parenthesis, the arrangement of adjacent phenol rings has been listed. This classification turns out to be insufficient for larger calixarenes, where additional degrees of freedom result from more variants of orientations of phenol rings with respect to their left and right neighbors. In order to characterize these calixarene conformers, Ugozzoli and Andreetti [27] proposed to represent a conformation of calix[*n*]arene through a list of *n* pairs of conformational parameters, defined as dihedral angles between a phenol ring and a plane defined by two C-C bonds originating from a linker carbon atom towards phenol groups. Formally these angles are calculated as dihedral angles $C_1C_2C_3C_4$ (ϕ) and $C_2C_3C_4C_5$ (χ), see Figure 1 for numbering of carbon atoms. Additionally, a "top side" of calixarene is defined in Ref. [27], and the order of (ϕ, χ) descriptors is set to be counterclockwise with respect to the top. For all cases considered in the present paper, the simplest definition of the top side applies, i.e., it is the side with the highest number of phenolic oxygen atoms. Some of these conformations are more stabilized by hydrogen bonds (H-bonds) between the hydroxy groups than others. Moreover, the cavity size and the rim properties differ significantly in various conformers. The conformations described above are in dynamic equilibrium, which is related to the possibility of rotation around the bonds bridging the individual phenol units. This rotation may be blocked by the introduction of large substituents into the aromatic rings, which will form a steric hinge within the upper edge of the calixarene ring (opposite to the hydroxyl groups) [28]. Other factors affecting the possibility of rotation of phenol units are: temperature and solvents. The availability of conformers in a finite temperature is governed by their energetic stability differences according to the Boltzmann law, while polar or nonpolar solvents can either preferably bind with the polar hydroxy groups of calixarenes or–on the opposite–stabilize conformers for which hydroxy groups can be hidden inside the hole creating the intramolecular H-bond network(s).

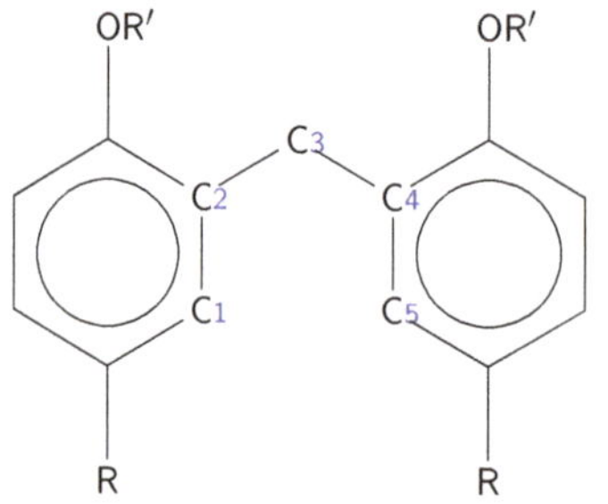

Figure 1. Dihedral angles between adjacent aryl groups of a calixarene. The first dihedral angle is defined through atoms 1–4, and the second one–through atoms 2–5.

Many theoretical and experimental studies exist for calixarenes, especially for the smallest calix[4]arenes, see, e.g., Refs. [29–31]. Studies of larger calixarenes, calix[6]arenes and calix[8]arenes as reported in Refs. [29,32,33] and [29,34,35], respectively, revealed that in the majority of cases, the cone conformation is the most stable. This finding can be explained by the creation of intramolecular H-bonds, as proposed by Gassoumi et al. [29] after a density-functional theory (DFT) study with several functionals for six- and eight-member unsubstituted calixarenes, while Furer et al. [32,34] arrived at the same conclusion for heavily substituted calix[*n*]arenes, $n = 6, 8$, with adamantane and *tert*-butyl serving as substituents.

Inclusion complexes with calixarenes were objects of several studies, too. For instance, Puchta et al. [36] examined the possibility of the binding of small molecules such as CH_3NH_2 and CH_3CN by calix[4]arenes by employing a simple DFT functional and a small basis set as an extension of the nuclear magnetic resonance (NMR) studies. They found that no proper inclusion complexes are created since the most favorable position of the guest molecules would be on the outer part of the edge. The binding energy was small (5 kcal/mol), but one should take into account that the utilized functional and basis set were not optimal for such an investigation. It should be also noted that Puchta et al. observed a proton transfer from the host species after binding of CH_3NH_2. In another study Galindo-Murillo et al. [6] examined the applicability of calix[n]arenes (where $n = 4, 5, 6, 8$) as hosts for 3-phenyl-1H-[1]benzofur[3,2-c]pyrasol (GTP). Their DFT calculations (with a better B97-D functional) showed that the most promising GTP carriers are the two largest calixarenes, which can be attributed to the existence of a large enough space inside the host molecule and to the presence of hydrogen bondings between the host and guest. The stabilizing role of a partial charge transfer (CT) between GTP and calixarene has also been noted in this work.

The usability of calix[4]rene and calix[4]resorcinarene and their derivatives to host a zoledronate acid molecule has been studied by Jang et al. [37] with DFT plus a dispersion correction (DFT+D), where the cone conformer of calixarene has been used. It turned out that the adduct contains a guest attached to the upper ring edge of the calixarene and that the inclusion of this guest causes only modest changes in the cavity of the host. If calixarene is modified with the sulfone groups or with phosphone ones, their cavities become notably wider, while the bottom remains mostly rigid, as it is stabilized by the intramolecular H-bonds. Further analysis of electronic structures shows that π–π interactions play a decisive role in the binding between the host and the guest. The number of H-bonds is important for the relative stability of similar complexes, e.g., the binding with calix[4]resorcinarene is stronger than that of the sulfonated calix[4]arene. In another study performed by Kryuchkova et al. [38], it has been found that calix[4]arenes modified by the R_2PO groups can ligate up to two cobalt complexes with the nitrate ions. The smallest calix[4]arenes were also studied as carriers of small gas molecules, such as H_2, O_2, N_2, H_2O, CO_2, NH_3, H_2S, N_2O, HCN, SO_2 [39], if the lower rim of the host has been methylated, either fully or in part. The calculations, performed with the DFT+D approach, revealed that one can modify the selectivity of calixarenes through their methylation. Very recently, a theoretical study on the docking of insulin on calixarenes have appeared, where the CHARMM force field has been applied to determine the preferred amino acids docking to substituted calixarenes [40]. Other investigations of docking on calixarenes have been performed, e.g., in Refs. [41,42].

General studies of the stability of calixarenes usually do not delve into the nuances of the nature of differences between various conformers and/or complexes with calixarenes. It is usually assumed in the literature that binding properties of calixarenes can be explained by the interactions of the CH-π, π-π, and ion-π types [2,43], but–to our knowledge–no systematic studies exist, which would explain the variety of properties of calixarene conformers and their complexes through the application of energy partitioning approaches in order to classify interaction types and to estimate their relative strength. Therefore, in this paper, we would like to fill this gap and present such partitioning analyses resulting from Symmetry-Adapted Perturbation Theory (SAPT) [44,45], Functional-group SAPT (F-SAPT) [46,47], and our recent modification of Systematic Molecular Fragmentation (SMF) [48], described in Refs. [49,50] and denoted as Symmetrized SMF (SSMF). Our study will include a systematic search for the H-bonds and other bonding types within the complexes. The recent redefinition of the H-bond is presented in Ref. [51], and numerous investigations concerning various types of H-bonds have been performed, including "unusual" ones (dihydrogen bonds, bonds including unusual hydrogen donors–such as, e.g., C–H, or unusual hydrogen acceptors–such as, e.g., benzene ring, or inverted H-bonds, etc.), see, e.g., Refs. [52–59]. Such investigations have been performed for other complexes with SAPT (see, e.g., Ref. [52]) or F-SAPT methods [46,47]. Additionally, applications of the SMF to intermolecular interaction energy partitionings or comparisons of F-SAPT and SMF approaches are missing in the literature.

We selected standard amino acids as potential guests and calix[6]arenes as hosts for our study because of the biochemical importance and a wide variety of binding sites of potential guests, which result from hydrophobic, polar, aliphatic, aromatic, etc. groups attached to the amino acid moiety, while the choice of larger calix[6]arenes instead of calix[4]arenes was dictated by the existence of a bigger cavity, in which larger molecules are able to fit. Therefore, the selected set of intermolecular complexes is expected to provide examples of noncovalent bonds of different types and strengths and to show the competition of various bonds under steric constraints, resulting from the existence of a cavity. An additional feature, which will be tracked in this study, is the confinement effect resulting in the enhancement of the attraction. Such cases, reported many times for endohedral complexes of fullerenes [60], appear if a guest molecule has an optimal size for a given cavity and can be attracted from many sides, while a larger molecule starts to feel a repulsion from the cavity wall, and finally a smaller molecule is attracted mostly by one side of the cavity or–if placed in the center–is attracted less strongly because of too large a distance between the guest and the closest atoms of the host. The calixarenes selected for this study are: unsubstituted calix[6]arene and hexa-*p-tert*-butylcalix[6]arene, where the latter has been added in order to examine how obstacles created by spatially large nonpolar groups modify host–guest interactions. The *tert*-butyl group is one of the smallest and one of the most popular groups of this type and it is frequently used to decorate calixarene molecules [61]. In the following, these two species will be abbreviated as CX and BCX, respectively.

In the literature, one finds reports on the use of calixarenes for recognition of amino acids [62–72]. The critical review of the available data enabled general observations on amino acid complexation by calixarenes. Usually, the ability to bind these small guests decreases in the order: calix[6]arene > calix[8]arene > calix[4]arene [70,73]. The binding process is controlled by hydrogen bonding, van der Waals, cation–π, and π–π stacking interactions [70]. The strength of the interaction depends strongly on the chemical character of the amino acid and the substituents introduced to the pristine calixarene scaffold. Stone et al. [73] studied the formation of complexes of calixarenes with amino acids with matrix-assisted laser desorption/ionization coupled with mass spectrometer (MALDI-MS) technique. They demonstrated that calix[6]arenes substituted by esters form stable complexes with amino acids which are bound mostly by cation–π interactions and hydrogen bondings. The native calix[6]arenes, on the contrary, do not form complexes stable enough in the gas phase to withstand a relatively energetic process of MALDI. Among the association complexes, one finds both: (i) outer-sphere ones, in which amino acids are located outside the pocket created by the phenyl rings (see, e.g., Ref. [66]); and (ii) typical host–guest structures, where the amino acid molecules are buried inside the calixarene cavity (see, e.g., results of Oshima et al. [74] for complexes between hydrophobic amino acids and calix[6]arenes modified by carboxylic groups). Due to the small size of the interacting moieties, often one calixarene species accommodates only one amino acid molecule, but the ratio increases while increasing the size of calixarene. Douteau-Guével and co-workers demonstrated by the microcalorimetry method that calix[6]arene sulfonates form weak 1:1 complexes with Lys and Arg in water [75]. The same stoichiometry was also determined for the host–guest type of binding of Trp by calix[6]arene carboxylic acid derivative [76]. The 1:1 and 1:2 complexes are found in the case of calix[8]arenes [77].

2. Methodology

2.1. Geometry Optimization–Calixarenes

The first step in the analysis of complexes is the selection of the most suitable conformers for the calixarene host. In order to facilitate this task, we performed a query in the Cambridge Structural Database (CSD) [78] for accessible calix[6]arene structures. This search resulted in three selected conformers: pinched-cone (CSD-VARGUR [79]) – found as the most stable in nonhydrogen-bonding solvents [80], 1,2,3-alternate (CSD-REGWIM [81]) – preferred in hydrogen-bonding solvents [80], and recently crystalized winged-cone one (CSD-REGWEI [81]). Starting geometries of these conformations were extracted from the corresponding crystal structures, and several starting geometries were generated with the Tinker program [82] by freezing all but dihedral angles of the hydroxyl groups. Then, the geometry optimization was performed with Density Functional Theory (DFT) using the B97-D3 functional [83], which is a popular choice for organic molecules. Electron repulsion integrals were treated within the Density Fitting (DF) approximation [84] and the def2-TZVP orbital basis set [85] was used with the corresponding default auxiliary DF basis. The resulting geometries are similar to the starting ones (no hydroxy group has switched from u to d or vice versa during the optimization), so the final conformers will be denoted as pc (from pinched), al (from alternate), and wc (from winged-cone), as the original structures from the CSD. The stability of these conformers for the CX case was tested by a modification of their geometries through a rotation of selected phenyl rings and reoptimization, which either led to the original conformer, or to local minima energetically higher than pc, al, or wc.

2.2. Geometry Optimization–Complexes

The search for the optimal geometries for complexes of twenty amino acids (for formulas and abbreviations see the Supplementary Information) with pc, wc, and al conformers of either calix[6]arene, or hexa-p-tert-butylcalix[6]arene has been performed in several steps. Firstly, the generation of approximate geometries of the most plausible host–guest conformers was performed with the AutoDock Vina program [86]. Secondly, the resulting geometries were used as starting points for the DFT optimization. Because of the high computational cost, the def2-SVP basis set was used for the search of local minima, and solely the lowest ones were reoptimized using the def2-TZVP basis set. It should be noted that for the arginine (Arg) case, all AutoDock Vina optimizations lead to the protonated (charged) version of the amino acid. Since the interaction energies between two ions would give a completely different interaction picture than for uncharged molecules, we decided to skip this amino acid from our test set. For further analysis, only the lowest minima for each calixarene conformer and amino acid were utilized.

Gaussian16 [87] was used for the geometry optimizations. The harmonic frequency analysis has been performed in order to verify that the stationary point is the minimum.

2.3. Energetics

In this part, we focus on the relative stability of pc, al, and wc conformers for both pristine calixarenes and complexes. The stability order can be established by comparing total energies, which are sums of the DFT+D electronic energy calculated at minimum and the nuclear contribution, given by zero-point vibrational energy (ZPVE). In practice, since these total energies have large absolute values and are similar to each other, it is the most useful to find the lowest conformer and to report relative differences of energies for the two remaining conformers. It should be emphasized that the energetic order of complexes of different calixarene conformers with the same amino acid, which results from this procedure, does not always correspond to the order of the interaction energies since a difference in conformers' energies contains additional deformation contributions for the interacting molecules. In particular, the calixarene's deformation may vary substantially for different complexes because of the large flexibility of calixarenes.

The selected complexes in their optimal geometries have been studied with the help of several methods in order to perform a detailed analysis of their energetics with special attention paid to energy partitioning. A method of choice for such a study is symmetry-adapted perturbation theory (SAPT) [44,88], which allows us to obtain not only intermolecular interaction energies but also their components, such as electrostatic, induction, dispersion, and their exchange counterparts, with a well-defined physical interpretation,

$$E_{int} = E^{(1)}_{elst} + E^{(1)}_{exch} + E^{(2)}_{ind} + E^{(2)}_{exch-ind} + E^{(2)}_{disp} + E^{(2)}_{exch-disp} + \delta E_{HF}. \tag{1}$$

In this study, we utilized the simplest SAPT variant, denoted as SAPT(HF) or alternatively as SAPT0 [89] SAPT0 is also a method of choice for further F-SAPT analysis.

Supermolecular interaction energies were obtained on the MP2 and Spin-Component-Scaled MP2 (SCS-MP2) [90] levels of theory with the jun-cc-pVDZ basis set [91–93]. Since the DF implementation of these methods was used, the corresponding default auxiliary basis sets [94–96] were applied in these calculations, which were performed with the Psi4 program [97]. In order to estimate the accuracy of the interaction energies with respect to the basis set, we performed the Complete-Basis-Set (CBS) extrapolation [98] for the correlation part of the interaction energy for one selected complex. For all supermolecular calculations, the Boys–Bernardi counterpoise correction was employed to avoid the basis-set superposition error [99].

Additionally, we studied the interaction energies with the help of a Symmetrized Systematic Molecular Fragmentation on the third level [49] with interaction energies between fragments calculated with the MP2 method (SSMF3-MP2) method. The SSMF arises from the SMF by Annihilation (SMFA) [100], and differs from the original one by treatment of branched molecules [49]. In the SMFA, one first defines molecular units, which are composed either of a single nonhydrogen atom, or atoms connected by multiple bonds, both with attached hydrogen atoms. Then, a molecule is fragmented into overlapping units, where broken single bonds are saturated with hydrogen atoms. Therefore, the molecule is represented as a sum of overlapping fragments with positive and negative weights, and the energy is calculated as a sum of the energies of overlapping fragments corrected with a sum of interaction energies of non-overlapping units [100–102], i.e., if the fragmentation of a molecule M into fragments G_i with weights w_i is performed, the total energy of M is expressed as

$$E(M) = \sum_i w_i E(G_i) + E_{nb}, \tag{2}$$

where $E(X)$ is the energy of the molecule X and E_{nb} is the (usually small) interaction energy of the non-overlapping parts. The size of fragments is given in terms of the number of units, and as a rule, one obtains fragments of size n and $n+1$ for the nth level of fragmentation. In the SMFA approach, the fragments which incorporate branches are bigger and can be formed of $n+2$ or more units, which makes the calculation more expensive computationally. In our recent publication [49] we proposed an alternative treatment of branching points, which keeps an immediate symmetry around branches (therefore the name) and–more importantly–preserves the number of units at cost of introducing fractional weights. Numerical tests show that the third fragmentation level (i.e., SSMF3) has sufficient accuracy for practical applications. We also studied the applicability of the SSMF3 approach for intermolecular interaction energies and found that the computationally expensive long-range correlation energy contribution to the interaction energy is well reproduced on the SSMF3 level [50].

The SMF approach may be also utilized to obtain the interaction energies. If one molecule (M) is large enough to undergo fragmentation, while the second one (A) is small and will be left unfragmented, then the interaction energy between M and A can be obtained by utilizing a *method* to calculate the interaction energy for each pair G_i with A, and then by applying the formula,

$$E_{int}^{method}(MA) = \sum_i w_i E_{int}^{method}(G_i A). \tag{3}$$

This is exactly the case of complexes with calixarenes, where the amino acids are left unfragmented, while the calixarene host is fragmented at the SSMF3 level [50]. Altogether 24 such fragments are generated for the CX case: twelve 6-methyl-2,2'-methanediyl-di-phenol, six 2,2'-methanediyl-di-phenol, and six 2,6-dimethylphenol molecules, so 24 (short) calculations of the interaction energies should be performed instead of one long calculation. The timing gain may be significant, especially for high-level electron-correlated theories, because of the steep scaling of such methods with respect to the molecular size, and since the calculations of fragments can be performed in parallel.

The SSMF3 partitioning has been performed by a homemade C++ program written by one of us (E.M.).

2.4. Molecular Properties

For pristine calix[6]arene, in addition to the total energy calculations, we performed computations of selected first-order and second-order properties (electric dipole moments and static dipole polarizabilities) on the HF and MP2 levels. For these properties, we applied Equation (2) with energies replaced by the corresponding property and neglecting the nonbonding contribution. The HF and MP2 properties for SSMF3 fragments were obtained analytically with Molpro [103] (all electrons were correlated in the MP2 method because of the limitations of the MP2 polarizabilities program in Molpro [104]). Properties for unfragmented calixarenes were obtained with the finite-field (FF) approach of a field strength 0.003 a.u. and by comparing the 3- and 5-point formula to control the accuracy of the FF procedure (1s orbitals of nonhydrogen atoms were frozen in the MP2 method to speed up the calculations after verification that this approximation has a very small effect on these properties). Because of the testing character of this investigation, we used a small cc-pVDZ basis, for which the second-order properties for the whole molecules can be obtained in a manageable computational time.

2.5. Energy Partitioning

The main workhorse for the energy partitioning in this work is the Functional-group SAPT (F-SAPT) method [46,47] and its intramolecular variant I-SAPT [105], which have been utilized to determine the inter- and intramolecular interactions between interesting groups of atoms. For this purpose, the partitioning on the SAPT0 level in the jun-cc-pVDZ basis set [93] was utilized. In the F-SAPT, both SAPT partitioning of the interaction energy into physically-sound components and the partitioning of these components into contributions, which can be attributed to the interactions between groups of atoms, are utilized, which gives us the opportunity to classify inter-group interactions not only in terms of their strength (i.e., by comparing the absolute value of the interaction energy) but also according to their dominant SAPT contribution. In practice, one usually examines electrostatics, first-order exchange, and effective induction and dispersion, which are obtained by adding up second-order induction, exchange-induction, and the δE_{HF} term, or second-order dispersion and exchange-dispersion, respectively. In order to facilitate the analysis, we developed a graph system [106], where groups of the molecule A and B are placed on the left and right edges of the graph and are connected with red (attractive) or blue (repulsive) lines, whose thickness is proportional to the absolute value of the component under study. It should be emphasized that both F- and I-SAPT energy partitionings should be treated with caution since such partitionings depend to some extent on the applied fragmentation approach. In the (F/I)-SAPT case, this fragmentation is orbital-based and therefore is dependent on the balanced description of various atoms and bonds within a given basis. It also contains contributions from the so-called "linkers", which appear at places where the bonds between groups are broken. The latter contributions are somewhat arbitrarily divided between adjacent groups in the reduced analysis; therefore, the analysis of tiny energy fluctuations for partitioned energies is of limited value. Numerous studies performed with either F-, or I-SAPT show [107–111] that many useful conclusions can be reached if one adheres to hints given in the original papers of Parrish et al.. Therefore, in this study, we apply a standard (for F-SAPT) jun-cc-pVDZ basis and the simplest SAPT0 rung of SAPT models, as advised in Ref. [46].

For purposes of the energy partitioning with F-SAPT, the selected functional groups for the case of calixarenes are: hydroxy (OH), phenyl (Ph), methylene (CH_2), and *tert*-butyl groups (in order to make an easier comparison of the calix[6]arene and hexa-*p*-*tert*-butylcalix[6]arene we separated the *para* hydrogen atom in the phenyl ring as a separate group–this hydrogen is replaced by the *tert*-butyl group in the latter molecule). The common functional groups of amino acids are: amino (NH_2) and carboxy (COOH) groups.

Depending on the amino acid type, other functional groups are singled out (see F-SAPT graphs for all the complexes under study in the Supplementary Information). For instance, for the simplest Gly, the only remaining group is the methylene group CH_2, while for the Phe amino acid, there are additionally the ring and CH groups. We have chosen not to fragment the carboxy group into C=O and OH; therefore, in some cases, the possible classification of the bonding involving this group as the H-bond (or not) will require a visual check.

For sake of conciseness, we define a calixarene "unit" consisting of a phenolic group (i.e., either a phenol group or *tert*-butyl-substituted phenol group) and a methylene bridge, and we will number these units as in Figure 2. If necessary, the respective groups within these units will be distinguished by adding a number behind the group (i.e., OH-k, Ph-k, CH_2-k, *tert*-butyl-k denote the hydroxy, phenyl, methylene, and *tert*-butyl groups of the kth calixarene unit, respectively). Analogously, we will differentiate the group X of an amino acid Y as X-Y, e.g., the COOH group of Gly will be denoted as COOH-Gly to avoid confusion.

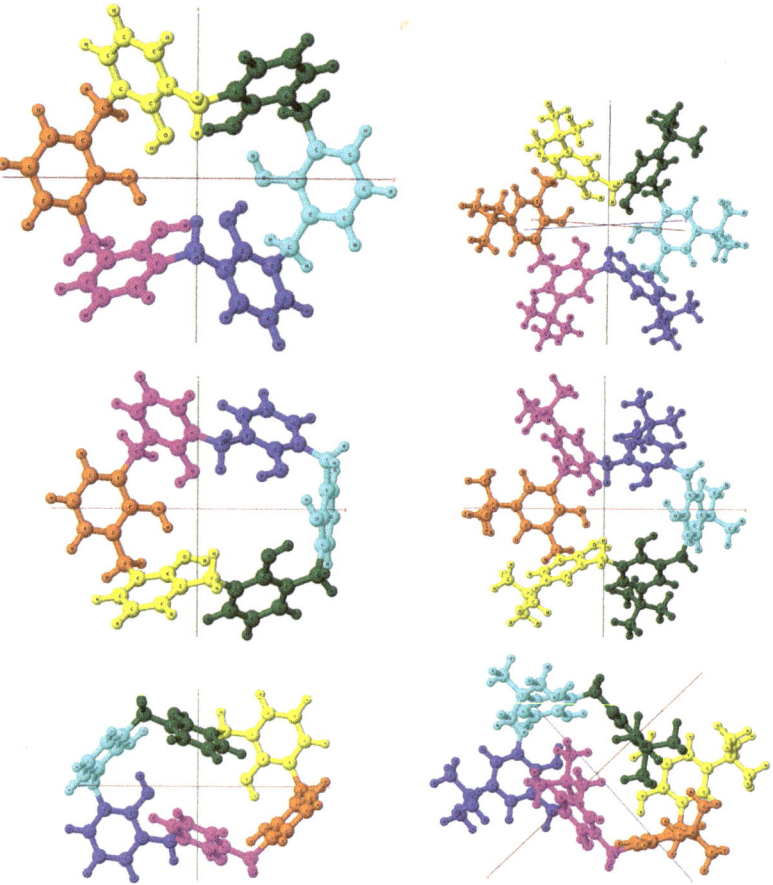

Figure 2. A numeration of calixarene groups for *pc*, *al*, and *wc* conformers (upper, middle, and lower rows, correspondingly). Colors correspond to the following numbers: orange–1, yellow–2, green–3, cyan –4, blue –5, violet– 6.

For the case of the SMF, the molecular units are selected automatically and grouped into fragments by the fragmentation procedure; therefore, in several cases, two or more interesting groups are placed together in one fragment. Nevertheless, by a difference analysis, one can often determine the main "culprit" for a strong interaction within the fragments. These analyses of SSMF3 pair interaction energies between fragments of molecules A and B give us a complementary view of the interaction energy partitioning.

The F-SAPT and I-SAPT partitionings, SAPT0 and most MP2 and SCS-MP2 calculations have been performed with the Psi4 program [112]. Some supermolecular calculations were performed with Molpro [103].

3. Results and Discussion

3.1. Pristine Calixarenes–Geometries

The characteristic feature of calixarenes is the existence of multiple intramolecular H-bonds, formed between its hydroxy groups. The creation of these bonds can be promoted or prohibited depending on the relative positions of the aryl rings in various conformers. To this end, two trimeric parts (i.e., consisting of three calixarene units), vaguely resembling semi-circles, can be distinguished in the structure of the *pc* conformer of calix[6]arene. All six hydroxy groups are on the same side of the backbone and because of a favorable orientation of semi-circles with respect to each other, each of these groups is involved in hydrogen binding as the H donor and the H acceptor, thus creating a ring-like structure of six H-bonds. Each one from two methylene bridges connecting trimeric subunits has one hydrogen atom pointing inside the macroring with a distance of 3.22 Å (3.02 Å) between these hydrogen atoms for the CX (BCX). Therefore, the cavity of the *pc* conformer is divided into two parts by this dihydrogen bridge and the top-side entrance is blocked by the second ring formed by the H-bonds. The distances between the oxygen atom of one hydroxy group and the hydrogen atom of the neighboring hydroxy group (denoted in the following as the H-bond distance) indicate that indeed all H-bonds formed in this case are strong–they are in range between 1.66 to 1.70 Å and–somewhat surprisingly–for the hexa-*p*-*tert*-butylcalix[6]arene case not all of them become larger, as one could expect (two become larger, while the remaining four–a bit shorter). For the H-bond the orientation of three atoms O-H $\cdots$ O should be close to 180° for optimal binding [51]. For a cyclic structure, as in the calixarene case, the H-bondings are always stressed in this aspect, and, e.g., for the *pc* conformer the H-bond angles are: 173, 165, 166° for the CX and 174, 164, 166° for BCX. Similar values of distances and angles in these both cases indicate that the *tert*-butyl substitution does not cause any significant distortion in calixarenes.

A similar semi-circle structure exists for the *al* conformer, but in this case, three adjacent hydroxy groups are placed upside down; therefore, only two H-bonds can be created on each side of the macrocycle. Moreover, in this case, one hydrogen atom from the methylene bridges connecting two trimeric semi-circles points to the interior of the molecule, but the H$\cdots$H distance is significantly larger and amounts to 3.70 Å for the CX and 3.58 Å for the BCX. Judging from the H-bond distances between oxygen and hydrogen atoms of the respective hydroxy groups, the *al* H-bonds should be weaker than for the *pc* case: these distances range between 1.75 and 1.81 Å, i.e., even the shortest distance is about 0.05 Å longer than for the *pc* conformer. From the H-bond angle values, which span from 166 to 168°, one can make a conclusion that only those more "stressed" H-bondings survive in the *al* conformer as compared to the *pc* one.

Finally, the *wc* conformer can be visualized by the flattening of two opposite phenyl rings in the *pc* conformer. As a result of this operation, the distance between two pairs of hydroxy groups becomes too large for an effective H-binding and only two triples of adjacent OH groups are still able to create two H-bonds each, similarly to the *al* case. The O$\cdots$H distances for the *wc* conformers range from 1.81 to 1.87 Å, i.e., are larger than in the *al* case, indicating that the formed H-bonds in the *wc* conformer are even weaker than in the *al* case. The addition of the *tert*-butyl groups does not have a significant impact on the lengths of the H-bonds, but it does affect the positioning of the phenol groups and the closest hydrogen atoms from the opposite methylene groups. Interestingly, in this case, the *tert*-butyl substitution results in an increase in the H-bond angles (168–170° for the BCX and 166–168° for CX). As can be expected, the distance between the closest carbon atoms of the opposite phenol rings (third and sixth) increases substantially for the BCX (from 3.82 to 5.28 Å). Additionally, some hydrogen atoms from different *tert*-butyl groups are exceptionally close to each other (2.28 Å).

For unsubstituted calix[6]arenes the *pc* and *wc* conformers have the C_2 symmetry axis, while no symmetry elements exist for the *al* conformer. If the molecular skeleton, i.e., the calixarene ring without hydrogen atoms attached to the hydroxy groups and–in the case of *tert*-butyl-substituted calixarenes–the ring without *tert*-butyl groups, is examined. Then, additional similarity features can be detected, although no exact symmetry elements can be named. For instance, for the *al* and *pc* conformers the lower and upper parts of the molecules, as seen in Figure 2, are quite similar to each other. A detailed examination of dihedral angles shows that this is indeed the case: most corresponding angles differ by a few degrees only, with the exception of one-two angles, where these differences are larger.

3.2. Pristine calixarenes–IR Spectra

The simulated harmonic IR spectra of three forms of calix[6]arene and hexa-*p*-*tert*-butylcalix[6]arene, together with a complete list of frequencies and intensities are presented in the Supplementary Information. Studies of smaller molecules containing the hydroxy groups, such as, e.g., studies of phenol and its complexes with water and ammonia [113] or a study of water dimer IR spectrum [114] with respect to basis sets, anharmonicity and couplings, etc., show that harmonic stretching frequencies are blue-shifted by 100–150 cm^{-1} in these cases, so one expects a similar shift of the calculated O–H stretching frequencies for calixarenes.

From IR spectra presented in Figures in the Supplementary Information one can see that there is a group of high-intensity frequencies above 3000 cm^{-1} for all the cases. The *pc* conformer is characterized by just one high-intensity composite peak, while for the remaining conformers, this peak is shifted to higher frequencies of about 3500 cm^{-1} and one finds an additional lower peak (or two peaks for *al*) at even higher energies. For the BCX case, an additional bunch of peaks appears at about 3000 cm^{-1}, which is well separated from the first group, and whose intensity is comparable to the previously discussed group for the *al* and *wc* conformers. Let us first focus on the highest vibrational frequencies, which turn out to be related to hydroxy group stretching. The six highest frequencies for the *wc* and *al* conformers correspond to various O-H stretching patterns, as can be expected for six OH groups. Since these frequencies span from about 3300 cm^{-1} up to about 3700 cm^{-1}, and the C-H stretching modes start from about 3100 cm^{-1}, the stretching modes for O-H and C-H do not mix. A different situation arises for the *pc* conformer. The highest frequency (3270 and 3280 cm^{-1} for CX and BCX, respectively) is much lower than for the *al* and *wc* conformers, and the examination of mode characters for consecutive frequencies reveals only the five highest ones are dominated by the O-H stretching. The "missing" sixth mode can be found at 3115 cm^{-1} for the BCX, separated from five OH-stretch frequencies mentioned above by several C-H stretching modes. The situation is even more complicated for the CX case, where instead of one missing mode dominated by the O-H stretch one finds three nearly degenerate modes of frequency 3099 cm^{-1}. They can be described as a simultaneous symmetric stretch of all present O-H bonds mixed with various patterns of C-H stretches, where the C-H bonds come from the phenyl groups. The replacement of the hydrogen atom in the *para* position by the *tert*-butyl group leads to a distortion of this mixing, which results in the absence of such mode combinations in the BCX case.

All the five highest frequencies of *pc* correspond to concerted stretching motions, i.e., involving simultaneously all six O-H bonds. They differ by the pattern of the motions, and, e.g., for the highest energetic mode there is–quite understandably – the alternate motion (when odd O-H bonds stretch then even ones shrink). More precisely, one can differentiate several patterns of these motions and–as usual in such cases–more nodes in the wave function mean higher energy. To this end, if we start counting the O-H bonds as in Figure 2, then the pattern corresponding to the highest frequency is the alternating one: (+,−,+,−,+,−), then the next highest is partially alternate (+,− −,++,+,− −,++), followed by more and more symmetrical patterns, such as (++,−,−,++,−,−), (++,−,− −,− −,+,++), and (++,++,+,− −,− −,−), whereby doubled plus or minus signs try to catch the increased amplitude of the motion. Finally, as discussed above, the expected totally symmetric case, i.e., (+,+,+,+,+,+), is mixed with C−H stretching motions for the CX case.

The lowering of spatial symmetry, as in the case of the *al* conformer, leads to a decoupling of the O-H stretch modes. Their highest two frequencies correspond to the stretching of a single O-H bond (for the OH-5 and OH-2 groups, respectively). The same two OH groups are involved in the two highest modes for the *wc* case. In all these cases, the high-frequency vibration corresponds to the OH group, which is involved in the H-bond as an electron donor only (i.e., through the oxygen ending), while the hydrogen atom does not participate in the H-bond because of the geometry hindrance. As a result, the O-H bond is not weakened by the H-bond and no red shift is observed in the IR spectrum. For the *wc* case there are two such high-energy (about 3700 cm^{-1}) vibrations, one corresponding to a simultaneous stretch and another–to an alternate stretch of the OH-2 and OH-5 groups. It should be noted that values of these frequencies are similar to water stretching frequencies [114], or to the stretching frequency of the hydroxy group in phenol [113]. However, for the *al* conformer (for both unsubstituted and substituted calixarenes) one of these high-energy vibrations becomes lower by approximately 100 cm^{-1}. This fact can be explained by a weak attractive interaction of the hydroxy group with the phenyl ring. Such interactions were reported and studied for the prototypical models of benzene with one or two water molecules in Ref. [115]. We will go back to this topic in Section 3.7.

Returning to the *pc* conformer of the CX, one can detect several characteristic IR peaks at lower frequencies. An examination of these cases shows that some of them belong to the O-H bending (either in-plane or out-of-plane with respect to the phenyl ring), while others are C-H bending from phenyl or breathing Kekule modes of phenyl rings. Similar peaks appear for the *al* and *wc* conformers, but they differ in intensities and exact positions of peaks, allowing us, in principle, to recognize which conformer has been measured.

For the hexa-*p*-*tert*-butylcalix[6]arene case, additional strong features around 3000 cm^{-1} should be attributed to multiple C-H stretching modes of methyl groups, which together form a *tert*-butyl group. Usually, several C-H bonds are simultaneously involved in these vibrations. Stretching frequencies of methylene bridges are of a similar energy range, so they contribute to these composite peaks as well.

3.3. Complexes with Amino Acids–Geometries

3.3.1. Dihedral Angles of Calixarenes

Before we analyze the binding (docking) sites for amino acids, let us first systematically analyze modifications of the calixarene macrocycle upon complexation with help of dihedral angles between selected carbon atoms between calixarene units. The values of the dihedral angles are defined through carbon atoms as indicated in Figure 1 together with the standard nomenclature for the considered calixarene conformers are listed in the Supplementary Information for both CX and BCX types. One should note that for each linking methylene group two angles are defined with the convention that the dihedral angle denoted as $RnaRm$ has two carbon atoms from the Rn ring, while $RnbRm$ has two carbon atoms from the Rm ring in its definition. The numbering for dihedral angles corresponds to the numeration presented in Figure 2.

Let us first analyze the pristine calixarenes. At the beginning, one should note that *pc*, *al*, and *wc* conformers can be distinguished by distinct sign patterns of the dihedral angles, which are: (± ∓ ± ± ∓±) for *pc*, (∓ = ∓± = ±) for *wc*, and (∓ ∓ ± ± ∓∓) for *al* (the ± symbol here denotes that first from two dihedral angles around a given methylene group is positive and the second negative, ∓–the opposite, while = denotes that both angles are negative). Secondly, because of the C_2 point-group symmetry within the *pc* and *wc* conformers for the CX, the second half of the dihedral angles is identical to the first half. No such exact symmetry exists in the *al* conformed; however, as noted above, there is an approximate correspondence between two parts of the *al* conformer and in an ideal case the kth angle from the table would correspond to the negative of the $(12 - k)$th angle. The same resemblances can be also found for the *pc* with the majority of differences of the order of a few degrees for the CX. For the hexa-*p*-*tert*-butylcalix[6]arene conformers, no exact point symmetry exists because of the presence of *tert*-butyl groups, but the same approximate resemblance is nevertheless preserved, confirming a small influence of the substitution on the macroring shape. It should be emphasized that no case has been found during geometry optimization where the complexation would modify the geometry of the calixarene to such an extent that the conformer became unrecognizable, although in a couple of cases the sign pattern is not preserved anymore.

In order to make the analysis of numerical values of the dihedral angles more complete, we compare them to two model molecules, which are also built from two phenol groups connected through the methylene linker. The geometry optimization of these molecules has been performed on the same level as for calixarenes. The simplest molecule of this type is diphenylmethane, for which the corresponding dihedral angles are equal to −62° and 118°. It turns out that there are indeed pairs ($RnaRm, RnbRm$), which resemble this pattern, but only for the *pc* conformer such a correspondence holds for all six pairs, while for the *al* and *wc* cases at least two angles are much smaller (by about 10–20°), so here other factors, such as hydroxy groups' interaction, should play a decisive role. In order to examine this issue in more detail, we likewise performed a geometry optimization for several conformers for the 2,2′-dihydroxydiphenylmethane molecule, which is the simplest molecule with two hydroxyphenyl groups connected with the methylene linker. Only one of these conformers has an intramolecular H-bond and the creation of this bond is gratified by the highest stability. The corresponding pair of dihedral angles for this H-bonded conformer is (−76°,101°), which corresponds quite accurately to all the pairs of the *pc* conformer and to four from six pairs of the *al* and *wc* conformers. The 2,2′-dihydroxydiphenylmethane conformer, which resembles the most the dihedral angles for the fifth and sixth ring of *al* and *wc* (and, because of symmetry, also for the second and third ring of *wc*) has two hydroxy groups separated by the CH_2 group, but they still point towards each other as an attempt to create an H-bond. It can be seen that although the distance between the corresponding oxygen and hydrogen is too large for effective creation of the H-bond (4.4 Å), it is still 0.3 Å smaller than for the *wc* conformer, i.e., one can say that these hydroxy groups are placed less optimally in the *wc* calixarene. One can also see this by a comparison of dihedral angles, which differ by as much as 22° when comparing *wc* to 2,2′-dihydroxydiphenylmethane. For the *al* case no such strains are observed. In order to fully describe the *al* conformer, one more local minimum of 2,2′-dihydroxydiphenylmethane should be used, which has the corresponding pair of angles equal to (−19°,117°).

Interestingly, the replacement of hydroxy groups by hydrogens and reoptimization of the resulting hydrocarbons leads to a complete distortion of semi-circles in the *al* and *pc* cases, while the *wc* conformer does not change so dramatically. Therefore, for the former two cases, the intramolecular H-bonds play a pivotal role in the stability of the molecule. Differences between dihedral angles and the corresponding angles for 2,2′-dihydroxydiphenylmethane can serve as an indicator of the intramolecular strain. One can see that for pristine calixarenes these differences are small for most angles, which supports the conclusion about the general stability of these conformers. Only for a few angles are these differences larger than 10°, such as, e.g., the angles between third and fourth (and symmetrically: first and sixth) rings for the *pc* case, first and second for the *al* and R2bR3 (and symmetrically: R5bR6) for the *wc* conformers of calix[6]arene. The substitution of hydrogens with the *tert*-butyl substituent leads to the largest strains for the *pc*-BCX case (two-digit differences for eight out of twelve angles, with the largest change of 28°). For *wc*-BCX conformer, the same R2xR3 (*x* means in the following both *a* and *b*) angles show differences of 10° and −13°, and the smallest differences with respect to the 2,2′-dihydroxydiphenylmethane conformers are found for the *al*-BCX conformer (−11° for R1bR2). The largest distortion for the *pc* case can be easily explained by the fact that in this case, all *tert*-butyl groups reside on the same bottom side of the calixarene macroring, so an adaptation of this ring should be performed to avoid repulsion between the *tert*-butyl groups.

The analysis of geometrical changes in calixarenes caused by the complexation can be performed at best by making a comparison between these angles for the pristine calixarenes and angles for calixarenes within complexes. On average, these changes are again quite small and become significantly larger only for a couple of angles (most often these are the angles R2bR3 and R5bR6 for *pc*, R2aR3 and R4bR5 for *al*, and R2bR3 and R5bR6 for *wc*). The presence of *tert*-butyl groups makes the calixarene backbone more rigid for the *wc* and *pc* conformers, since the average change for the BCX is smaller than for the CX. This rigidity can be explained by the fact that the *tert*-butyl groups cannot be easily moved in space because they start to overlap, and this problem is especially severe when all these groups reside on the same side of calix[6]arene (i.e., for the *pc* and to some extent for the *wc* cases). For several amino acids, the distortion of one or more dihedral angles is quite significant, which indicates a larger geometry modification of the calixarene and possibly larger deformation energy. Especially for complexes with the *wc* conformer of CX for the majority of cases, one angle is modified by more than 30°. The only amino acids for which there is no such large distortion are: AspH, Cys, Gln, GluH, Lys, and Pro. For the *al* conformer, such a large distortion occurs for two cases only (HisE and Pro), while other large distortions are at most 20–22° (for Cys, Gln, Gly, Ile, and Thr). For the *pc* conformer changes are even smaller: only for two cases the largest discrepancy amounts to 20° (for Asn and GluH) and is smaller for the remaining cases. A comparison to the BCX complexes leads to a conclusion that the largest differences for the *wc* conformers are much smaller than for the unsubstituted counterpart–only for two cases the difference is larger than 20° (HisE and Trp). For the *al*-BCX conformer, there are two cases of distortion larger or close to 30° (Tyr and Phe). Finally, for the *pc* case there is one case of a change close to 20° (for HisE), while the other largest discrepancies are much smaller (about 15° for several cases, but usually–a one-digit number). Therefore, the overall conclusion is that the substitution with *tert*-butyl groups hinders geometry modifications under complexation in the majority of complexes.

3.3.2. Binding Sites

An analysis of complexes of calix[6]arenes with amino acids reveals that several preferred binding sites can be identified. The most straightforward situation arises for the *wc* conformer, where for all forty cases (twenty amino acids and two calixarenes), the preferred binding site is on the top of the molecule, i.e., where all the hydroxy groups reside. For this conformer the hole is smaller because of the flattened shape–one should note that two phenyl rings (third and sixth according to Figure 2) are placed approximately in a position of a "shifted sandwich", which turns out to be the optimal one for two benzene molecules [116].

For the *al* conformer, two preferred binding sites can be identified, which both have a form of a cavity because of a specific orientation of calixarene units for this conformer: in comparison to the *pc* case three units are turned upside down. The most popular cavity is the cavity created by the third, fourth, and fifth units in the case of CX, but in some cases, the second cavity created by the remaining rings is used by amino acid guests. The latter situation occurs for: Phe-CX, Phe-BCX, Trp-CX, and Tyr-BCX. It should be noted that for all these cases, the amino acid contains an aryl group.

For the *pc* conformer, three docking places can be identified. One of these places is the top of the cone, while the other–the bottom cavity. If a larger amino acid occupies the cone, its residual part is usually placed along the groove made of units two and three, i.e., between two semi-circles, and in some cases, the groove becomes the only docking site (see the complex with Tyr) without attaching to the hydroxy-rich cone part at all. In the latter case, the guest molecule is, as a rule, shifted from the center of the cavity and resides on one side of the semi-circle, but in several cases (for larger molecules) both cavities are used. Somewhat surprisingly, the top position is preferred for the calix[6]arene host, while inclusion complexes are more common for the hexa-*p-tert*-butylcalix[6]arene, in spite of space obstacles on the calixarene rim for the latter case. Only Ile, Lys, Pro, and Val amino acids are docked in the cavity for the *pc*-CX. In all these cases one side of the cavity, created by units third to fifth, is used. Because of this position of the guest molecule, the structure of the six-member H-bond ring is often unharmed by the complexation, which can be also seen by the O$\cdots$H distance in the H-bonds, which remains within the range 1.63–1.7 Å after the complexation. The distance between the closest hydrogen atoms from the methylene linkers (CH$_2$-5 and CH$_2$-2) become larger by 0.2–0.3 Å for Lys and Val, which are the largest amino acids from this set, while it is enlarged by only 0.1 Å for Ile and becomes 0.1 Å smaller for the most compact Pro guest. For the hexa-*p-tert*-butylcalix[6]arene host, a majority of amino acids choose the cavity as the preferred binding site and seven only are attached to the hydroxy-decorated top (Asn, Cys, Gly, HisD, HisE, Ser, and Thr).

The Cartesian geometries of all the complexes are listed in the Supplementary Information.

3.3.3. Example: Complexes with Ala

Let us focus on changes in hydroxy-group bond lengths in selected complexes. An exemplary case is shown for the complexes with Ala in Figure 3. For the *wc* and *al* cases, there is one O-H bond longer than 1 Å (for both *al*-CX and *al*-BCX, and for *wc*-CX) and in both cases, this O-H bond seems to be weakened by the formation of the H-bond with the NH$_2$ group of amino acids. For the *al*-BCX there is also a second O-H bond longer than 1 Å, and again one can see that this elongation results from the involvement of the network of the H-bonds (this time with the second OH group). The O$\cdots$H distance, in this case, is 0.1 Å smaller than usual. No O-H bond is significantly elongated in the *wc*-BCX, although the OH-3 group does interact with the amino group of Ala. One can see, however, that the H$\cdots$N bond is longer than usual for this intermolecular H-bond. Interestingly, there is another intermolecular H-bond in this case which involves the hydroxy group of Ala and the O-6 atom of calixarene. For the *pc*-CX one has an H-bond with the amino group from Ala, too, but in this case, the H$\cdots$N binding is by about 0.2 Å longer than for the *al* or *wc* cases. One should keep in mind that for the pristine *pc* conformer there is a network of six H-bonds that virtually close up the top part of the calixarene, but the Ala amino acid is able to break this network for the CX case. Interestingly, on the other side of the calixarene cavity, there is one H-bond which becomes shorter by about 0.1 Å in comparison to the uncomplexed case (to 1.6 Å), and the involved O-H bond becomes somewhat elongated (to 1.0 Å). Finally, for the case of the *pc*-BCX the network of the H-bonds remains unharmed by the interaction with Ala. In this case, the amino group is not directly involved in the interaction with calixarene.

When considering all the cases, it turns out that for all the *wc*-CX complexes there is one hydroxy group that becomes significantly elongated after the complexation (e.g., to 1.02 Å for Ala or 1.04 Å for Asn). This length modification should be, in principle, detectable in the IR spectrum. The elongated O-H bond points to the nitrogen atom from the amino acid group, therefore it creates an intermolecular H-bond with the hydroxy group donating its hydrogen. The N···H distance for this H-bond is quite small (1.6 Å) as for the intermolecular interaction, indicating its strong character.

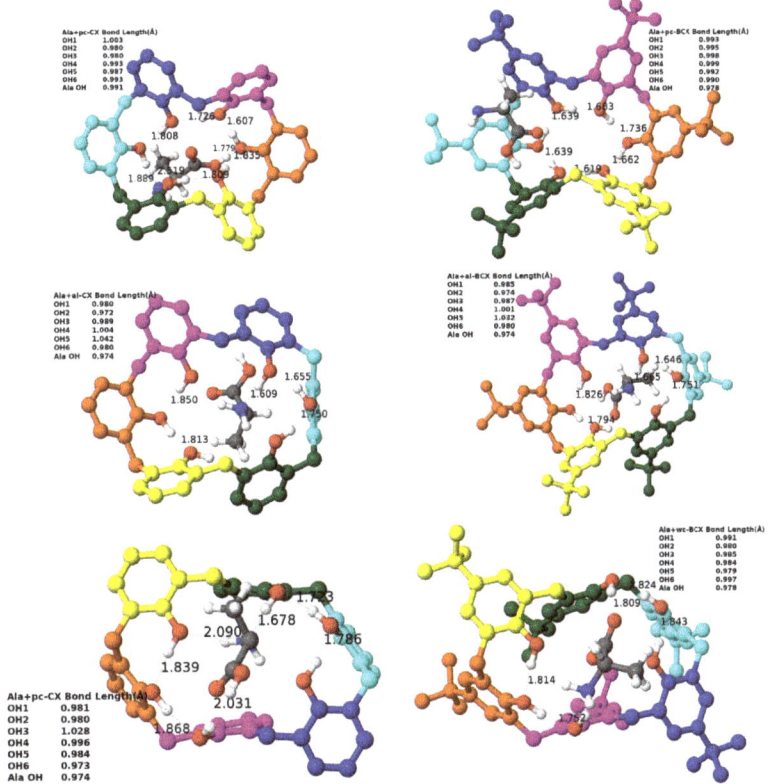

Figure 3. Selected bond lengths for O-H and H-bonds for the case of complexes with Ala. Hydrogen atoms of calixarenes are removed for a better view.

3.4. Stability of Complexes with Amino Acids

In this study, we are mostly interested in energy partitioning and relative energies; therefore, we do not aim at achieving very accurate interaction energies. Nevertheless, we selected one case (a complex of *pc*-CX with Gly), for which we calculated the interaction energy in two basis sets from a modified Dunning series for the SAPT0, MP2, and SCS-MP2 theories. One of these basis sets, jun-cc-pVDZ, has been used for all other complexes, while the larger one–jun-cc-pVTZ–has been used for the purpose of the CBS study. As expected, the main part of the basis set unsaturation comes from the electron-correlation part of the interaction energy. A difference in the HF interaction energy between both basis sets amounts to 0.7 mH only, while the net dispersion from SAPT0 ($E_{disp}^{(20)} + E_{exch-disp}^{(20)}$) is equal to 4.8 mH, a quantity similar to the correlation part of the MP2 and SCS-MP2 interaction energies, where these differences are equal to 5.0 and 4.1 mH, respectively. If a popular inverse cubic extrapolation formula [98],

$$E_L = E_\infty + \frac{A}{L^3}, \tag{4}$$

is applied to the correlation part of the interaction energy (with L equal to 2 and 3 for jun-cc-pVDZ and jun-cc-pVTZ, respectively), then the estimated correlation energy decreases by another 1.7 mH for the SCS-MP2 case, so (taking the HF interaction energy on a larger basis) the estimate of the CBS limit of the SCS-MP2 interaction energy for this complex is −21.5 mH, which should be compared with −19.1 mH and −15.6 mH for the jun-cc-pVTZ and jun-cc-pVDZ basis sets, respectively. Therefore, the energetic results in the jun-cc-pVDZ basis set are rather semiquantitative and as a rule of thumb it should be rescaled by approximately $\frac{4}{3}$ to make the estimate for the CBS interaction energy.

The relative stability of both empty calixarenes for three considered conformations and complexes of these conformations with amino acids calculated from their total energies with respect to the lowest conformation are presented in Table 1 and in Figure 4. For empty calixarenes, the *pc* conformer is the most stable for both the CX and BCX cases, which can be explained by two more H-bonds stabilizing the macrocycle structure in comparison to the *wc* and *al* cases. For the *al* and *wc* conformers, the folding of the calixarene macrocycle prevents the creation of H-bonds between the following hydroxy pairs: first with sixth and fourth with third. The order of the *wc* and *al* conformers differs for the CX and BCX cases–in the former case, the *al* conformer has the lower energy, while the opposite is true for the BCX case. It should also be noted that the *wc*-BCX has the energy higher only by 2.4 mH than the *pc*-BCX, while both *al*-CX and *wc*-CX lie over 12 mH higher than the *pc*-CX. Therefore, at standard conditions, one can expect a sizable (about 7%) contribution of the *wc* conformer for the hexa-*p*-*tert*-butylcalix[6]arene, while for the case of calix[6]arene there is only one dominant conformer (*pc*).

The energetic order of complexes of *pc*, *al*, and *wc* calixarenes with amino acids is in most cases different than the pristine series. The energetic sequence is preserved for amino acids: Leu, Phe, Pro, Trp, and Val interacting with CX and only Met, Phe, and Tyr interacting with BCX. If the energetic difference between the highest and the lowest conformers of empty calixarenes with the analogous energetic span of complexes are compared, one can find that this difference becomes smaller (within 2.0 mH or less) for CX interacting with Ala, Gln, GluH, HisD, and Tyr, and for BCX interacting with Ala, Met, and Phe. For these complexes, all types of conformers are accessible under standard conditions. On the other hand, the energy span between conformers becomes larger than for pristine conformers only for one case (Asn) for the CX case and for Asn, AspH, Cys, Gly, HistD, Lys, Ser, and Trp for the BCX one. However, it should be noted that the energy span between conformers of the empty BCX (10.7 mH) is smaller than for the CX (15.8 mH). In general, the addition of large *tert*-butyl groups seems to reduce energetic differences in both pristine calix[6]arenes and their complexes. In several cases, the complexation diminishes the energetic differences for two from three conformers. If access to a specific conformer is desired, the special cases, for which one conformer has much lower energy than the two remaining ones are the most promising in view of potential applications. For complexes with CX especially, often the *al*-CX complexes become energetically more favorable than the *pc* ones. The exceptionally large difference between the *al*-CX complex and *pc* or *wc* ones occurs for the Asn amino acid (16 mH). Other cases include: Gly (8 mH), Ser (10 mH), and Thr (6 mH). The *wc*-CX complexes have the lowest energy for three cases only, but in all of these cases, the differences are very small (2 mH and smaller). The *pc*-CX remains the lowest one for a couple of cases, but only for the complex with Val energetic differences larger than 2.5 mH.

Contrary to the complexes with CX, it is the *wc*-BCX conformer that becomes the lowest energetically for the majority of complexes with amino acids. The situation where the other two conformers lie higher than 4 mH occurs for AspH, Gln, GluH, Gly, HisD, HisE, Ile, Leu, Thr, Trp, and Val. In several cases, two conformers (*wc* and *al*) are almost isoenergetic and lower than *pc*, such as in the case of Asn, Lys, and Ser. Finally, for the Phe amino acid, the differences between all three conformers are less than 2 mH. It should be noted that the *al*-BCX conformer does not become the preferred one for all but one case, which is the complex with Cys. However, even in this case, the next lowest conformer is the *wc* conformer.

Summarizing, the complexation with amino acids tends to modify the energetic order of both CX and BCX conformers, and in several cases, these differences are quite substantial. The extreme differences are seen more often for the unsubstituted calix[6]arene case. One should mention that one important potential application resulting from such a change of energetic order of *pc*, *al*, or *wc* conformers is a possibility of accessing, e.g., the *al*-type conformer for the purpose of chemical reactions with calix[6]arene, which would otherwise be dominated by the *pc* conformer.

Table 1. The relative electronic and total energy of both empty calixarenes in three considered conformations, and complexes of these conformations with amino acids with respect to the lowest conformation energy. Energies in millihartree.

Amino Acid	CX Electronic Energy			CX Total Energy			BCX Electronic Energy			BCX Total Energy		
	al-CX	*pc*-CX	*wc*-CX	*al*-CX	*pc*-CX	*wc*-CX	*al*-BCX	*pc*-BCX	*wc*-BCX	*al*-BCX	*pc*-BCX	*wc*-BCX
-	13.2	0.0	15.8	12.0	0.0	14.9	10.7	0.0	1.7	9.7	0.0	2.4
Ala	2.7	2.4	0.0	2.0	1.5	0.0	3.5	3.6	0.0	1.6	1.5	0.0
Asn	0.0	16.1	15.7	0.0	16.3	17.5	1.8	14.4	0.0	0.0	12.0	0.0
AspH	0.0	4.1	8.7	0.0	4.2	9.4	3.9	11.1	0.0	3.5	11.0	0.0
Cys	0.0	4.7	6.0	0.0	4.3	6.3	0.0	11.2	0.2	0.0	10.0	1.8
Gln	0.0	1.6	1.8	0.0	0.1	1.9	4.2	6.9	0.0	2.9	6.4	0.0
GluH	0.0	1.1	1.1	0.0	1.5	1.7	9.9	10.2	0.0	9.1	9.2	0.0
Gly	0.0	8.7	7.0	0.0	8.3	8.0	7.6	17.0	0.0	6.0	14.7	0.0
HisD	2.0	2.4	0.0	0.9	1.6	0.0	5.1	11.9	0.0	4.5	9.9	0.0
HisE	0.0	3.5	4.8	0.0	4.1	5.6	5.2	4.8	0.0	2.5	2.9	0.0
Ile	2.9	0.0	1.7	3.6	0.0	2.5	8.8	7.9	0.0	7.2	6.5	0.0
Leu	0.8	0.0	13.7	0.3	0.0	13.5	9.8	6.6	0.0	7.9	6.1	0.0
Lys	0.0	1.0	1.6	0.0	2.5	2.2	1.6	12.6	0.0	0.0	13.3	0.0
Met	0.0	5.9	4.8	0.0	5.1	4.6	2.4	1.1	0.0	0.4	0.0	0.2
Phe	1.9	0.0	8.6	2.5	0.0	8.8	1.6	0.1	0.0	1.1	0.0	0.3
Pro	2.1	0.0	9.5	0.7	0.0	9.4	4.5	0.0	3.8	4.2	0.0	4.8
Ser	0.0	11.8	9.4	0.0	9.8	9.8	2.1	13.5	0.0	0.0	10.4	0.0
Thr	0.0	7.6	11.2	0.0	5.9	11.2	4.4	11.0	0.0	3.1	8.1	0.0
Trp	0.1	0.0	3.7	0.8	0.0	4.5	13.7	8.2	0.0	10.8	6.7	0.0
Tyr	2.7	2.2	0.0	1.9	1.2	0.0	8.6	0.0	5.7	6.9	0.0	5.3
Val	7.9	0.0	8.4	7.3	0.0	8.8	7.4	5.4	0.0	5.6	3.9	0.0

3.5. Complexes with Amino Acids–Interaction Energies

Interaction energies of the calixarene complexes with amino acids are presented for SAPT0, MP2, and SCS-MP2 methods in Figure 5. We treat the SCS-MP2 results as the most accurate since the scaling of the MP2 same- and opposite-spin energies, which make up the SCS-MP2 model, is devised [117,118] to reproduce the CCSD(T) correlation energy. The MP2 and SAPT0 results are consistently lower than SCS-MP2, but the interaction energies for all three methods follow the same global trend, as can be seen in Figure 5, where for a given conformer, the lines connecting interaction energies for various amino acids obtained with the same method look similar for SAPT0, MP2, and SCS-MP2. In particular, the *pc* conformer is the least prone to making a strong binding with amino acids, while the remaining conformers provide more attractive contributions (with a slight advantage of *al*). Since the F-SAPT and I-SAPT partitioning schemes utilize the preceding SAPT0 calculations, the consistency of SAPT0 and SCS-MP2 results should be especially emphasized in view of the reliability of further F/I-SAPT analyses. It is also interesting to note that on average the

BCX complexes have lower interaction energies than the CX ones. A general analysis of the SAPT contributions shows that indeed the mean SAPT0 interaction energy for the BCX complexes is about 6 mH lower than this for the CX complexes, and the main contribution to this difference is due to the effective dispersion (10 mH lower), while the more repulsive first-order exchange contribution partially counterweights the additional attractive effect coming from electron correlation.

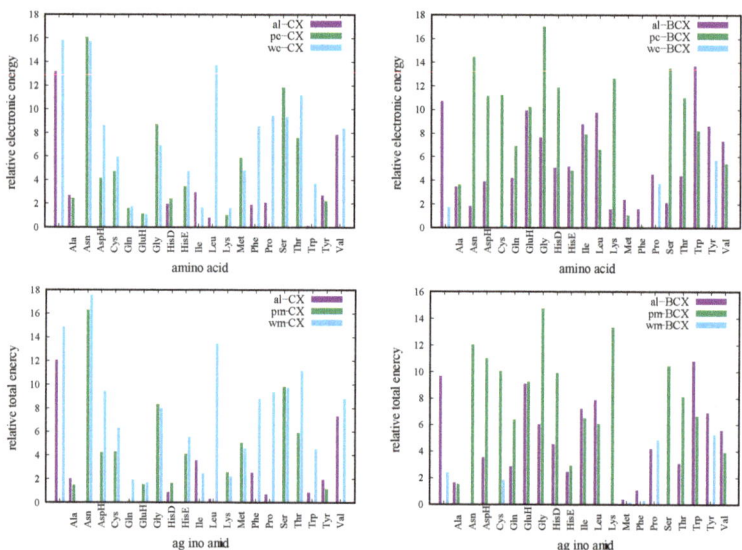

Figure 4. Relative total and electronic energies for complexes with three conformers of calix[6]arene and hexa-*p-tert*-butylcalix[6]arene. Energies are in millihartree.

The analysis of Figure 5 leads to the conclusion that the largest attraction appears for the Asn and Gln amino acids for the *al*-CX conformer and for Asn for the *al*-BCX, and for Ala and (again) Asn for the *pc*-CX and *pc*-BCX, respectively. On the other hand, in the case of the *wc* conformer, it is the HisD amino acid, which shows the most attractive interaction for the CX, while the Lys molecule exhibits the largest attraction for the BCX.

The comparison of the SSMF3 interaction energies with the standard (unfragmented) ones, which is presented in Table 2 for the calix[6]arene (the analogous table for hexa-*p-tert*-butylcalix[6]arene has been moved to the Supplementary Information) shows that the SSMF3 approach (see Equation (3)) quite accurately reproduces interaction energies for investigated complexes. The first study of the applicability of SSMF3 for the interaction energies [50] revealed that the electron-correlation contribution is well reproduced by the SSMF3, while the accuracy of the HF part depends on the distance between the interacting molecules. The present study supports these conclusions, i.e., the electron-correlation parts of the SCS-MP2 interaction energies for complexes with the CX are reproduced with a mean percent error of 1.7%, with the largest absolute error of 3.7%. The electrostatic and first-order exchange SAPT0 contributions are obtained with good accuracy (the mean percent errors of 0.6% and 0.2%, and maximum ones 2.7% and 0.8%), while the most problematic component is the second-order induction (the mean percent error of 3.2%, but the maximum error is as large as 18%). One should note, however, that large percent errors for induction appear for those cases, for which this component is small with respect to other components (the largest error appears for the *pc*-CX with Met, where the second-order induction and dispersion contribute with -8.6 mH and -26.9 mH, respectively). The second-order dispersion, i.e., the pure electron-correlation effect, is reproduced quite accurately for all the cases (the mean and maximum errors of

1.2% and 2.0%, respectively). For the hexa-*p-tert*-butylcalix[6]arene complexes, similar error ranges occur, e.g., the maximum absolute errors for the electrostatic and first-order exchange are equal to 2.3% and 0.9%, and again the highest error arises for the second-order induction, but anyway, the maximum error is more than twice smaller (7.9%). The second-order SAPT0 dispersion is reproduced with a stable mean error of about 2% and the maximum error of 2.8%. The accuracy of the electron-correlated contributions from the MP2 and SCS-MP2 theories are of the same range as for the dispersion. Therefore, the comparison of the SSMF3-SAPT0 and SAPT0 results, as well as SSMF3-(SCS-MP2) and SCS-MP2 allows us to make a conclusion that in the region of a minimum of the intermolecular potential energy hypersurface the SMF-based methods give quite accurate results. Note that in the case of large percentage errors for the HF interaction energy, the energy itself is close to zero. This conclusion is important in view of our perspective applications of SSMF3 in the analysis of energy partitioning among functional groups.

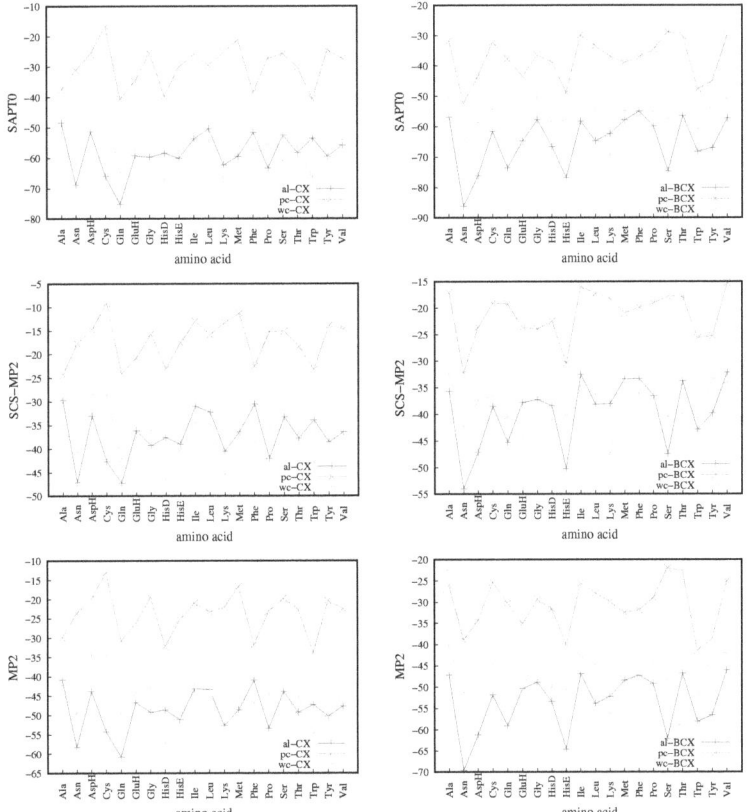

Figure 5. Trends in interaction energies for complexes with three conformers of calix[6]arene and hexa-*p-tert*-butylcalix[6]arene. Energies are in millihartree.

Table 2. The SAPT0 energy components, SAPT0 and HF interaction energies, MP2 and SCS-MP2 electron-correlated parts of the interaction energies for CX + amino acid complexes. The percent errors of the SSMF3 fragmentation scheme are given in parentheses. Energies in millihartree.

	$E_{elst}^{(10)}$	$E_{exch}^{(10)}$	$E_{ind,resp}^{(20)}$	$E_{exch-ind,resp}^{(20)}$	$E_{disp}^{(20)}$	$E_{exch-disp}^{(20)}$	E_{int}^{HF}	E_{tot}^{SAPT0}	$E_{corr,int}^{MP2}$	$E_{corr,int}^{SCS-MP2}$
al–CX-Ala	−66.5 (−0.3)	97.9 (0.6)	−46.2 (1.4)	29.4 (0.7)	−56.6 (−1.4)	8.3 (−2.9)	−0.1 (−423.9)	−48.4 (−1.9)	−40.7 (−1.0)	−29.6 (−1.1)
al–CX-Asn	−86.9 (0.4)	110.4 (0.1)	−58.9 (−1.1)	35.0 (−1.4)	−57.0 (−0.9)	8.5 (−1.3)	−20.3 (−3.8)	−68.8 (−1.7)	−37.9 (−1.4)	−26.8 (−1.7)
al–CX-AspH	−67.5 (−0.3)	94.5 (0.3)	−46.7 (−1.0)	29.9 (−1.9)	−55.3 (−1.1)	7.6 (−1.4)	−3.7 (−24.7)	−51.3 (−2.8)	−40.1 (−1.4)	−29.2 (−1.5)
al–CX-Cys	−84.5 (0.4)	104.7 (0.1)	−51.6 (−0.4)	31.4 (−1.1)	−58.8 (−1.5)	8.6 (−2.7)	−15.8 (−0.8)	−66.0 (−1.1)	−38.4 (−1.8)	−26.9 (−2.2)
al–CX-Gln	−102.6 (0.0)	128.1 (0.6)	−59.0 (−0.3)	36.0 (−0.5)	−68.9 (−0.9)	10.1 (−1.4)	−16.5 (−4.8)	−75.3 (−1.7)	−44.4 (−1.3)	−30.8 (−1.6)
al–CX-GluH	−59.4 (1.2)	66.9 (0.1)	−30.4 (0.5)	18.7 (0.0)	−54.0 (−1.8)	6.5 (−3.5)	−11.8 (7.1)	−59.2 (0.2)	−34.9 (−2.0)	−24.3 (−2.7)
al–CX-Gly	−74.7 (0.5)	91.3 (0.2)	−46.8 (−0.4)	28.2 (−1.2)	−50.6 (−1.6)	7.5 (−2.7)	−16.6 (−1.2)	−59.7 (−1.3)	−32.6 (−1.4)	−22.7 (−1.6)
al–CX-HisD	−57.1 (1.7)	67.4 (0.2)	−30.6 (2.7)	18.3 (0.3)	−54.6 (−1.4)	6.4 (−3.2)	−10.2 (18.7)	−58.4 (2.3)	−38.4 (−1.1)	−27.4 (−1.4)
al–CX-HisE	−67.6 (−0.1)	96.5 (0.1)	−49.0 (−1.4)	30.6 (−0.8)	−63.7 (−0.8)	8.7 (−0.8)	−5.2 (−26.7)	−60.1 (−3.1)	−45.9 (−1.3)	−33.8 (−1.4)
al–CX-Ile	−56.1 (−0.3)	76.0 (0.1)	−32.6 (0.9)	21.5 (−0.6)	−61.6 (−1.5)	7.6 (−3.3)	0.4 (−73.8)	−53.6 (−0.7)	−43.6 (−1.5)	−31.4 (−1.9)
al–CX-Leu	−58.3 (−0.1)	83.7 (0.3)	−39.0 (−1.1)	25.0 (−1.7)	−57.6 (−1.2)	7.5 (−2.1)	−0.3 (−267.8)	−50.4 (−2.7)	−43.1 (−1.5)	−31.9 (−1.6)
al–CX-Lys	−75.7 (0.2)	102.7 (0.4)	−48.3 (−0.6)	29.8 (−1.5)	−63.0 (−0.9)	8.7 (−2.0)	−7.9 (−11.3)	−62.2 (−2.1)	−44.7 (−1.1)	−32.5 (−1.3)
al–CX-Met	−58.3 (1.9)	74.6 (0.1)	−33.7 (1.5)	21.2 (−0.4)	−62.2 (−1.7)	7.6 (−3.5)	−4.9 (−38.4)	−59.5 (1.8)	−43.7 (−1.1)	−31.5 (−1.4)
al–CX-Phe	−52.8 (−0.2)	68.5 (−0.1)	−29.5 (−0.4)	18.2 (−0.4)	−53.5 (−1.0)	6.4 (−2.0)	−4.5 (−11.2)	−51.6 (−1.8)	−36.4 (−1.5)	−26.0 (−1.7)
al–CX-Pro	−85.3 (−0.1)	115.5 (0.3)	−59.8 (−1.5)	36.2 (−1.1)	−58.3 (−1.1)	8.8 (−1.7)	−13.8 (−11.9)	−63.3 (−3.4)	−39.5 (−1.2)	−28.2 (−1.4)
al–CX-Ser	−61.7 (−0.4)	74.4 (0.4)	−31.1 (1.6)	19.8 (0.8)	−53.1 (−1.8)	6.8 (−4.3)	−6.2 (3.1)	−52.5 (−0.9)	−37.6 (−1.4)	−27.0 (−1.7)
al–CX-Thr	−62.4 (−0.2)	72.3 (0.2)	−29.8 (1.2)	18.9 (0.2)	−57.1 (−1.7)	7.1 (−3.6)	−8.3 (3.8)	−58.3 (−0.7)	−41.0 (−1.4)	−29.5 (−1.7)
al–CX-Trp	−53.4 (−0.3)	79.1 (0.3)	−34.7 (−0.4)	24.3 (−0.7)	−67.7 (−0.7)	8.3 (−1.4)	5.7 (5.4)	−53.6 (−1.2)	−53.0 (−0.9)	−39.6 (−1.0)
al–CX-Tyr	−60.8 (−0.1)	73.3 (0.4)	−31.7 (−0.4)	20.1 (0.8)	−60.1 (−1.6)	7.4 (−3.5)	−6.9 (9.7)	−59.6 (−0.3)	−43.4 (−1.5)	−31.7 (−1.8)
al–CX-Val	−60.1 (−0.8)	78.8 (0.2)	−36.6 (−0.2)	22.9 (−1.7)	−57.7 (−1.2)	7.3 (−2.9)	−5.4 (−11.2)	−55.9 (−2.0)	−42.3 (−1.7)	−31.0 (−2.0)
pc–CX-Ala	−55.1 (−0.2)	62.7 (0.1)	−29.5 (2.9)	18.2 (0.0)	−30.1 (−1.4)	4.6 (−1.9)	−12.0 (2.8)	−37.6 (0.0)	−18.0 (−1.4)	−12.3 (−1.8)
pc–CX-Asn	−34.1 (1.2)	37.7 (0.0)	−14.4 (13.3)	10.1 (2.1)	−31.2 (−2.0)	3.4 (−2.6)	−3.3 (65.5)	−31.1 (5.3)	−20.3 (−2.0)	−14.4 (−2.6)
pc–CX-AspH	−26.6 (2.1)	32.8 (0.4)	−13.5 (13.9)	8.7 (2.7)	−26.9 (−1.4)	2.8 (−1.8)	−1.4 (147.3)	−25.6 (7.0)	−18.3 (−1.6)	−13.2 (−2.1)
pc–CX-Cys	−14.6 (2.7)	19.6 (0.3)	−7.0 (14.8)	5.0 (1.9)	−19.7 (−1.1)	1.8 (−1.0)	1.4 (−95.5)	−16.6 (6.7)	−14.4 (−1.1)	−10.5 (−1.3)
pc–CX-Gln	−50.1 (0.2)	57.3 (0.3)	−26.1 (6.5)	16.2 (1.3)	−35.4 (−1.5)	4.5 (−1.8)	−9.7 (15.8)	−40.6 (2.7)	−21.1 (−1.2)	−14.4 (−1.7)
pc–CX-GluH	−41.8 (0.6)	45.6 (0.2)	−20.7 (9.7)	13.3 (2.0)	−29.8 (−1.6)	3.6 (−2.1)	−8.5 (25.2)	−34.7 (5.0)	−17.8 (−2.0)	−12.3 (−2.6)
pc–CX-Gly	−33.4 (0.8)	36.0 (0.0)	−15.9 (6.8)	9.7 (0.9)	−20.1 (−1.4)	2.6 (−1.9)	−8.0 (15.3)	−25.5 (3.9)	−11.4 (−1.3)	−7.7 (−1.9)
pc–CX-HisD	−39.8 (0.6)	52.5 (0.1)	−22.6 (7.0)	16.3 (1.2)	−47.0 (−1.1)	5.4 (−1.1)	2.0 (−84.7)	−39.6 (3.2)	−34.1 (−0.9)	−25.1 (−1.1)
pc–CX-HisE	−24.9 (0.2)	36.2 (0.3)	−15.2 (7.0)	11.8 (1.6)	−39.4 (−0.6)	4.3 (−0.1)	5.0 (−18.9)	−30.1 (2.3)	−30.3 (−0.7)	−22.6 (−0.7)
pc–CX-Ile	−22.2 (2.4)	39.8 (0.8)	−14.7 (4.1)	11.4 (1.2)	−41.2 (−1.3)	4.4 (−1.8)	11.1 (−6.8)	−25.7 (1.1)	−32.1 (−1.6)	−24.1 (−1.8)
pc–CX-Leu	−31.0 (1.3)	43.8 (0.3)	−16.1 (10.4)	11.8 (1.7)	−38.5 (−1.0)	4.1 (−1.4)	4.9 (−40.4)	−29.5 (5.5)	−28.1 (−0.8)	−20.7 (−1.0)
pc–CX-Lys	−19.3 (0.3)	38.9 (0.2)	−12.4 (3.2)	9.4 (0.5)	−43.2 (−1.0)	4.5 (−1.4)	13.4 (−3.5)	−25.2 (0.4)	−35.4 (−0.9)	−26.7 (−1.0)
pc–CX-Met	−17.4 (2.1)	25.6 (0.5)	−8.6 (17.7)	6.0 (2.2)	−26.9 (−0.9)	2.3 (−0.2)	3.5 (−49.2)	−21.0 (7.1)	−20.1 (−1.6)	−14.8 (−1.9)
pc–CX-Phe	−39.9 (0.9)	56.2 (0.0)	−24.9 (5.8)	18.4 (0.9)	−49.1 (−1.0)	6.0 (−1.7)	4.6 (−36.2)	−38.5 (3.3)	−36.4 (−0.8)	−27.1 (−1.0)
pc–CX-Pro	−20.1 (0.0)	34.2 (0.2)	−12.8 (3.4)	9.6 (0.7)	−39.0 (−0.9)	4.1 (−0.9)	7.7 (−4.9)	−27.2 (0.3)	−30.8 (−1.1)	−22.9 (−1.2)
pc–CX-Ser	−30.0 (1.4)	34.7 (0.3)	−14.4 (11.5)	9.6 (2.0)	−24.8 (−1.5)	2.8 (−1.8)	−3.5 (57.0)	−25.6 (6.6)	−16.2 (−1.3)	−11.5 (−1.7)
pc–CX-Thr	−37.9 (0.7)	39.6 (−0.3)	−17.7 (4.3)	10.5 (0.2)	−22.5 (−1.3)	2.8 (−1.7)	−10.4 (11.0)	−30.1 (3.0)	−12.1 (−1.4)	−8.0 (−2.0)
pc–CX-Trp	−35.8 (1.0)	54.5 (0.0)	−22.2 (7.7)	16.9 (1.1)	−55.7 (−0.9)	6.2 (−1.1)	8.9 (−22.9)	−40.6 (3.9)	−42.7 (−0.7)	−32.1 (−0.8)
pc–CX-Tyr	−17.3 (1.8)	27.4 (0.1)	−10.8 (4.6)	7.0 (1.2)	−31.4 (−0.8)	3.1 (−0.7)	4.0 (−18.4)	−24.3 (2.1)	−24.3 (−0.5)	−17.9 (−0.5)
pc–CX-Val	−23.1 (0.7)	37.5 (0.5)	−13.0 (4.1)	9.6 (1.4)	−39.5 (−1.1)	4.2 (−1.3)	8.0 (−5.8)	−27.3 (0.3)	−30.4 (−1.1)	−22.6 (−1.3)

176

Table 2. Cont.

	$E^{(10)}_{elst}$	$E^{(10)}_{exch}$	$E^{(20)}_{ind,resp}$	$E^{(20)}_{exch-ind,resp}$	$E^{(20)}_{disp}$	$E^{(20)}_{exch-disp}$	E^{HF}_{int}	E^{SAPT0}_{tot}	$E^{MP2}_{corr,int}$	$E^{SCS-MP2}_{corr,int}$
wc–CX-Ala	−72.9 (−0.1)	80.5 (0.1)	−39.3 (−1.3)	23.0 (−1.9)	−36.4 (−1.2)	5.7 (−3.0)	−20.9 (−0.4)	−51.7 (−0.7)	−21.3 (−1.3)	−14.2 (−1.6)
wc–CX-Asn	−92.1 (−0.1)	109.0 (0.2)	−55.3 (−0.6)	31.5 (−1.4)	−45.4 (−1.0)	7.1 (−2.5)	−23.8 (−1.0)	−62.1 (−0.8)	−25.6 (−1.3)	−16.9 (−1.8)
wc–CX-AspH	−49.7 (0.6)	48.1 (−0.3)	−22.4 (−0.8)	12.5 (−1.2)	−30.0 (−1.2)	3.6 (−2.3)	−17.6 (1.6)	−44.1 (0.0)	−16.7 (−2.0)	−10.8 (−2.8)
wc–CX-Cys	−54.9 (−0.3)	58.0 (−0.3)	−27.8 (−0.2)	16.0 (−0.7)	−30.9 (−1.3)	4.1 (−3.2)	−17.7 (−1.4)	−44.4 (−1.2)	−16.7 (−2.0)	−10.8 (−2.6)
wc–CX-Gln	−83.6 (−0.2)	85.2 (−0.4)	−41.1 (0.1)	22.2 (−1.0)	−39.3 (−1.3)	5.7 (−2.9)	−31.4 (0.7)	−65.0 (−0.2)	−18.1 (−2.0)	−10.8 (−2.9)
wc–CX-GluH	−71.5 (0.2)	73.2 (−0.4)	−37.1 (−0.9)	19.9 (−0.9)	−34.0 (−1.3)	4.8 (−3.1)	−28.0 (−0.7)	−57.2 (−0.9)	−16.0 (−2.6)	−9.6 (−3.7)
wc–CX-Gly	−74.2 (0.0)	80.0 (0.2)	−39.5 (−1.2)	23.0 (−1.8)	−34.7 (−1.3)	5.6 (−3.3)	−22.9 (−0.4)	−52.0 (−0.7)	−19.8 (−1.5)	−13.0 (−1.9)
wc–CX-HisD	−89.0 (0.2)	104.4 (0.1)	−52.6 (−0.5)	31.5 (−1.2)	−52.9 (−1.0)	7.9 (−2.4)	−22.0 (0.0)	−67.1 (−0.5)	−34.9 (−1.1)	−24.6 (−1.3)
wc–CX-HisE	−71.0 (0.1)	83.3 (0.1)	−40.8 (−0.6)	25.9 (−1.2)	−51.2 (−1.0)	7.1 (−2.0)	−13.7 (1.4)	−57.7 (−0.3)	−36.3 (−0.7)	−26.4 (−0.8)
wc–CX-Ile	−68.1 (0.0)	75.6 (0.0)	−35.0 (−1.0)	20.5 (−1.8)	−39.4 (−1.1)	5.5 (−2.7)	−17.7 (0.7)	−51.6 (−0.3)	−24.6 (−1.2)	−16.8 (−1.5)
wc–CX-Leu	−73.8 (0.1)	86.4 (−0.1)	−42.5 (−0.2)	25.2 (−1.1)	−43.6 (−1.2)	6.4 (−2.2)	−17.0 (2.1)	−54.2 (0.0)	−26.3 (−1.4)	−17.7 (−1.7)
wc–CX-Lys	−92.1 (0.2)	105.1 (−0.3)	−55.9 (−0.6)	32.2 (−0.9)	−44.8 (−1.4)	7.1 (−2.9)	−29.3 (0.1)	−66.9 (−0.5)	−24.7 (−2.1)	−16.1 (−2.8)
wc–CX-Met	−77.5 (−0.1)	90.1 (0.1)	−43.9 (−0.8)	25.6 (−1.9)	−42.4 (−1.1)	6.3 (−2.8)	−19.7 (−0.1)	−55.8 (−0.6)	−25.7 (−1.1)	−17.4 (−1.4)
wc–CX-Phe	−78.8 (0.1)	88.2 (0.1)	−43.6 (−1.2)	25.3 (−2.1)	−41.3 (−1.2)	6.2 (−3.1)	−22.4 (0.5)	−57.4 (−0.4)	−24.4 (−1.3)	−16.4 (−1.7)
wc–CX-Pro	−70.7 (0.9)	81.1 (−0.1)	−39.8 (−0.1)	24.7 (−0.6)	−46.7 (−1.6)	6.6 (−3.1)	−15.4 (6.1)	−55.5 (0.7)	−29.0 (−1.6)	−20.0 (−2.0)
wc–CX-Ser	−81.1 (0.0)	84.1 (−0.1)	−41.5 (−0.4)	23.3 (−1.9)	−39.4 (−1.0)	5.9 (−2.8)	−27.3 (1.4)	−60.8 (0.3)	−22.0 (−1.3)	−14.2 (−1.7)
wc–CX-Thr	−77.1 (0.3)	85.9 (0.0)	−40.0 (−0.2)	23.7 (−1.9)	−43.5 (−1.3)	6.3 (−3.0)	−19.3 (3.8)	−56.5 (0.6)	−25.8 (−1.7)	−17.1 (−2.2)
wc–CX-Trp	−79.9 (0.2)	93.8 (−0.1)	−43.3 (0.1)	27.4 (−1.6)	−60.4 (−1.1)	8.0 (−2.6)	−13.9 (5.8)	−66.3 (0.5)	−40.8 (−1.3)	−28.9 (−1.6)
wc–CX-Tyr	−72.2 (0.1)	83.8 (−0.1)	−40.2 (−0.7)	26.7 (−1.3)	−53.7 (−0.7)	7.5 (−1.7)	−12.7 (1.5)	−58.9 (−0.1)	−37.1 (−0.7)	−26.6 (−0.8)
wc–CX-Val	−71.5 (0.2)	78.8 (−0.1)	−37.3 (−1.1)	21.8 (−2.0)	−40.3 (−1.1)	5.7 (−2.8)	−19.4 (1.2)	−54.0 (−0.1)	−24.6 (−1.3)	−16.7 (−1.6)

3.6. Complexes with Amino Acids–IR Spectra

The simulated IR spectra of complexes of both calixarenes with selected amino acids can be found in the Supplementary Information.

The examination of vibrational modes of complexes of selected amino acids with calixarenes reveals many interesting features, with a general conclusion that the IR spectra in many cases can serve as a fingerprint of a complex. Especially useful in this aspect is the high-energy part of the spectrum, where the X-H stretch modes (either within one isolated bond or combined from several simultaneous motions within such bonds) undergo various modifications depending on their involvement in noncovalent interactions. In the following, as an exemplary case, we present only a short discussion of the IR spectra for the *wc* conformer.

A characteristic feature of all complexes with the *wc* conformer is a strong line at about 2500–2700 cm^{-1}, which corresponds to the O-H stretching motion for the hydroxy group involved in a strong H-bond with the nitrogen atom from the amino acid group. One can see that the elongation of this bond is well correlated with the red shift of the frequency, as can be expected.

When one compares the region of 3000 cm^{-1} and higher, significant differences in intensities and/or emergence or line shifting can be noticed, which represent a combined result of additional motions of the N-H stretching and–in some cases–O-H stretching from the amino acid side. In particular, there are instances where the N-H stretch becomes the highest energetic line, such as, e.g., in *wc*-CX with Asn, *wc*-CX with Leu, or *wc*-CX with Lys.

The stretch within the COOH group appears for the *wc* case as a well-separated line at about 1700 cm^{-1}. In some cases, there is a second line in this region, such as for the Asn guest, which corresponds to the stretch of the CONH$_2$ group. In both cases, the main stretching involves the double C-O bond, but this motion is concerted with the neighboring OH or NH$_2$ groups.

3.7. Pristine Calixarenes–Energy Partitioning

The calix[6]arenes contain six hydroxy groups, which can be oriented in such a way that one hydroxy group serves as a hydrogen donor, and the second one–as a hydrogen acceptor. Additionally, some hydroxy groups play both roles simultaneously. As already noted, for the *pc* conformer all six groups are connected in a way that allows the creation of six H-bonds, while for the *al* and *wc* cases four H-bonds are present. According to Figure 2 in the *al* conformer, the OH-6 group is connected (by its hydrogen ending) to OH-1, which in turn interacts with OH-2 on the left semi-circle of calixarene, and on the right semi-circle, the OH-3 group donates its hydrogen to OH-4, which in turn passes its hydrogen to OH-5. Similarly, for the *wc* conformer two chains of H-bonds are: OH-5 to OH-4 to OH-3 and OH-2 to OH-1 to OH-6. In Table 3, we list interaction energies between hydroxy groups of all six cases of calixarenes obtained from the I-SAPT method. (Note that I-SAPT calculations for the interaction between hydroxy groups from calixarene have also been performed for selected complexes, see the Supplementary Information).

Table 3. I-SAPT interaction energies between hydroxy groups for empty calixarenes. The upper triangle presents the BCX case, the bottom triangle–the CX case, energy values for the *al*, *pc*, and *wc* conformers, respectively, are separated by a dash. Energies in millihartree.

	OH-1	OH-2	OH-3	OH-4	OH-5	OH-6
OH-1	-	−10.9/−12.3/−9.3	−0.5/−1.6/0.0	−0.4/−1.0/0.0	−0.7/−3.2/−0.3	−11.4/−15.1/−9.7
OH-2	−11.8/−12.6/−9.6	-	−0.1/−14.3/−1.2	−0.2/−3.2/−0.3	0.4 /2.1/0.0	−4.0 /−5.7/−2.7
OH-3	−0.3/−1.5/−0.3	−0.1/−13.9/−1.1	-	−11.6/−15.1/−9.6	−2.9/−5.7/−2.7	−0.3 /−1.5 /1.2
OH-4	−0.4/−0.9/0.0	−0.6/−3.2/0.1	−11.2/−15.2/−9.2	-	−10.5/−12.3/−9.3	−0.3/−1.6 /0.0
OH-5	−0.3/−3.2/0.1	−0.4/ 2.2/1.4	−4.0/−5.9/−2.7	−11.7/−12.6/−9.6	-	−1.8/−14.3/−1.2
OH-6	−11.2/−15.3/−9.2	−3.0/−5.9/−2.7	0.4 /−1.2/0.0	−0.8/−1.5/−0.3	−1.8/−13.8/−1.1	-

The results in Table 3 indicate that the intramolecular OH-OH interactions strictly correspond to the H-bond pattern, i.e., for the *pc* conformer there are six and for the remaining conformers–four interaction energies, which are negative and below 9 mH. A more detailed analysis shows that the largest absolute value of these energies appears for the *pc* case (15 mH), followed by *al* (12 mH), while the energies for the *wc* conformer (9 mH) have the smallest absolute value. This order strictly corresponds to the increasing O$\cdots$H distance. It should be noted that the absolute values of interaction energies between the second neighbors (such as OH-3 and OH-5) are larger than for the opposite groups, in full agreement with chemical intuition. An analysis of SAPT components (not shown) reveals that the second-neighbor interaction has a practically pure electrostatic character, while the interaction of the adjacent hydroxy groups contains similarly important contributions from electrostatics, induction, dispersion, and the exchange counterparts. Although a common explanation of the highest stability of the *pc* conformer is the existence of two additional intramolecular H-bonds, a simple addition of the interaction energies, mentioned above, predicts much higher stability than found from the total energies' differences, see Section 3.4. In a fact, much more factors should be accounted for, among which secondary interactions, such as those involving phenyl groups, etc. should play a significant role.

Hydroxy groups in calixarenes reside in a tight neighborhood of other groups, from which the largest phenyl groups are of the highest importance. Since neighboring hydroxy and phenyl groups are placed in different relative orientations, interactions of various types can be obtained. We performed a detailed analysis of I-SAPT components of the hydroxy–phenyl interactions for the *pc* and *al* conformers, and found that the majority of these pairs interact electrostatically. Additionally, the closeness of the H or O ending of the hydroxy group allows us to predict the sign of the electrostatic interaction. For instance, OH groups with the H-ending placed closer to a neighboring Ph group form as a rule the repelling pair (e.g., OH-3 and Ph-4 with the energy of 9 mH, OH-4 and Ph-5 with the energy of 8 mH, OH-5 and Ph-6 with the energy of 2.5 mH, OH-6 and Ph-1 with the energy of 9 mH, and OH-1 and Ph-2 with the energy of 8 mH, for the *al*-CX). If the O-ending is closer to a neighboring Ph group, an attraction pair is formed (e.g., OH-4 and Ph-3 with the energy of -7 mH, OH-5 and Ph-4 with the energy of -8 mH, and OH-2 and Ph-1 with the energy of -7.5 mH for the *al*-CX). The same picture has been found for the *pc*-CX case, where the electrostatic attraction between pairs: OH-6 and Ph-1 with the energy of -9 mH, OH-1 and Ph-2 with the energy of -6 mH, OH-5 and Ph-6 with the energy of -8 mH can be found, while the electrostatic repulsion of the pairs: OH-6 and Ph-5 with the energy of 11 mH, OH-5 and Ph-4 with the energy of 11 mH, OH-1 and Ph-6 with the energy of 10 mH (plus C_2 point-group counterparts) reduce the stability of these conformers.

This rule of thumb does not work for neighboring pairs from the first and second semi-circle for the *al* conformers (both CX and BCX), i.e., for OH-6 and Ph-5 (+4 mH for CX and BCX) and OH-2 and Ph-3 (-4.5 mH for CX and -5 mH for BCX). The latter pair is different from all other neighboring hydroxy–phenyl pairs, since in this case the I-SAPT partitioning shows that all energy components are of equal importance and the resulting attraction is a result of a subtle balance of attractive and repulsive components of similar absolute values. Relatively large exchange components signify that electron clouds of both groups significantly overlap, which allows us to identify a weak secondary H-bond of the π-type [119,120]. It is especially worth noting that this noncovalent bonding facilitates the elongation of the OH-2 bond towards Ph-3 during the oscillation. We have already noticed in the study of the IR spectra that the frequency corresponding to the OH-2 stretch for the *al* conformers is lower than the OH-5 counterpart on the opposite site of the calixarene, so the behavior of the I-SAPT interaction energies clarifies the mechanism of this red shift.

3.8. Pristine Calixarenes–Molecular Properties

Since in a further discussion the F-SAPT and SSMF3 methods will be employed for the analysis of intermolecular interactions, it is interesting to explore how the latter method reproduces first- and second-order molecular properties, such as electric dipole moments and dipole polarizabilities. As discussed in the previous section, the HF and MP2 methods were selected for this test. For the CX case, the HF (MP2) dipole length is reproduced with the error of 6.6, −7.7, −1.9% (−13.5, −18, −16.5%) for the *pc*, *al*, and *wc* conformers, respectively, i.e., this property is reproduced with medium quality. However, the HF (MP2) average polarizability is reproduced with percent errors of 0.6, 1.4, and 3.2% (0.6, 1.5, 2.9%) for these conformers, respectively, which is a much better result. The average polarizability is a tensor invariant, but it is also interesting to examine off-diagonal terms in a given coordinate system since these terms are much smaller than diagonal terms and are more sensitive to applied approximations (see, e.g., the investigation of the influence of local approximations for static polarizabilities [104]). In particular, it is interesting to see whether the balancing of the fragmentation terms with plus and minus signs is accurate enough to reproduce zero or close-to-zero off-diagonal terms for the unfragmented molecule from possibly large (in terms of absolute values) off-diagonal terms for fragments. It turns out that this is indeed the case and, e.g., a small term $\alpha_{xz} = -1.9$ a.u. for the *al* conformer is reproduced as −1.3 a.u. as a summation of terms, which are individually as large as about ±30 a.u. Therefore, one can conclude that the molecular fragmentation model is useful for a semiqualitative reproduction of first-order molecular properties, while it behaves better for second-order properties if the third level of fragmentation is employed. It should be noted that nonbonded parts of the property were omitted here for the sake of simplicity. A full comparison of the dipole moment and polarizability components is presented in the Supplementary Information.

3.9. Complexes with Amino Acids–Energy Partitioning with SSMF and F-SAPT

The energy partitioning in the SMF approaches is a byproduct of the SMF procedure and sometimes two interesting groups are placed in the same fragments, and thus, impossible to separate. However, in many other cases, they are sometimes attributed to different fragments, and then by the elimination analysis one can establish which molecular group in a given molecule is responsible for the highest interaction strength. The systematic way of performing such an analysis consists in a selection of those interaction energies between the calixarene fragment and the amino acid (see Equation (3)), which have the highest absolute values, and the following investigation of how the removal or addition of a neighboring unit influences the interaction energy with the amino acid.

3.9.1. Special Case: SSMF3 and F-SAPT Partitioning Analyses for Complexes of CX···Gly

The SAPT0 interaction energies of 24 pairs (fragment···Gly), resulting from the SSMF3 fragmentation of the CX, are presented in Table 4 for *al*, *pc*, and *wc* conformers of CX, while the corresponding F-SAPT partitionings are presented in Figure 6. One immediately sees that the resulting interaction energies are completely different for these three cases. Let us analyze these differences in more detail, starting from the *pc* conformer. In this case, the SSMF3-SAPT0 fragmentation energies with the largest absolute values correspond to fragments 7, 8, and 9 (see Figure 7). All these fragments contain the same hydroxy group OH-4, which donates its oxygen atom to a strong H-bond with the hydroxy group of Gly (the 1.78 Å distance between the H-ending of the glycine hydroxy group and the oxygen atom from the OH-4). Removal or addition of the methyl group does not have any influence on this energy, which is equal to −18.5 mH. However, it would be incorrect to attribute this energy to the isolated H-bond since the removal of the phenyl ring with the attached OH-3 group reduces the attraction by as much as 5 mH (see fragment 10). The weaker attraction cannot be simply explained as a lack of the attraction between Gly and Ph-3 or OH-3, because the fragment 11, which possesses the same OH-4 group bound to Gly but additionally has the phenyl ring with the OH-5 group on another side, reduces the attraction with Gly by another 5 mH. Therefore, the relative position of the second unit is crucial. One can presume that for fragments 7 to 9 the interaction of OH-4 with OH-3 plays a role since according to the I-SAPT partitioning (see Table 3) there is a strong intramolecular H-bond between these two groups. From geometrical considerations and from a more detailed analysis of induction components, one can see that the OH-4 group donates the H atom while the OH-3 group is the H acceptor, which results in shifting more negative charge to oxygen in OH-4 and making it a better H acceptor for the COOH-Gly. An opposite situation arises for the fragment 11, where the OH-4 group accepts the hydrogen atom from the OH-5 group, what makes the O-4 atom less negative, leading to a weakening of the attraction between the OH-4 and the OH-Gly group. The SSMF3 fragmentation is not detailed enough to directly separate contributions from phenyl ring, methylene, or hydroxy groups, but the F-SAPT partitioning of the complex reveals that indeed (i) the attraction of the carboxy group of Gly with the OH-4 is the strongest one (−10 mH), (ii) there is an additional attraction of COOH-Gly to the OH-3 and Ph-3 groups (−2 mH), which explains why fragment 10, deprived of those groups, shows a weaker attraction, and (iii) there is a relatively strong repulsion between COOH and OH-5 (+4 mH), which explains a reduction in attraction for fragment 11. It should be noted that the attraction between the carboxy group of Gly and the OH-4 group is a net result of a balance between several components of similar importance, such as electrostatics, induction, dispersion and exchange. A significant exchange term indicates that electron clouds of these two groups overlap as in the case of covalent bonds, but the magnitude, which is smaller than for typical covalent bonds, classifies this bond as noncovalent. Therefore, the nonzero exchange and other SAPT components signify that the H-bond should exist between some atoms of COOH-Gly and OH-4 (note that according to the IUPAC criteria atoms participating in H-bonds should be close to each other so that the distance between them was smaller than the sum of atomic Van der Waals radii [121]). If the SSMF3 contributions of a range of about −9 mH are analyzed in the same manner (not shown in the figure), a secondary binding is revealed, which can be attributed to the interaction between NH_2-Gly and the OH-1 groups with the amino group being the H donor. As it could be guessed from a relatively large distance between the hydrogen of NH_2 and oxygen of the OH-1 (2.23 Å) and from the value of the H···OH-1 angle (143°), which is quite different from the full angle, this interaction should be quite weak, but nonetheless, it is still composed of nonzero polarization and exchange contribution, which sum up to −4 mH according to F-SAPT, therefore we can still classify it as an H-bond.

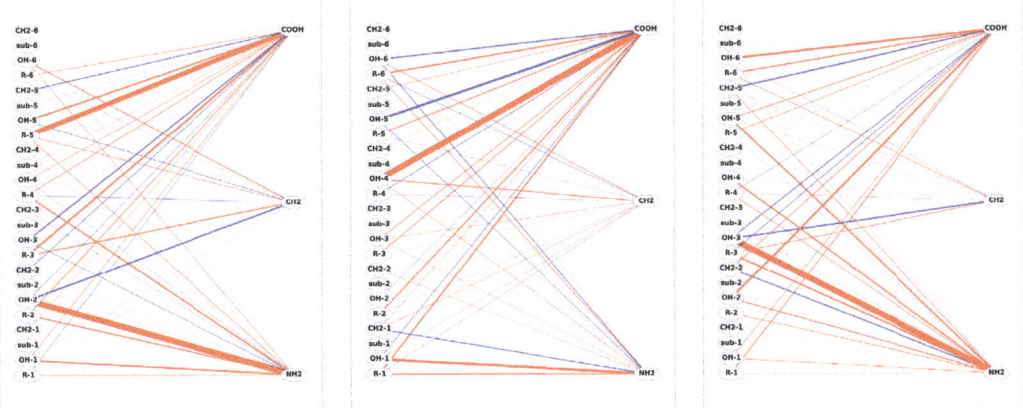

Figure 6. The F-SAPT interaction energy graph for the Gly amino acid interacting with the *al*-CX (leftmost graph), *pc*-CX (middle graph), and *wc*-CX (rightmost graph). Calixarene functional groups are depicted on the left and those for Gly–on the right. The red (blue) lines denote attraction (repulsion), and their thickness is proportional to the interaction strength.

Table 4. The total SAPT0 interaction energies of unfragmented and the SSMF3 fragments for all CX conformers + Gly complex are presented. Energies are in millihartree.

	Weight	al-CX	pc-CX	wc-CX
No Fragmentation		−59.7	−25.5	−52.0
Fragment #1	1	−25.3	−8.7	−9.5
Fragment #2	−1	−18.1	−1.7	−6.1
Fragment #3	1	−22.6	−4.8	−24.3
Fragment #4	−1	−22.6	−4.7	−24.1
Fragment #5	1	−22.7	−4.7	−24.2
Fragment #6	−1	−4.3	−2.7	−16.3
Fragment #7	1	−9.6	−18.7	−22.5
Fragment #8	−1	−6.9	−18.5	−23.1
Fragment #9	1	−7.3	−18.6	−23.0
Fragment #10	−1	−6.5	−13.3	−3.5
Fragment #11	1	−25.5	−9.7	−10.0
Fragment #12	−1	−25.0	−9.4	−9.8
Fragment #13	1	−26.4	−9.7	−9.5
Fragment #14	−1	−19.4	2.5	−5.4
Fragment #15	1	−21.8	2.7	−16.1
Fragment #16	−1	−21.1	3.1	−16.0
Fragment #17	1	−21.2	1.9	−16.2
Fragment #18	−1	−3.6	−0.9	−10.8
Fragment #19	1	−7.8	−8.6	−14.4
Fragment #20	−1	−7.2	−8.7	−14.5
Fragment #21	1	−7.4	−8.7	−14.5
Fragment #22	−1	−4.0	−7.8	−2.9
Fragment #23	1	−24.5	−9.4	−10.9
Fragment #24	−1	−24.6	−8.6	−10.8
Sum		−58.9	−26.5	−51.6

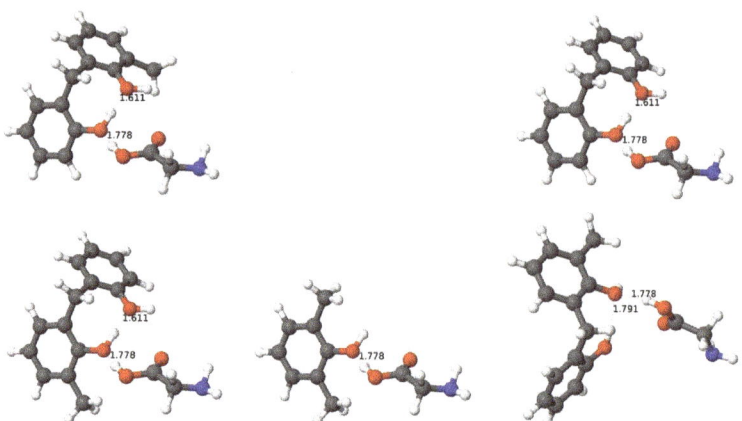

Figure 7. The fragments 7 to 11 generated from the SSMF3 partitioning of the *pc*-CX interacting with Gly. The numeration goes row-wise from top left to bottom right. Selected H-bond distances (in Å) are shown in order to identify their placement in the original calixarene.

For the *al*-CX··· Gly complex, there exists much more important terms since the Gly molecule resides in the *al* cavity, and in Figure 8, we present only those fragments, which are the most relevant to the discussion below. The important fragments can be segregated into those corresponding to the binding to the NH$_2$-Gly and COOH-Gly sites and–contrary to the *pc* case–these two sets are of similar importance. The most attractive contributions for the first and second sites come from fragments 1 and 13 and amount to −25 and −26 mH, respectively. Fragments 1 to 3 have an H-bond between the OH-2 and NH$_2$ groups, while fragments 13 to 15–between OH-5 and COOH. Similarly, as in the *pc*-CX case above, the removal of the phenyl ring with the hydroxy group, serving as a hydrogen donor in the intramolecular H-bond, reduces the interaction energy by about 7 mH (fragments 2 and 14). However, contrary to the *pc*-CX case, the addition of the $n+1$th calixarene unit (number 3 for the fragment 3 and number 6 for the fragment 15) does not lead to a further reduction in attraction. Quite the opposite, a small rise of the attraction (by 3-4 mH) is observed in comparison to fragments 2 and 14. The geometry analysis shows that these additional phenyl and hydroxy groups are more twisted in comparison to the *pc* case, so that the creation of the H-bond between OH-2 and OH-3 or between OH-5 and OH-6 is prevented, and such bonds would weaken the negative character of the O-2 and O-5 atoms. What remains to be explained is the increase in the attraction for fragment 3 in comparison to fragment 2 (and for fragment 15 in comparison to 14). Since these fragments differ by other phenyl plus hydroxy groups, one (or both) of these groups should be responsible for this phenomenon. Because of the limitations of the SSMF3 partitioning, the explanation of this fact should be postponed till the F-SAPT analysis is made. Now let us move to a complementary view of the F-SAPT partitioning. As expected, there is a strong (-14 mH) binding between the OH-2 and NH$_2$ groups. Surprisingly, the interaction between OH-5 and COOH amounts to -3.5 mH only, but the carboxy group is strongly attracted to the Ph-5 group (-11.5 mH). All F-SAPT components, including the exchange one, are significant for the COOH$\cdots$Ph-5 interaction; therefore, the F-SAPT analysis reveals the existence of the untypical H$\cdots\pi$ H-bond. Such interactions were reported, e.g., in molecules containing aromatic rings with those with the S-H bond [122]. Another candidate for the H$\cdots\pi$ H-bond would be the H atom from the NH$_2$ group interacting with the Ph-3 group. However, in this case, the total interaction energy is close to zero, which is a result of a perfect cancellation of several contributions of similar magnitude (the electrostatics of -3 mH and dispersion of -4 mH are counterbalanced by the large exchange component). The question remains why fragments 2 and 3 differ in attraction by as much as 4 mH. A perusal of the F-SAPT partitioning table reveals that this difference is due to the electrostatic attraction (-4 mH) between the Ph-3 and COOH groups. Finally, a similar difference for fragments 14 and 15 can be explained by the net attraction to the OH-6 group (-2.5 mH).

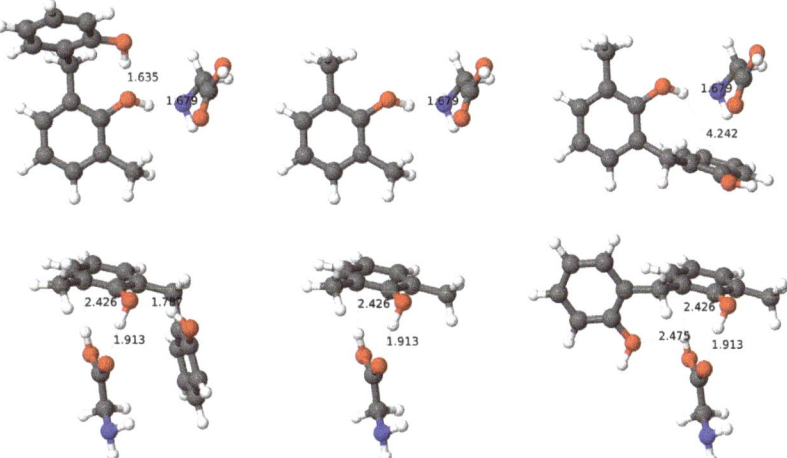

Figure 8. The fragments 1 to 3 and 13 to 15 generated from the SSMF3 partitioning of the *al*-CX interacting with Gly. The numeration goes row-wise from top left to bottom right.

The most important contributions from the SSMF3 partitioning for the *wc*-CX$\cdots$Gly complex provide interaction energies of about -22 to -24 mH (see two representative fragments 3 and 7 in Figure 9) and have in common the H-bond between the OH-3 and NH$_2$ with the latter group as the H acceptor (the H$\cdots$N bond of 1.67 Å). As in the previous cases, this energy cannot be attributed to this one H-bond only, since, e.g., fragment 6 containing this H-bond attracts the Gly molecule weaker by 8 mH. The geometry considerations allow us to point to a possible additional interaction with the OH-2 group for fragment 3 since the O···H distance in this case is equal to 2.10 Å, that is, the amino group of Gly donates its hydrogen to create a weak H-bond. The mechanism of an increased attraction in case of the fragment 7 is different: here the intramolecular H-bond between the OH-3 and OH-4 groups makes the O-3 oxygen less negative, thus allowing the H-3 to interact more strongly with the nitrogen from the amino group. Not shown in the figure are fragments 15 to 17, for which two interactions: between the amino group and OH-5 (the NH_2 as H donor) and between the carboxy group and OH-6 (COOH as the H acceptor) can be guessed from the geometry considerations. The F-SAPT partitioning confirms these predictions. First, a strong interaction between the OH-3 and NH_2 groups has been obtained with this method (−13 mH). There are also three weak H-bonds (again recognized by the significance of all SAPT components, including exchange): between OH-6 and COOH (−4 mH), OH-5 and NH_2 (−3 mH), and OH-2 and NH_2 (−2 mH). The latter interaction is too weak to explain a difference between the attraction from fragments 3 and 6, but the F-SAPT provides the additional strong attraction of a purely electrostatic type between the OH-2 and carboxy group (−4 mH) to fill this gap.

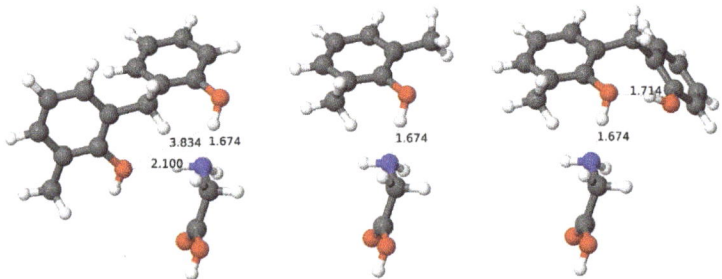

Figure 9. The fragments 3, 6, and 7 generated from the SSMF3 partitioning of the *wc*-CX interacting with Gly. The numeration goes from left to right.

Summarizing, all three cases of the interaction with the simplest Gly amino acid provide different mechanisms for secondary stabilizing interactions, which would be difficult to elucidate just from the analysis of the total energies.

3.9.2. Special Case: SSMF3 and F-SAPT Partitioning Analyses for Complexes of BCX···Gly

The SSMF3 model for hexa-*p-tert*-butylcalix[6]arene produces as many as 150 fragments; therefore, the tables analogous to Table 4 were shifted to the Supplementary Information. The SAPT0 interaction energies for complexes al-BCX···Gly and wc-BCX···Gly are close to those for the CX counterpart (−59.7 mH and −57.7 mH for *al*, −52.0 mH and −50.0 mH for *wc*, for CX and BCX, respectively), but the for *pc*-BCX the interaction energy is 11 mH lower than for the unsubstituted case. Naively, one could assume that it is the attraction with the *tert*-butyl groups, which makes the total interaction energy more negative, but the sum of interaction energies between fragments made from *tert*-butyl with Gly gives a negligible contribution. Therefore, the influence of the *tert*-butyl substituents is more subtle–their presence leads to geometry modifications, which, in turn, allow for a better arrangement of Gly on top of the *pc*-BCX. Similarly, as in the unsubstituted case, there is a set of interaction energies of about −19 mH each, which contain the same type of an H-bond between the carboxyl group of Gly and the OH-1 group (note that for the pristine calixarene OH-1 and OH-4 groups etc. are equivalent, see Figure 2). Again, the presence of the calixarene unit 2 causes an enhancement of the electronegative character of the O-1 atom through the intramolecular H-bond with the OH-2 group, which explains its weaker attraction to Gly (to −13 mH for fragments with the unit 2 removed). The first significant difference appears for numerical values of interaction energies for fragments with the unit added on the other side (the unit 6) of the unit 1. Although the intramolecular H-bond between OH-1 and OH-6 is created, as in the CX case, it weakens the attraction to Gly by 2 mH only, while for the CX case this change amounts to 5 mH. Still, this difference alone does not explain the 11 mH gap between the total interaction energies for the CX and BCX for this conformer, and other differing factors should be looked for. It turns out that the missing difference can be found from the examination of fragments, which contain units 4 and 3 of BCX and have interaction energies of −14 to −15 mH. In these fragments, one H-bond between the H atom from the OH-4 group and the NH$_2$ group of Gly can be found, while the neighboring OH-3 group forms the intramolecular H-bond acting as the H donor, i.e., it enhances the negative character of the O-4 atom. The latter explanation is confirmed by comparison with fragments without the OH-3 group, for which the interaction energy changes to −10 mH. Summarizing, in comparison to the CX case, the interaction with the NH$_2$ group is much stronger and it is this interaction which according to the SSMF3 partitioning is responsible for the stronger attractive force for the *pc*-BCX in comparison to the unsubstituted calixarene. It is also worth noting that the direction of the primary H-bond for the NH$_2$ group changes, i.e., this group becomes the H acceptor for the BCX case. The F-SAPT partitioning confirms these findings: there is still a strong bond of −15 mH between the COOH and OH-1 groups, but the attraction of the OH-4 to the amino group is stronger (−15 mH) than for the CX case. This new intermolecular H-bond causes a distortion of one from the intramolecular H-bonds in the BCX, what can be also observed from the pattern of I-SAPT interaction energies between the hydroxy groups, where instead of one strong attraction of −15 mH between OH-1 and OH-6 only a weak one of −4 mH remains.

The *al*-BCX···Gly complex is of the inclusion type, and the Gly molecule resides in the cavity created by units 3, 4, and 5 of the BCX, as for the *al*-CX counterpart. The fragments which give the largest attractive SSMF3 contributions (−18 mH) contain the phenyl and hydroxy groups from units 1 and 2, and–similarly as for the *al*-CX case – one of them donates its H atom to the NH$_2$-Gly group forming an H-bond, and another one enhances this interaction through a creation of the intramolecular H-bond with the O-2 atom. This effect can be estimated as about 6 mH, based on the energy of the fragment without the unit 1 (−12 mH). There is another group of fragments with no analogs for the CX case, which gives rise to contributions of about −15 mH. It turns out that they all contain the H-bond between NH$_2$ (the H donor) and the OH-6 group (H···O distance of 2.005 Å) and additionally possess the phenyl and hydroxy groups from the neighboring unit 5. The lack of these two groups leads to a strong decrease in attraction (to −8 mH), but for the *al*-BCX geometry, the intramolecular H-bond between the OH-5 and OH-6 groups cannot exist; therefore, the reason for a large attraction in the former case is the direct interaction of the phenyl and hydroxy groups of unit 5 with Gly. The latter conclusion is confirmed by the interaction energy of about −9 mH for fragments, which contain the Ph-5 and OH-5 groups, but no other groups of these types. It should be noted that a relatively large distance between Gly and such fragments suggests that the interaction should be of an electrostatic type since electrostatic contributions are known to be long-ranged. There are also fragments containing the phenyl and hydroxy groups of the unit 4, which attract Gly with the strength of −10 mH. It is evident from the geometry analysis that the OH-4 group cannot effectively participate in any H-bond with Gly, but since the carboxy group is positioned quite close to the Ph-4 (the closest distance between the hydrogen atom of COOH and the carbon atom is about 2.5 Å), one can predict a formation of an unusual H-bond between this hydrogen and the π cloud of the phenyl ring. Note that this distribution of interaction strengths is different from the *al*-CX case, where contributions as large as −26 mH are present, so–surprisingly–such close total interaction energies result from the summation of contributions of a partially different origin. The F-SAPT partitioning confirms the existence of strong bonding between the OH-2 and NH$_2$ groups (−21 mH), as in the CX case. However, contrary to the CX case, the carboxy group is attracted mostly not to the Ph-5 group, but to a closer Ph-4 group with a remarkable strength of −24 mH. Nevertheless, in both cases, the existence of the H-π bond can be postulated based on the position of the hydrogen atom and on the analysis of SAPT components. Another feature of the binding pattern is the existence of a secondary H-bond, in which the NH_2 group donates a hydrogen atom. This H-bond can be identified based on the analysis of the interaction between the NH_2 and OH-6 (−5 mH, with about +10 mH from the exchange component). The strong attraction from unit 5 found in the SSMF3 partitioning is also reproduced here as a strong electrostatic-dominated interaction between the COOH and Ph-5 (−10 mH). It should be emphasized that the F-SAPT and SSMF3 partitionings for the *al*-CX and *al*-BCX complexes with Gly reveal that the Gly molecule is attracted by the cavity from several sites with similar strength. The geometry analysis shows that this relatively small molecule seems to fit well into the small cavity of the *al* conformer; therefore, these complexes represent examples of the enhancement of the interaction due to a confinement effect.

The main features of the complex of Gly with the *wc*-BCX are similar to the *wc*-CX case. The first set of interaction energies of about −23.5 mH corresponds to fragments containing the phenyl groups plus hydroxy groups from units 5 and 6. The OH-5 group serves as an H donor, and the OH-6–as an H acceptor for two H-bonds with the NH_2-Gly. Since the fragment without the OH-5 group, but with the OH-6 group remaining, has the interaction energy reduced to −16 mH, the strength of the second H bond can be estimated as about 7 mH. The addition of the phenyl and OH groups from the unit 1 leads to the increased attraction (−21 mH), which can be explained by the intramolecular H-bond creation between the OH-6 and OH-1 groups, which increases the electronegative character of the O-6 atom (this H-bond is also seen from the I-SAPT analysis). The next sets of interaction energies correspond to the fragments containing either both phenyl and hydroxy groups from units 2 and 3 (energies of −15.5 mH), or having groups from unit 3 only, which reduces the interaction to −10 mH. Since the OH-3 group donates its H atom to the oxygen from the carbonyl group of Gly (the H$\cdots$O distance of 1.99 Å) and the OH-2 group is a hydrogen acceptor for the NH_2 group; therefore, two H-bonds are present here, and the H-bond between the OH-2 and amino groups can be estimated as about 5.5 mH from a difference analysis. The F-SAPT partitioning results are in line with these findings. Firstly, the most important interaction of −19 mH exists between the amino and OH-6 groups. The carboxy group forms a weaker bond with the OH-3 group with the strength of −7 mH. The analysis of SAPT components confirms that these two bonds are H-bonds. The amino group is also connected with the groups OH-1 and OH-2, but from these two pairs, the first one is dominated by electrostatics, while the second again represents an H-bond. It is interesting to note that the distance between the corresponding hydrogen atoms of the amino group and oxygen atoms of the hydroxy groups is only 0.15 Å longer (2.22 *versus* 2.04 Å) for the electrostatic driven interaction and differences in the total interaction energies are also not very large (−4 mH *versus* −6 mH). Nevertheless, in the second case, one has a significant first-order exchange component of 7 mH, which is counterbalanced by other components, from which the electrostatics gives the most negative contribution (−9 mH). It should be noted that for the CX case, both these interactions were of the H-bond type.

3.9.3. Special Cases: Analysis of *pc*- CX Inclusion Complexes

Let us use the F-SAPT partitioning to examine in detail the only four cases, for which the *pc* conformer of CX prefers to form an inclusion complex with an amino acid (Ile, Lys, Val, and Pro). Since the dispersion energy can be only negative and its magnitude is correlated with the number of electrons, it can be used to locate electron-rich environments. Indeed, for all four cases, one can identify pairs of functional groups (one from CX and one from an amino acid), which are close to each other and have a large number of electrons. In particular, a perusal of the dispersion graphs (see the Supplementary Information) allows us to instantly locate the guest molecule into one from two parts of the *pc*-conformer cavity.

(Note that if the dispersion component is significant, then usually other SAPT contributions are nonzero and we deal with a potential candidate for a noncovalent bonding, such as the H-bond.) The next components to examine are the electrostatic and induction energies. One can see that indeed, depending on a relative orientation, some pairs attract each other more strongly than it would be possible with dispersion only. However, the analysis of these three components only is not sufficient to guess the total interaction strength, since many of these components are quenched by short-range first-order exchange contribution. One should emphasize again that the presence of nonzero exchange components together with the polarization (induction and dispersion) components show that we have to deal with a noncovalent bond.

Let us first consider the pc-CX$\cdots$Ile case. The F-SAPT reveals here a strong attraction between the OH-2 group of the calixarene and the COOH group of Ile, which can be explained by the creation of a weak H-bond (the O$\cdots$H distance of 2.0 Å). The F-SAPT estimation of this bond strength can be given by the interaction energy of the OH-2 and COOH groups and it amounts to -9 mH, which should be compared with -10 mH for the intramolecular attraction of two adjacent hydroxy groups in calixarene. This interaction is not the only one in this complex. Moreover, the Ph-6 ring is attracted to the carboxy group, but it is also strongly repelled by the neighboring CH group. It is also interesting that the chain of six intra-calixarene H-bonds is not destroyed in spite of a contribution of OH-2 to another H-bond, which can be seen in the examination of the interaction between the hydroxy groups of calixarene with help of I-SAPT (there are still six interaction energies between these groups of value below -10 mH).

For the complex of pc-CX with Pro the F-SAPT partitioning of dispersion energies reveals that the ring of Pro is strongly attracted by the calixarene phenyl rings on one part of the macrocycle (with Ph-4, Ph-3, and Ph-5), which agrees with the fact that the Pro amino acid occupies just one side of the calixarene cavity. For the interaction with Ph-4 and Ph-5, there is some attractive electrostatic component, too. However, the major part of the interaction comes from the attraction of the COOH group of Pro with Ph-3, which has a predominantly electrostatic origin. It turns out that a strong attractive dispersion interaction of the Pro-ring with three phenyl rings is to a large extent counterweighted by a repulsive exchange interaction, leaving a weak net attraction for these three pairs. It is also interesting to note that practically no interaction exists between Pro and the hydroxy groups of calixarene, in spite of the fact that both Pro fragments are polar and close to at least the OH-3, OH-4, and OH-5 groups. Apparently, the sextet of H-bonds makes the hydroxy group less polar, so they are not involved even in a long-range electrostatic interaction (it should be also noted that the H-bonds' distances range between 1.63 to 1.69 Å, i.e., they do not elongate upon the complexation with Pro).

In the complex with the Lys molecule, there are three Ph groups, which are mostly responsible for the net attraction. From the Lys side, all major polar groups contribute to the interaction and it is impossible to select one or two major interacting pairs. One can note that one from two NH_2 groups of Lys is involved in a secondary bond with the nearest Ph-2 ring. This interaction of -3.2 mH is composed of significant attractive electrostatic and dispersion components and is quenched by the exchange. Interestingly, the interaction of another amino group with the same ring is of a very similar value (-3.7 mH), but is predominantly of the electrostatic character. The hydroxy groups of calixarene do not contribute to these interactions in a significant way (what can be also seen by practically fixed lengths of H-bonds similar to the empty calixarene). It turns out that the COOH and NH_2 groups of Lys connect through an intramolecular H-bond, which apparently depolarizes this molecule to some extent and makes it a weaker target for calixarene.

Finally, for the complex with Val again the net attraction comes from the same three Ph groups, but in this case, one can pinpoint the COOH group as the main partner on the Val side. The interaction between the COOH group and Ph-1, Ph-2, and Ph-6 is of mostly electrostatic character. There are several cases where the exchange contribution is relatively large, but, e.g., for the largest exchange contribution amounting to $+5$ mH (the Ph-1 with the amino group) the electrostatic one is close to zero and after adding the dispersion component a small net interaction energy below 1 mH remains. The alkyl skeleton gives together a weak attraction of −1 mH, but the closest CH group is oriented in a way that produces repulsive electrostatics. Therefore, all potential candidates for a formation of weak noncovalent bonds are excluded in this way and altogether the Val@*pc*-CX complex can be described as the electrostatically bound complex.

3.9.4. Types of Noncovalent Bonding in Calixarene-Amino Acid Complexes

As shown in Sections 3.9.1–3.9.3, the F-SAPT analysis is very useful as a tool for the identification of various types of noncovalent bonds. A comprehensive study of all 120 complexes reveals many more interesting bindings, among which the various sorts of H-bonds form an important group. The IUPAC classification of H-bonds [51,54,121] provides the following H-bond criteria, which may be useful to browse through the present data: *(i)* their energy (63–167 kJ · mol^{-1} for strong, 17–36 kJ · mol^{-1} for medium, and less than 17 kJ · mol^{-1} for weak hydrogen bonds); *(ii)* a large contribution to the electrostatic interaction (with a nonnegligible dispersion), and geometry constraints; such as *(iii)* a preferable alignment of a hydrogen donor, a hydrogen atom, and a hydrogen acceptor and; *(iv)* distances between interacting atoms, which should be significantly smaller than the sum of the Van der Waals radii. The SAPT approach has been applied many times for studying H-bonds, such as, e.g., in the comprehensive study of the behavior of various SAPT components for the water dimer [123], where the near-linear preference of the H-O··· H atoms has been explained by a different angle dependence of first- and second-order SAPT energy components, with the prevailing importance of the second-order induction and dispersion. Therefore, from the SAPT point of view, one should examine not only the pure electrostatic contribution as a fingerprint of the H-bond but also the exchange component, which indicates how large the overlap of the electron clouds of the hydrogen is, the hydrogen donor and the hydrogen acceptor (which is indirectly implied in the H-bond classification by the requirement of a small enough distance between atoms).

The interaction types for all pairs of groups (i.e., 24 groups for a calixarene and a differing number of groups for amino acids) can be first classified according to their total interaction energy E_{int}. According to the IUPAC criteria (64–24 mH–strong, 24–6.5 mH–medium, below 6.5 mH–weak H-bonds), almost all bondings present in calixarenes belong to medium or weak types (note, however, that this criterion is of a limited value here, as the F-SAPT energies are calculated between functional groups containing more atoms than just three involved in an H-bond). The strength of the pair interaction does not imply automatically the H-bond character, although most of the strongest pairs do contain typical H-bonds, i.e., these interactions occur between hydroxy groups of calixarenes and, e.g., hydroxy groups (usually from the carboxy group) or amino groups of amino acids.

The interaction type can be established through the analysis of weights of SAPT components with respect to the interaction energy, i.e., by examining the ratios: $|E^{(1)}_{elst}/E_{int}|$, $|E^{(1)}_{exch}/E_{int}|$, $|E^{(2)}_{ind,eff}/E_{int}|$, and $|E^{(2)}_{disp,eff}/E_{int}|$, where the "eff" subscript denotes effective components as described in Section 2.5. In particular, a large contribution of the exchange energy signifies that both functional groups are close enough for the electron exchange and are a necessary prerequisite of a true bonding (of course, it could also signify a repulsion). A large exchange contribution usually implies that the remaining ratios are nonnegligible. If the only significant contribution is the electrostatic energy, then the functional groups do not share the electron cloud and their attraction or repulsion can be described classically *via* the Coulomb law. The selection of a cutoff value for such ratios is to a large extent arbitrary, and we assumed that if a given component accounts for at least 30% of the interaction energy, the contribution from this component is significant.

The F-SAPT partitioning for selected pairs is presented in Table 5, while tables for all pairs are shifted to the Supplementary Information. A representative of a typical moderately strong H-bond can be characterized by high ratios for electrostatic, exchange, and induction– usually above 1.0 for both first-order components and above 0.5 for the induction and by a smaller contribution from the dispersion (below 0.3). The contribution from the dispersion rises for special cases, such as Pro or His, where a ring with a heteroatom (nitrogen) with the attached hydrogen serves as a hydrogen donor, but additionally, the ring itself interacts with the calixarene hydroxy group, which occurs to a large extent through the dispersion interaction. It should be noted that in such cases, the first three ratios remain high, similar to other H-bonds discussed so far.

Table 5. The F-SAPT partitioning for selected pairs. Energies are in millihartree. The numbers in parenthesis are ratios with respect to total interaction energies between pairs, as defined in Section 3.9.4.

Complex	Group A	Group B	E_{elst} (Ratio)	E_{exch} (Ratio)	$E_{ind,eff}$ (Ratio)	$E_{disp,eff}$ (Ratio)	E_{total}	Bond Type
al-CX-Asn	OH-2	NH$_2$-1	−53.9 (1.8)	60.7 (2.0)	−29.0 (1.0)	−7.5 (0.3)	−29.7	typical H-bond
al-BCX-Lys	OH-2	NH$_2$-1	−48.3 (1.6)	47.9 (1.6)	−23.1 (0.8)	−6.5 (0.2)	−29.9	typical H-bond
al-CX-Pro	OH-2	Ring	−44.7 (1.7)	54.0 (2.1)	−26.2 (1.0)	−8.7 (0.3)	−25.5	typical H-bond+disp
wc-CX-Pro	OH-3	Ring	−26.8 (1.8)	29.1 (2.0)	−11.2 (0.8)	−5.8 (0.4)	−14.6	typical H-bond+disp
wc-BCX-Pro	OH-6	Ring	−25.7 (1.8)	28.4 (2.0)	−11.0 (0.8)	−5.6 (0.4)	−13.9	typical H-bond+disp
wc-CX-Leu	Ph-6	COOH	−19.7 (0.9)	7.5 (0.3)	−4.7 (0.2)	−4.8 (0.2)	−21.7	H···π
wc-CX-Pro	Ph-6	COOH	−16.6 (0.9)	6.2 (0.3)	−3.8 (0.2)	−4.5 (0.2)	−18.6	H···π
al-BCX-Ser	Ph-5	COOH	−19.4 (0.9)	10.2 (0.5)	−5.8 (0.3)	−5.8 (0.3)	−20.8	H···π
pc-BCX-Tyr	Ph-4	OH	−10.2 (1.0)	5.5 (0.6)	−2.5 (0.3)	−2.8 (0.3)	−9.9	H···π
pc-CX-HisE	Ph-6	Ring	−11.9 (0.9)	10.1 (0.8)	−2.7 (0.2)	−8.5 (0.7)	−13.0	disp
wc-CX-HisE	Ph-2	Ring	−11.8 (0.9)	7.5 (0.6)	−1.7 (0.1)	−7.2 (0.5)	−13.2	disp
pc-BCX-HisE	Ph-5	Ring	−13.4 (0.9)	10.6 (0.7)	−3.5 (0.2)	−8.8 (0.6)	−15.1	disp
wc-BCX-HisE	Ph-5	Ring	−13.9 (0.9)	10.0 (0.7)	−2.3 (0.2)	−9.0 (0.6)	−15.1	disp
wc-BCX-Trp	Ph-5	Ring	−12.1 (0.8)	10.7 (0.7)	−2.6 (0.2)	−10.8 (0.7)	−14.8	disp
pc-CX-Tyr	Ph-3	COOH	−15.7 (0.9)	1.2 (0.1)	−1.4 (0.1)	−2.2 (0.1)	−18.1	elst
al-CX-HisD	Ph-4	COOH	−11.7 (0.8)	0.3 (0.0)	−2.3 (0.2)	−1.5 (0.1)	−15.1	elst
al-BCX-Gly	Ph-5	COOH	−7.7 (0.7)	0.8 (0.1)	−1.4 (0.1)	−2.1 (0.2)	−10.5	elst
pc-BCX-GluH	Ph-3	COOH-1	−15.4 (0.8)	2.3 (0.1)	−2.2 (0.1)	−3.2 (0.2)	−18.5	elst

Apart from such cases, there are also several moderately strong bondings, which are characterized by a different ratio pattern. Usually, they arise from the interaction of a phenyl ring of calixarene with a polar group belonging to an amino acid. They are characterized by a lower ratio for the exchange (much lower than 1, but not less than 0.3), the still high ratio for electrostatics, and usually by smaller ratios for induction and dispersion (about 0.2–0.3).

An examination of the relative position of the calixarene phenyl ring in question and the hydrogen atom from the polar group of the amino acid reveals that such cases correspond to a partial donation of the hydrogen atom to the ring, i.e., here, the untypical H-bond is detected (H··· π bond), where the phenyl ring serves as a hydrogen acceptor. One should add that usually an elongation of the H–X bond in comparison to the isolated amino acid is found in such cases, similarly to typical H-bonds.

The next characteristic pattern occurs for the interaction between a calixarene phenyl ring and a ring of an amino acid (e.g., for His). For such interactions there is still a high electrostatics ratio, the exchange ratio is higher than for the H··· π case, but lower than for a typical H-bond (usually about 0.5), and finally, a large dispersion contribution amounts to at least 0.5. Summarizing, in comparison to all other cases, these pairs are to a large extent dispersion-bound, especially if one takes into account that the electrostatics and first-order exchange contributions partially cancel each other.

One should add that even among quite strong interactions one could find electrostatic-driven ones, especially for the interaction of a calixarene ring with a polar group of the amino acid; therefore, the ratio check of components is usually necessary in order to determine the interaction type. In the case of calixarenes, the macrocyclic structure represents an obstacle in relaxing the geometry after the complexation. In some cases, this means that apart from the dominant attractive pairs, some other pairs exist, for which the interaction energy becomes zero or is even positive (repulsive). In the interaction table, the largest such contributions amount to 7–8 mH. Among them, there are pairs that repel each other electrostatically, but also those that are close enough to have a large exchange contribution.

The sorting of F-SAPT total energies for pair interactions reveals in many cases two or more pairs with interaction energies of a similar value (which one can set arbitrarily for, e.g., no more than 50% difference). Among such cases, both inclusion and outer complexes can be found. For the inclusion complexes, this analysis helps to identify the confinement (encapsulation) effects, i.e., those cases where an amino acid occupies the calixarene cavity and utilizes its active groups to bind with several groups of the host. One such example has already been presented in Sections 3.9.1–3.9.2, where the *al* conformers of both CX and BCX use both polar groups of Gly to encapsulate it effectively in the cavity. More such complexes can be found, e.g., for the *pc*-CX case for complexes with Val, Pro, and Ile and for the *pc*-BCX one–with Pro, GluH, Ala, and AspH. In all these cases, the guest molecule fits into one of two cavities created by three calixarene units and the dihydrogen bridge. Other complexes with the confinement effect are: complexes with Ile, Val, Tyr, Thr, and Ser for the *al*-CX, and complexes with Val, Trp, Thr, Ser, Pro, Met, HisD, Gln, AspH, and Asn for the *al*-BCX. It should be noted that in the case of partial confinement, which we have in the case of calixarenes, the cage (or better saying: half-cage) has more flexibility to adapt itself for a particular guest than an inherently rigid full cage, such as in fullerenes. In all these cases, the guest molecules are attached in the cavity by at least two comparable interactions, which are usually (but not always) the proper H-bonds.

The same analysis for the outer complexes reveals that a general classification into: either cases for which one molecule has a dominating group interacting with at least two groups of another molecule, or cases for which each molecule has at least two such groups. The first scheme corresponds to the pattern $A_1 \leftrightarrow B_1 \wedge A_1 \leftrightarrow B_2$, and the second–to $A_1 \leftrightarrow B_1 \wedge A_2 \leftrightarrow B_2$, where A_i and B_i are functional groups of molecules A and B. The first case can be named, similarly as in the ligand theory, as di- (or in general: poly-) dentate complex, while the second case–as di- (or in general: poly-) site complex. The di-dentate complexes are the most common. Among several examples, one can name the complex of the *pc*-CX with Gln, where the Gln molecule uses its carboxy group to make two H-bonds with the OH-1 and OH-6 groups (of similar strength of about −9 mH), or with Tyr, for which the carboxy group binds electrostatically with the Ph-3 and Ph-2 groups. Yet another example of this type is the *pc*-BCX complex with Asn, where the carboxy group forms two H-bonds with two hydroxy groups of calixarene, with the interaction energies of −13 and −11 mH.

The poly-site complexes are, e.g., complexes of the *pc*-CX with Ala, where two H-bonds are created: between the OH-4 and NH_2 groups and between the OH-1 and COOH groups with the interaction energies of −17 mH and −15 mH, respectively, or with AspH, where again two opposite hydroxy groups (OH-3 and OH-6) form H-bonds with two AspH carboxy groups. For the *wc*-CX case, the complex with AspH is particularly interesting, since the examination of the SAPT energy decomposition allows us to identify two H-bonds as for the *pc* case, but additionally both COOH groups interact electrostatically with other groups of calixarene with total energies of the same magnitude. Other *wc*-CX examples of poly-site binding are: (*i*) Gln, where two large groups (COOH and CONH) each form two H-bonds with hydroxy groups of calixarene (the strongest two interactions amount to −18 and −17 mH); (*ii*) HisE, where one typical H-bond is formed between the OH-3 and NH_2 groups, but additionally a strong dispersion-dominated interaction between two rings (the Ph-2 group with the HisE ring) occurs (the first interaction amounts to −20 mH, and the second–to −13 mH); (*iii*) Leu or Lys, where two H-bonds are formed: one with help of the COOH, and the second–with the NH_2 group.

There is also a number of poly-site complexes for the hexa-*p*-tert-butylcalix[6]arene case, such as, e.g., already discussed *pc*-BCX· · ·Gly case, see Section 3.9.2. For the *wc*-BCX complex with HisE an analogous situation appears as for the CX case with very similar interaction energies of −18 and −15 mH for the H-bond and the dispersion-dominated bond, respectively. The same pair of interactions exists for the complex with Trp, where apart from a typical H-bond between the OH-6 and NH_2 groups there exists a dispersion-dominated interaction between the Ph-5 and Trp rings (of energies of −17 and −15 mH, respectively). Other interesting cases are: *(i)* the complex with Thr, where apart from a typical H-bond between the OH-3 and NH_2, the Ph-6 and carboxy groups attract each other electrostatically (both bonds of strength −16 mH) and; *(ii)* a similar pair of H-bond and electrostatic interactions in the complex with Pro, where the nitrogen from the Pro ring forms the H-bond, while the COOH group is attracted by the phenyl ring (with the interaction energies of −14 and −15 mH, respectively), or; *(iii)* the interaction with Met, where again pairs OH-6 with NH_2 and Ph-3 with COOH form the H-bond and the electrostatic-dominated interaction of the same strength of −14 mH. There are also cases of typical H-bonds of similar strength, such as in the complex with Ser (the OH-6 with NH_2 and the OH-3 with OH-Ser groups with interaction energies of −19 and −13 mH, respectively), with Lys (interaction energies of about −20 mH) or with GluH (−19 mH).

Many more similar examples can be found after the analysis of the F-SAPT interaction energy partitioning, listed in the Supplementary Information.

4. Conclusions

Three calix[6]arene and hexa-*p-tert*-butylcalix[6]arene conformers and their most stable complexes with amino acids were studied with DFT+D, SSMF, MP2, SCS-MP2, SAPT0, and F-SAPT methods.

The *pc* (pinched-cone) conformer is the most stable for the pristine calix[6]arene and hexa-*p-tert*-butylcalix[6]arene. A ring of six H-bonds strongly stabilizes the calixarene pinched-cone shape, with a strength of the H-bond of about −15 mH, according to the I-SAPT analysis. This ring partially closes the top of the calixarene molecule. No significant differences between calix[6]arene and hexa-*p-tert*-butylcalix[6]arene stability of the *pc* conformer have been found. The *wc* (winged-cone) and *al* (alternate) conformers each have two triples of hydroxy groups connected by H-bonds, which are weaker than for the *pc* case (−10 mH and −12 mH).

The energetic order of complexes with the same amino acid usually differs from the stability order of empty calixarene conformers. The interaction energy order is different too, which can be attributed to deformations, which strongly depend on the conformer and on the amino acid type. The absolute values of interaction energies of the *pc* conformers are about two times smaller than those for the *wc* and *al* ones, indicating that because of the more stable structure the *pc* conformer is less prone to interact with polar molecules such as amino acids. No molecule was able to penetrate inside the *wc* conformer, while for other conformers both inclusion and outer complexes are found as the lowest complex conformers.

Systematic molecular fragmentation is proven to reproduce the interaction energy of the complexes with good accuracy. Especially the dispersion energy (i.e., electron-correlated part of the interaction energy) is well suited to be treated by the SSMF3 approach. The examination of individual contributions to the interaction energy calculated *via* the SSMF3 method is useful to select those parts of the calixarene molecule, which are of major importance for the interaction. By a difference analysis of two or more such fragments, more information can be obtained about the nature of the interaction, such as, e.g., about the cooperative or anti-cooperative interaction of the major binding site with neighboring functional groups or about a subtle dependence of the sign of such effects on the relative position of these groups.

A partition of the SAPT interaction energy into components and – simultaneously–into contributions from pairs of functional groups with help of the F-SAPT approach, gives a plethora of information about the strength and nature of interactions between these groups. Depending on the relative importance of the electrostatic, first-order exchange, and effective induction and dispersion contributions with respect to the total interaction energy of the pair, we propose a preliminary classification of the interactions into: (i) typical H-bonds; (ii) H-bonds with a dispersion flavor; (iii) H-bonds with an aryl ring as the hydrogen acceptor; (iv) dispersion-dominated, and finally; (v) electrostatic-dominated interactions, depending on ratios of SAPT components with respect to the total interaction energy between these two groups. The characteristic feature of H-bonds is a high ratio of first-order exchange, which is a prerequisite of the effective overlap of the electron clouds.

After pair interactions are classified according to their strength and origin, one can single out cases, for which two or more such interactions are of similar importance. Numerous such cases arise in this study because of multiple candidates for binding sites for both calixarenes and amino acids. Especially interesting among them are those inclusion complexes for which the confinement effect plays a role in stabilization.

Finally, smaller-than-expected stability of some complexes can be explained by the existence of pair interactions that are weakly bound or which repel each other because of an inadequate alignment of the interacting pair.

The picture of the binding between the studied calixarenes and the amino acids is not simple and unequivocal. Therefore, it was not possible to find a clear tendency in the studied set of complexes. However, the analysis of data reveals that irrespectively of the type of the amino acid its NH_2 group and, to a lesser extent also the COOH group, participate in the calixarene-amino acid binding and that they can form H-bonds of various strengths and geometries, as described in more detail above. Further, in the case of His, a large contribution comes from the interaction of its imidazole ring. This ring can participate both in the H-bond formation through its nitrogen atoms, as well as in the π-π stacking interactions. The latter is also very well developed and plays a decisive role in the case of the binding of Phe. Surprisingly, the indole ring of Trp contributes less to the overall binding. The amino acids possessing the OH group (Ser, Thr, and Tyr) involve this entity to form relatively strong H-bonds with the calixarenes hosts. Finally, the contribution of the SH entity is also seen in the case of the Cys complexes.

This study shows that the I- and F-SAPT approaches are very useful to elucidate the nature of inter- and intramolecular noncovalent bindings. Additionally, they can be used to refine explanations or other phenomena, such as the lowering of the IR frequency for the *al*-CX, which–as revealed by the I-SAPT analysis–resulted from the noncovalent attractive interaction between hydroxy and phenyl groups of different calixarene units.

Supplementary Materials: The following are available at https://www.mdpi.com/article/10.3390/molecules27227938/s1, Figure S1: Amino acids as guest molecules used in this study, Figures S2–S9: IR spectra of calixarenes, Figures S10–105: The F-SAPT partitioning graphs, Tables S1 and S2: Dihedral angles for complexes of calixarenes, Table S3: I-SAPT interaction energies al-CX, Tables S4–S9: A comparison of the SSMF3 fragmentation scheme with the unfragmented results for supermolecular and SAPT0 interaction energies for amino acid and calixarene complex, Table S10: The SAPT0 energy components, SAPT0 and HF interaction energies, MP2 and SCS-MP2 electron-correlated parts of the interaction energies for BCX+amino acid complexes, Table S11: The electronic and total energy of both empty calixarenes in three considered conformations, and complexes of these conformations with amino acids, Tables S12–S17: Percent errors of the SSMF3 fragmentation scheme in respect to unfragmented results for amino acid+calixarene complex, Tables S18–S23: A complete list of frequencies and intensities of calixarenes, Tables S24–S26: The total SAPT0 interaction energies of unfragmented and the SSMF3 fragments for BCX+Gly complex, Table S27: The accuracy of the SSMF3-HF(b) and SSMF3-MP2(b) electric dipole moments and dipole static polarizabilities for three conformers of calix[6]arene, Table S28: The F-SAPT partitioning for selected CX+amino acid pairs with their classification, Table S29: The F-SAPT partitioning for selected BCX+amino acid pairs with their classification, Table S30: Interaction of hydroxy groups of the BCX calixarene with and without the presence of Gly as calculated with the I-SAPT method. FSAPT_h-sub.ods, F-SAPT partitioning for complexes with calix[6]arene; FSAPTz_b-sub.ods, F-SAPT partitioning for complexes with hexa-*p-tert*-butylcalix[6]arene; xyz_files.zip, Cartesian coordinates of studied systems.

Author Contributions: Conceptualization, T.K. and D.R.-Z.; methodology, M.C., T.K. and E.M.; software (SSMF program), E.M.; validation, T.K. and D.R.-Z.; formal analysis, T.K., M.C., E.M.; investigation, E.M., M.C., T.K.; writing—original draft preparation, M.C., T.K., D.R.-Z.; writing—review and editing, T.K., D.R.-Z., E.M. and M.C.; visualization, E.M., T.K., M.C.; supervision, T.K.; project administration, T.K.; resources, T.K.; funding acquisition, T.K. All authors have read and agreed to the published version of the manuscript.

Funding: This research was funded by the National Science Centre of Poland through grant 2017/27/B/ST4/02699. This research was supported by PCSS infrastructure and PRACE-6IP 823767, MNISW DIR/WK/2016/18.

Data Availability Statement: Data is contained within the article or supplementary material

Conflicts of Interest: The authors declare no conflict of interest.

Abbreviations

The following abbreviations are used in this manuscript:

al	1,2,3-alternate
BCX	hexa-*p-tert*-butylcalix[6]arene
CBS	Complete Basis Set
CCSD(T)	coupled cluster singles and doubles with perturbative triples correction
CSD	Cambridge Structural Database
CX	calix[6]arene
DF	density fitting
DFT	Density-Functional Theory
DFT+D	Density-Functional Theory with dispersion correction
FF	Finite-field
F-SAPT	Functional-group SAPT
I-SAPT	Intramolecular SAPT
HF	Hartree-Fock
MP2	Møller-Plesset theory to the second order
pc	pinched-cone
SAPT	Symmetry-Adapted Perturbation Theory
SMFA	Systematic Molecular Fragmentation by Annihilation
SSMF	Symmetrized Systematic Molecular Fragmentation
SCS-MP2	spin-component-scaled MP2
wc	winged-cone
ZPVE	zero-point vibrational energy

References

1. Da Silva, E.; Lazar, A.; Coleman, A. Biopharmaceutical applications of calixarenes. *J. Drug Deliv. Sci. Technol.* **2004**, *14*, 3–20. [CrossRef]
2. Guo, D.S.; Liu, Y. Supramolecular chemistry of p-sulfonatocalix[n]arenes and its biological applications. *Accounts Chem. Res.* **2014**, *47*, 1925–1934. [CrossRef]
3. Yousaf, A.; Hamid, S.A.; Bunnori, N.M.; Ishola, A.A. Applications of calixarenes in cancer chemotherapy: Facts and perspectives. *Drug Des. Dev. Ther.* **2015**, *9*, 2831–2838. [CrossRef]
4. Deraedt, C.; Astruc, D. Supramolecular nanoreactors for catalysis. *Coord. Chem. Rev.* **2016**, *324*, 106–122. [CrossRef]
5. Mammino, L. Bowl-shaped structures from acylphloroglucinols: An ab initio and DFT study. *Mol. Phys.* **2017**, *115*, 2254–2266. [CrossRef]
6. Galindo-Murillo, R.; Olmedo-Romero, A.; Cruz-Flores, E.; Petrar, P.; Kunsagi-Mate, S.; Barroso-Flores, J. Calix[n]arene-based drug carriers: A DFT study of their electronic interactions with a chemotherapeutic agent used against leukemia. *Comput. Theor. Chem.* **2014**, *1035*, 84–91. [CrossRef]
7. An, L.; Wang, J.W.; Liu, J.D.; Zhao, Z.M.; Song, Y.J. Design, Preparation, and Characterization of Novel Calix[4]arene Bioactive Carrier for Antitumor Drug Delivery. *Front. Chem.* **2019**, *7*, 732. [CrossRef] [PubMed]
8. Feng, H.T.; Li, Y.; Duan, X.; Wang, X.; Qi, C.; Lam, J.W.Y.; Ding, D.; Tang, B.Z. Substitution Activated Precise Phototheranostics through Supramolecular Assembly of AIEgen and Calixarene. *J. Am. Chem. Soc.* **2020**, *142*, 15966–15974. [CrossRef]
9. Cacciapaglia, R.; Stefano, S.D.; Mandolini, L.; Salvio, R. Reactivity of carbonyl and phosphoryl groups at calixarenes. *Supramol. Chem.* **2013**, *25*, 537–554. [CrossRef]

10. Rebilly, J.N.; Reinaud, O. Calixarenes and resorcinarenes as scaffolds for supramolecular metallo-enzyme mimicry. *Supramol. Chem.* **2014**, *26*, 454–479. [CrossRef]
11. Hennrich, G.; Murillo, M.T.; Prados, P.; Song, K.; Asselberghs, I.; Clays, K.; Persoons, A.; Benet-Buchholz, J.; de Mendoza, J. Tetraalkynyl calix[4]arenes with advanced NLO properties. *Chem. Commun.* **2005**, 2747–2749. [CrossRef]
12. Pichierri, F. Cs+–π interactions and the design of macrocycles for the capture of environmental radiocesium (Cs-137): DFT, QTAIM, and CSD studies. *Theor. Chem. Accounts* **2018**, *137*, 118. [CrossRef]
13. Chen, X.; Häkkinen, H. Protected but Accessible: Oxygen Activation by a Calixarene-Stabilized Undecagold Cluster. *J. Am. Chem. Soc.* **2013**, *135*, 12944–12947. [CrossRef]
14. Ocak, Ü.; Ocak, M.; Bartsch, R.A. Calixarenes with dansyl groups as potential chemosensors. *Inorganica Chim. Acta* **2012**, *381*, 44–57. [CrossRef]
15. Deska, M.; Dondela, B.; Sliwa, W. Selected applications of calixarene derivatives. *Arkivoc* **2015**, *2015*, 393–416. [CrossRef]
16. Kumar, S.; Chawla, S.; Zou, M.C. Calixarenes based materials for gas sensing applications: A review. *J. Incl. Phenom. Macrocycl. Chem.* **2017**, *88*, 129–158. [CrossRef]
17. Harris, S.J. Calixarene-Based Compounds Having Antibacterial, Antifungal, Anticancer-Hiv Activity. Patent WO 95/19974, 27 July 1995.
18. Coleman, W.A.; Baggetto, L.G.; Lazar, A.N.; Michaud, M.H.; Magnard, S. Calixarene Derivatives as Anticancer Agent. U.S. Patent US2010/0056482A1, 4 March 2010.
19. Naseer, M.M.; Ahmed, M.; Hameed, S. Functionalized calix[4]arenes as potential therapeutic agents. *Chem. Biol. Drug Des.* **2017**, *89*, 243–256. [CrossRef]
20. Pur, F.N.; Dilmaghani, K.A. Calixpenams: Synthesis, characterization, and biological evaluation of penicillins V and X clustered by calixarene scaffold. *Turk. J. Chem.* **2014**, *38*, 288–296. [CrossRef]
21. Ben Salem, A.; Sautrey, G.; Fontanay, S.; Duval, R.E.; Regnouf-de Vains, J.B. Molecular drug-organiser: Synthesis, characterization and biological evaluation of penicillin V and/or nalidixic acid calixarene-based podands. *Bioorg. Med. Chem.* **2011**, *19*, 7534–7540. [CrossRef]
22. Casnati, A.; Fabbi, M.; Pelizzi, N.; Pochini, A.; Sansone, F.; Ungaro, R.; Di Modugno, E.; Tarzia, G. Synthesis, antimicrobial activity and binding properties of calix[4]arene based vancomycin mimics. *Bioorg. Med. Chem. Lett.* **1996**, *6*, 2699–2704. [CrossRef]
23. Boukerb, A.M.; Rousset, A.; Galanos, N.; Méar, J.B.; Thépaut, M.; Grandjean, T.; Gillon, E.; Cecioni, S.; Abderrahmen, C.; Faure, K.; et al. Antiadhesive Properties of Glycoclusters against Pseudomonas aeruginosa Lung Infection. *J. Med. Chem.* **2014**, *57*, 10275–10289. [CrossRef] [PubMed]
24. Mourer, M.; Dibama, H.M.; Fontanay, S.; Grare, M.; Duval, R.E.; Finance, C.; Regnouf-de Vains, J.B. p-Guanidinoethyl calixarene and parent phenol derivatives exhibiting antibacterial activities. Synthesis and biological evaluation. *Bioorg. Med. Chem.* **2009**, *17*, 5496–5509. [CrossRef]
25. Shurpik, D.N.; Padnya, P.L.; Stoikov, I.I.; Cragg, P.J. Antimicrobial Activity of Calixarenes and Related Macrocycles. *Molecules* **2020**, *25*, 5145. [CrossRef]
26. Neagu, M.; Ion, R.M.; Manda, G.; Constantin, C.; Radu, E.; Cristu, Z. Antitumoral Effect of Calixarenes in Experimental Photodynamic Therapy with K562 Tumor Cell Line. *Rom. J. Biochem.* **2010**, *47*, 17–35.
27. Ugozzoli, F.; Andreetti, G.D. Symbolic Representation of the Molecular Conformation of Calixarenes. *J. Incl. Phenom. Mol. Recognit. Chem.* **1992**, *13*, 337–348. [CrossRef]
28. Matthews, S.E.; Cecioni, S.; O'Brien, J.E.; MacDonald, C.J.; Hughes, D.L.; Jones, G.A.; Ashworth, S.H.; Vidal, S. Fixing the Conformation of Calix[4]arenes: When Are Three Carbons Not Enough? *Chem. A Eur. J.* **2018**, *24*, 4436–4444. [CrossRef]
29. Gassoumi, B.; Chaabene, M.; Ghalla, H.; Chaabane, R.B. Physicochemical properties of the three cavity form of calix[n = 4, 6, 8] aren molecules: DFT investigation. *Theor. Chem. Accounts* **2019**, *138*, 58. [CrossRef]
30. Kumar, S.; Kaur, J.; Verma, A.; Mukesh; Kumar, A.; Dominic, S. Influence of polyether chain on the non-covalent interactions and stability of the conformers of calix[4]crown ethers. *J. Incl. Phenom. Macrocycl. Chem.* **2018**, *91*, 81–93. [CrossRef]
31. Özkinali, S.; Karayel, A. Synthesis, characterization, conformational equilibrium and intramolecular hydrogen bond analysis of Novel Azocalix[4]arenes including acryloyl moiety using DFT studies. *J. Mol. Struct.* **2019**, *1176*, 303–313. [CrossRef]
32. Furer, V.L.; Potapova, L.I.; Vatsouro, I.M.; Kovalev, V.V.; Shokova, E.A.; Kovalenko, V.I. Investigation of the conformation and hydrogen bonds in adamantylthiacalix[4]arene by IR spectroscopy and DFT. *J. Mol. Struct.* **2018**, *1171*, 207–213. [CrossRef]
33. Malinska, M. Insights into molecular recognition from the crystal structures of p-tert-butylcalix[6]arene complexed with different solvents. *IUCrJ* **2022**, *9*, 55–64. [CrossRef]
34. Furer, V.L.; Potapova, L.I.; Vatsouro, I.M.; Kovalev, V.V.; Shokova, E.A.; Kovalenko, V.I. Study of conformation and hydrogen bonds in the p-1-adamantylcalix[8]arene by IR spectroscopy and DFT. *J. Incl. Phenom. Macrocycl. Chem.* **2019**, *95*, 63–71. [CrossRef]
35. Kieliszek, A.; Malinska, M. Conformations of p-tert-Butylcalix[8]arene in Solvated Crystal Structures. *Cryst. Growth Des.* **2021**, *21*, 6862–6871. [CrossRef]
36. Puchta, R.; Clark, T.; Bauer, W. The formation of endo-complexes between calixarenes and amines–a reinvestigation. *J. Mol. Model.* **2006**, *12*, 739–747. [CrossRef] [PubMed]
37. Jang, Y.M.; Yu, C.J.; Kim, J.S.; Kim, S.U. Ab initio design of drug carriers for zoledronate guest molecule using phosphonated and sulfonated calix[4]arene and calix[4]resorcinarene host molecules. *J. Mater. Sci.* **2018**, *53*, 5125–5139. [CrossRef]

38. Kryuchkova, N.A.; Kostin, G.A.; Korotaev, E.V.; Kalinkin, A.V. XPS and quantum chemical investigation of electronic structure of Co complexes with calix[4]arenes modified by R2PO groups in upper or lower rim. *J. Electron Spectrosc. Relat. Phenom.* **2018**, *229*, 114–123. [CrossRef]
39. Murphy, P.; Dalgarno, S.J.; Paterson, M.J. Systematic Study of the Effect of Lower-Rim Methylation on Small Guest Binding within the Host Cavity of Calix[4]arene. *J. Phys. Chem. A* **2017**, *121*, 7986–7992. [CrossRef]
40. Sharafdini, R.; Mosaddeghi, H. Inhibition of Insulin Amyloid Fibrillation by Salvianolic Acids and Calix[n]arenes: Molecular Docking Insight. *J. Comput. Biophys. Chem.* **2021**, *20*, 539–555. [CrossRef]
41. Shinde, M.N.; Barooah, N.; Bhasikuttan, A.C.; Mohanty, J. Inhibition and disintegration of insulin amyloid fibrils: A facile supramolecular strategy with p-sulfonatocalixarenes. *Chem. Commun.* **2016**, *52*, 2992–2995. [CrossRef]
42. Zhao, H.; Yang, X.H.; Pan, Y.C.; Tian, H.W.; Hu, X.Y.; Guo, D.S. Inhibition of insulin fibrillation by amphiphilic sulfonatocalixarene. *Chin. Chem. Lett.* **2020**, *31*, 1873–1876. [CrossRef]
43. Böhmer, V. Calixarenes, Macrocycles with (Almost) Unlimited Possibilities. *Angew. Chem. Int. Ed. Engl.* **1995**, *34*, 713–745. [CrossRef]
44. Jeziorski, B.; Moszynski, R.; Szalewicz, K. Perturbation Theory Approach to Intermolecular Potential Energy Surfaces of van der Waals Complexes. *Chem. Rev.* **1994**, *94*, 1887–1930. [CrossRef]
45. Szalewicz, K. Symmetry-Adapted Perturbation Theory of Intermolecular Forces. *Wiley Interdiscip. Rev. Comput. Mol. Sci.* **2012**, *2*, 254–272. [CrossRef]
46. Parrish, R.M.; Sherrill, C.D. Spatial assignment of symmetry adapted perturbation theory interaction energy components: The atomic SAPT partition. *J. Chem. Phys.* **2014**, *141*, 044115. [CrossRef] [PubMed]
47. Parrish, R.M.; Parker, T.M.; Sherrill, C.D. Chemical Assignment of Symmetry-Adapted Perturbation Theory Interaction Energy Components: The Functional-Group SAPT Partition. *J. Chem. Theory Comput.* **2014**, *10*, 4417–4431. [CrossRef]
48. Collins, M.A.; Bettens, R.P. Energy-Based Molecular Fragmentation Methods. *Chem. Rev.* **2015**, *115*, 5607–5642. [CrossRef]
49. Masoumifeshani, E.; Korona, T. Symmetrized Systematic Molecular Fragmentation Model and its Application for Molecular Properties. *Comput. Theor. Chem.* **2021**, *1202*, 113303. [CrossRef]
50. Masoumifeshani, E.; Chojecki, M.; Korona, T. Electronic Correlation Contribution to the Intermolecular Interaction Energy from Symmetrized Systematic Molecular Fragmentation Model. *Comput. Theor. Chem.* **2022**, *1211*, 113684. [CrossRef]
51. Arunan, E.; Desiraju, G.R.; Klein, R.A.; Sadlej, J.; Scheiner, S.; Alkorta, I.; Clary, D.C.; Crabtree, R.H.; Dannenberg, J.J.; Hobza, P.; et al. Definition of the hydrogen bond (IUPAC Recommendations 2011). *Pure Appl. Chem.* **2011**, *83*, 1637–1641. [CrossRef]
52. Izgorodina, E.I.; MacFarlane, D.R. Nature of Hydrogen Bonding in Charged Hydrogen-Bonded Complexes and Imidazolium-Based Ionic Liquids. *J. Phys. Chem. B* **2011**, *115*, 14659–14667. [CrossRef]
53. Grabowski, S.J. What Is the Covalency of Hydrogen Bonding? *Chem. Rev.* **2011**, *111*, 2597–2625. [CrossRef] [PubMed]
54. Ilnicka, A.; Sadlej, J. Inverse hydrogen bond: Theoretical investigation on the nature of interaction and spectroscopic properties. *Struct. Chem.* **2012**, *23*, 1323–1332. [CrossRef]
55. Smaga, A.; Sadlej, J. Computational study on interaction energy changes during double proton transfer process. *Comput. Theor. Chem.* **2012**, *998*, 120–128. Non-covalent interactions and hydrogen bonding: Commonalities and differences. [CrossRef]
56. Jabłoński, M. Theoretical insight into the nature of the intermolecular charge-inverted hydrogen bond. *Comput. Theor. Chem.* **2012**, *998*, 39–45. [CrossRef]
57. Gallardo, A.; Fanfrlík, J.; Hobza, P.; Jelinek, P. Nature of Binding in Planar Halogen-Benzene Assemblies and Their Possible Visualization in Scanning Probe Microscopy. *J. Phys. Chem. C* **2019**, *123*, 8379–8386. [CrossRef]
58. Varadwaj, P.R.; Varadwaj, A.; Marques, H.M.; Yamashita, K. Significance of hydrogen bonding and other noncovalent interactions in determining octahedral tilting in the $CH_3NH_3PbI_3$ hybrid organic-inorganic halide perovskite solar cell semiconductor. *Sci. Rep.* **2019**, *9*, 50. [CrossRef]
59. Jabłoński, M. Ten years of charge-inverted hydrogen bonds. *Struct. Chem.* **2020**, *31*, 61–80. [CrossRef]
60. Korona, T.; Dodziuk, H. Small Molecules in C_{60} and C_{70}: Which Complexes Could Be Stabilized? *J. Chem. Theory Comput.* **2011**, *7*, 1476–1483. [CrossRef]
61. Pan, Y.C.; Hu, X.Y.; Guo, D.S. Biomedical Applications of Calixarenes: State of the Art and Perspectives. *Angew. Chem. Int. Ed.* **2021**, *60*, 2768–2794. [CrossRef]
62. Arena, G.; Contino, A.; Gulino, F.G.; Magrì, A.; Sansone, F.; Sciotto, D.; Ungaro, R. Complexation of native L-α-aminoacids by water soluble calix[4]arenes. *Tetrahedron Lett.* **1999**, *40*, 1597–1600. [CrossRef]
63. Douteau-Guével, N.; Coleman, A.W.; Morel, J.P.; Morel-Desrosiers, N. Complexation of basic amino acids by water-soluble calixarene sulphonates as a study of the possible mechanisms of recognition of calixarene sulphonates by proteins. *J. Phys. Org. Chem.* **1998**, *11*, 693–696. [CrossRef]
64. Giuliani, M.; Morbioli, I.; Sansone, F.; Casnati, A. Moulding calixarenes for biomacromolecule targeting. *Chem. Commun.* **2015**, *51*, 14140–14159. [CrossRef] [PubMed]
65. Antipin, I.S.; Stoikov, I.I.; Pinkhassik, E.M.; Fitseva, N.A.; Stibor, I.; Konovalov, A.I. Calix[4]arene based α-aminophosphonates: Novel carriers for zwitterionic amino acids transport. *Tetrahedron Lett.* **1997**, *38*, 5865–5868. [CrossRef]
66. Selkti, M.; Coleman, A.W.; Nicolis, I.; Douteau-Guével, N.; Villain, F.; Tomas, A.; de Rango, C. The first example of a substrate spanning the calix[4]arene bilayer: The solid state complex of sulfonatocalix[4]arene with lysine. *Chem. Commun.* **2000**, 161–162. [CrossRef]

67. Buschmann, H.J.; Mutihac, L.; Jansen, K. Complexation of some amine compounds by macrocyclic receptors. *J. Incl. Phenom. Macrocycl. Chem.* **2001**, *39*, 1–11. [CrossRef]
68. Atwood, J.L.; Ness, T.; Nichols, P.J.; Raston, C.L. Confinement of Amino Acids in Tetra-p-Sulfonated Calix[4]arene Bilayers. *Cryst. Growth Des.* **2002**, *2*, 171–176. [CrossRef]
69. Hassen, W.M.; Martelet, C.; Davis, F.; Higson, S.P.; Abdelghani, A.; Helali, S.; Jaffrezic-Renault, N. Calix[4]arene based molecules for amino-acid detection. *Sens. Actuators B Chem.* **2007**, *124*, 38–45. [CrossRef]
70. Mutihac, L.; Lee, J.H.; Kim, J.S.; Vicens, J. Recognition of amino acids by functionalized calixarenes. *Chem. Soc. Rev.* **2011**, *40*, 2777–2796. [CrossRef]
71. Español, E.S.; Villamil, M.M. Calixarenes: Generalities and Their Role in Improving the Solubility, Biocompatibility, Stability, Bioavailability, Detection, and Transport of Biomolecules. *Biomolecules* **2019**, *9*, 90. [CrossRef]
72. Parikh, J.; Bhatt, K.; Modi, K.; Patel, N.; Desai, A.; Kumar, S.; Mohan, B. A versatile enrichment of functionalized calixarene as a facile sensor for amino acids. *Luminescence* **2022**, *37*, 370–390. [CrossRef]
73. Stone, M.M.; Franz, A.H.; Lebrilla, C.B. Non-covalent calixarene–amino acid complexes formed by MALDI-MS. *J. Am. Soc. Mass Spectrom.* **2002**, *13*, 964–974. [CrossRef]
74. Oshima, T.; Inoue, K.; Furusaki, S.; Goto, M. Liquid membrane transport of amino acids by a calix[6]arene carboxylic acid derivative. *J. Membr. Sci.* **2003**, *217*, 87–97. [CrossRef]
75. Douteau-Guével, N.; Coleman, A.W.; Morel, J.P.; Morel-Desrosiers, N. Complexation of the basic amino acids lysine and arginine by three sulfonatocalix[n]arenes (n = 4, 6 and 8) in water: Microcalorimetric determination of the Gibbs energies, enthalpies and entropies of complexation. *J. Chem. Soc. Perkin Trans.* **1999**, *2*, 629–634. [CrossRef]
76. Oshima, T.; Goto, M.; Furusaki, S. Extraction Behavior of Amino Acids by Calix[6]arene Carboxylic Acid Derivatives. *J. Incl. Phenom. Macrocycl. Chem.* **2002**, *43*, 77–86. [CrossRef]
77. Martins, J.N.; Lima, J.C.; Basílio, N. Selective Recognition of Amino Acids and Peptides by Small Supramolecular Receptors. *Molecules* **2021**, *26*, 106. [CrossRef]
78. Groom, C.R.; Bruno, I.J.; Lightfoot, M.P.; Ward, S.C. The Cambridge Structural Database. *Acta Crystallogr. Sect. Struct. Sci. Cryst. Eng. Mater.* **2016**, *72*, 171–179. [CrossRef]
79. Dale, S.H.; Elsegood, M.R.; Redshaw, C. Polymorphism and Pseudopolymorphism in Calixarenes: Acetonitrile Clathrates of p-But-Calix[n]arenes (n= 6 and 8). *CrystEngComm* **2003**, *5*, 368–373. [CrossRef]
80. Wolfgong, W.J.; Talafuse, L.K.; Smith, J.M.; Adams, M.J.; Adeogba, F.; Valenzuela, M.; Rodriguez, E.; Contreras, K.; Carter, D.M.; Bacchus, A.; et al. The Influence of Solvent of Crystallization upon the Solid-State Conformation of Calix[6]Arenes. *Supramol. Chem.* **1996**, *7*, 67–78. [CrossRef]
81. Martins, F.T.; De Freitas Oliveira, B.G.; Sarotti, A.M.; De Fátima, Â. Winged-Cone Conformation in Hexa-p-tert-Butylcalix[6]Arene Driven by the Unusually Strong Guest Encapsulation. *ACS Omega* **2017**, *2*, 5315–5323. [CrossRef]
82. Rackers, J.A.; Wang, Z.; Lu, C.; Laury, M.L.; Lagardère, L.; Schnieders, M.J.; Piquemal, J.P.; Ren, P.; Ponder, J.W. Tinker 8: Software Tools for Molecular Design. *J. Chem. Theory Comput.* **2018**, *14*, 5273–5289. [CrossRef]
83. Grimme, S. Semiempirical GGA-Type Density Functional Constructed with a Long-Range Dispersion Correction. *J. Comput. Chem.* **2006**, *27*, 1787–1799. [CrossRef] [PubMed]
84. Eichkorn, K.; Treutler, O.; Öhm, H.; Häser, M.; Ahlrichs, R. Auxiliary Basis Sets to Approximate Coulomb Potentials. *Chem. Phys. Lett.* **1995**, *240*, 283–290. [CrossRef]
85. Weigend, F.; Ahlrichs, R. Balanced Basis Sets of Split Valence, Triple Zeta Valence and Quadruple Zeta Valence Quality for H to Rn: Design and Assessment of Accuracy. *Phys. Chem. Chem. Phys.* **2005**, *7*, 3297–3305. [CrossRef] [PubMed]
86. Trott, O.; Olson, A.J. AutoDock Vina: Improving the Speed and Accuracy of Docking with a New Scoring Function, Efficient Optimization, and Multithreading. *J. Comput. Chem.* **2010**, *31*, 455–461. [CrossRef]
87. Frisch, M.J.; Trucks, G.W.; Schlegel, H.B.; Scuseria, G.E.; Robb, M.A.; Cheeseman, J.R.; Scalmani, G.; Barone, V.; Petersson, G.A.; Nakatsuji, H.; et al. *Gaussian 16 Revision B.01*; Gaussian Inc.: Wallingford, CT, USA, 2016.
88. Garcia, J.; Podeszwa, R.; Szalewicz, K. SAPT codes for calculations of intermolecular interaction energies. *J. Chem. Phys.* **2020**, *152*, 184109. [CrossRef]
89. Parker, T.M.; Burns, L.A.; Parrish, R.M.; Ryno, A.G.; Sherrill, C.D. Levels of symmetry adapted perturbation theory (SAPT). I. Efficiency and performance for interaction energies. *J. Chem. Phys.* **2014**, *140*, 094106. [CrossRef]
90. Grimme, S. Improved second-order Møller–Plesset perturbation theory by separate scaling of parallel- and antiparallel-spin pair correlation energies. *J. Chem. Phys.* **2003**, *118*, 9095–9102. [CrossRef]
91. Kendall, R.A.; Dunning, T.H.; Harrison, R.J. Electron affinities of the first-row atoms revisited. Systematic basis sets and wave functions. *J. Chem. Phys.* **1992**, *96*, 6796–6806. [CrossRef]
92. Dunning, T.H. Gaussian basis sets for use in correlated molecular calculations. I. The atoms boron through neon and hydrogen. *J. Chem. Phys.* **1989**, *90*, 1007–1023. [CrossRef]
93. Papajak, E.; Zheng, J.; Xu, X.; Leverentz, H.R.; Truhlar, D.G. Perspectives on Basis Sets Beautiful: Seasonal Plantings of Diffuse Basis Functions. *J. Chem. Theory Comput.* **2011**, *7*, 3027–3034. [CrossRef]
94. Weigend, F.; Köhn, A.; Hättig, C. Efficient Use of the Correlation Consistent Basis Sets in Resolution of the Identity MP2 Calculations. *J. Chem. Phys.* **2002**, *116*, 3175–3183. [CrossRef]

95. Weigend, F. A fully direct RI-HF algorithm: Implementation, optimised auxiliary basis sets, demonstration of accuracy and efficiency. *Phys. Chem. Chem. Phys.* **2002**, *4*, 4285–4291. [CrossRef]
96. Available online: http://xxx.lanl.gov/abs/https://github.com/psi4/psi4/blob/master/psi4/share/psi4/basis/aug-cc-pvdz-jkfit.gbs (accessed on 22 December 2021).
97. Parrish, R.M.; Burns, L.A.; Smith, D.G.; Simmonett, A.C.; DePrince, A.E.; Hohenstein, E.G.; Bozkaya, U.; Sokolov, A.Y.; Di Remigio, R.; Richard, R.M.; et al. Psi4 1.1: An Open-Source Electronic Structure Program Emphasizing Automation, Advanced Libraries, and Interoperability. *J. Chem. Theory Comput.* **2017**, *13*, 3185–3197. [CrossRef] [PubMed]
98. Helgaker, T.; Klopper, W.; Koch, H.; Noga, J. Basis-Set Convergence of Correlated Calculations on Water. *J. Chem. Phys.* **1997**, *106*, 9639–9646. [CrossRef]
99. Boys, S.; Bernardi, F. The calculation of small molecular interactions by the differences of separate total energies. Some procedures with reduced errors. *Mol. Phys.* **1970**, *19*, 553–566. [CrossRef]
100. Collins, M.A. Systematic fragmentation of large molecules by annihilation. *Phys. Chem. Chem. Phys.* **2012**, *14*, 7744–7751. [CrossRef]
101. Collins, M.A.; Deev, V.A. Accuracy and efficiency of electronic energies from systematic molecular fragmentation. *J. Chem. Phys.* **2006**, *125*, 104104. [CrossRef]
102. Addicoat, M.A.; Collins, M.A. Accurate treatment of nonbonded interactions within systematic molecular fragmentation. *J. Chem. Phys.* **2009**, *131*, 104103. [CrossRef]
103. Werner, H.J.; Knowles, P.J.; Knizia, G.; Manby, F.R.; Schütz, M.; Celani, P.; Györffy, W.; Kats, D.; Korona, T.; Lindh, R.; et al. MOLPRO, version 2020.1, a Package of ab Initio Programs. Available online: https://www.molpro.net (accessed on 15 September 2022).
104. Korona, T.; Pflüger, K.; Werner, H.J. The effect of local approximations in coupled-cluster wave functions on dipole moments and static dipole polarisabilities. *Phys. Chem. Chem. Phys.* **2004**, *6*, 2059–2065. [CrossRef]
105. Parrish, R.M.; Gonthier, J.F.; Corminbœuf, C.; Sherrill, C.D. Communication: Practical intramolecular symmetry adapted perturbation theory via Hartree-Fock embedding. *J. Chem. Phys.* **2015**, *143*, 051103. [CrossRef]
106. Chojecki, M.; Rutkowska-Zbik, D.; Korona, T. On the applicability of functional-group symmetry-adapted perturbation theory and other partitioning models for chiral recognition - the case of popular drug molecules interacting with chiral phases. *Phys. Chem. Chem. Phys.* **2019**, *21*, 22491–22510. [CrossRef] [PubMed]
107. Sirianni, D.A.; Zhu, X.; Sitkoff, D.F.; Cheney, D.L.; Sherrill, C.D. The influence of a solvent environment on direct non-covalent interactions between two molecules: A symmetry-adapted perturbation theory study of polarization tuning of π-π interactions by water. *J. Chem. Phys.* **2022**, *156*, 194306. [CrossRef] [PubMed]
108. Masumian, E.; Daniel Boese, A. Intramolecular resonance-assisted hydrogen bonds: Insights from symmetry adapted perturbation theory. *Chem. Phys.* **2022**, *557*, 111474. [CrossRef]
109. Muchowska, K.B.; Pascoe, D.J.; Borsley, S.; Smolyar, I.V.; Mati, I.K.; Adam, C.; Nichol, G.S.; Ling, K.B.; Cockroft, S.L. Reconciling Electrostatic and n → π^* Orbital Contributions in Carbonyl Interactions. *Angew. Chem. Int. Ed.* **2020**, *59*, 14602–14608. [CrossRef]
110. Li, C.; Liu, X.; Han, Y.; Guo, Q.; Yang, W.; Liu, Q.; Song, B.; Zheng, X.; Tao, S. Ultra-stable anti-counterfeiting materials inspired by water stains. *Cell Rep. Phys. Sci.* **2021**, *2*, 100571. [CrossRef]
111. Cukras, J.; Sadlej, J. Towards quantum-chemical modeling of the activity of anesthetic compounds. *Int. J. Mol. Sci.* **2021**, *22*, 9272. [CrossRef]
112. Smith, D.G.A.; Burns, L.A.; Simmonett, A.C.; Parrish, R.M.; Schieber, M.C.; Galvelis, R.; Kraus, P.; Kruse, H.; Di Remigio, R.; Alenaizan, A.; et al. PSI4 1.4: Open-source software for high-throughput quantum chemistry. *J. Chem. Phys.* **2020**, *152*, 184108. [CrossRef]
113. Petković, M. O-H Stretch in Phenol and Its Hydrogen-Bonded Complexes: Band Position and Relaxation Pathways. *J. Phys. Chem. A* **2012**, *116*, 364–371. [CrossRef]
114. Kalescky, R.; Zou, W.; Kraka, E.; Cremer, D. Local vibrational modes of the water dimer-Comparison of theory and experiment. *Chem. Phys. Lett.* **2012**, *554*, 243–247. [CrossRef]
115. Malenov, D.P.; Janjić, G.V.; Veljković, D.Z.; Zarić, S.D. Mutual influence of parallel, CH/O, OH/π and lone pair/π interactions in water/benzene/water system. *Comput. Theor. Chem.* **2013**, *1018*, 59–65. [CrossRef]
116. Podeszwa, R.; Bukowski, R.; Szalewicz, K. Potential Energy Surface for the Benzene Dimer and Perturbational Analysis of π-π Interactions. *J. Phys. Chem. A* **2006**, *110*, 10345–10354. [CrossRef] [PubMed]
117. Fink, R.F. Why does MP2 work? *J. Chem. Phys.* **2016**, *145*, 184101. [CrossRef] [PubMed]
118. Szabados, A. Theoretical interpretation of Grimme's spin-component-scaled second order Møller-Plesset theory. *J. Chem. Phys.* **2006**, *125*, 214105. [CrossRef] [PubMed]
119. Nekoei, A.R.; Vatanparast, M. π−Hydrogen bonding and aromaticity: A systematic interplay study. *Phys. Chem. Chem. Phys.* **2019**, *21*, 623–630. [CrossRef] [PubMed]
120. Bandyopadhyay, I.; Lee, H.M.; Kim, K.S. Phenol vs Water Molecule Interacting with Various Molecules: σ-type, π-type, and χ-type Hydrogen Bonds, Interaction Energies, and Their Energy Components. *J. Phys. Chem. A* **2005**, *109*, 1720–1728. [CrossRef] [PubMed]
121. Arunan, E.; Desiraju, G.R.; Klein, R.A.; Sadlej, J.; Scheiner, S.; Alkorta, I.; Clary, D.C.; Crabtree, R.H.; Dannenberg, J.J.; Hobza, P.; et al. Defining the hydrogen bond: An account (IUPAC Technical Report). *Pure Appl. Chem.* **2011**, *83*, 1619–1636. [CrossRef]

122. Forbes, C.R.; Sinha, S.K.; Ganguly, H.K.; Bai, S.; Yap, G.P.A.; Patel, S.; Zondlo, N.J. Insights into Thiol-Aromatic Interactions: A Stereoelectronic Basis for S-H/π Interactions. *J. Am. Chem. Soc.* **2017**, *139*, 1842–1855. [CrossRef]
123. Tafipolsky, M. Challenging dogmas: Hydrogen bond revisited. *J. Phys. Chem. A* **2016**, *120*, 4550–4559. [CrossRef]

Article

Exploring the Dynamical Nature of Intermolecular Hydrogen Bonds in Benzamide, Quinoline and Benzoic Acid Derivatives

Kamil Wojtkowiak and Aneta Jezierska *

Faculty of Chemistry, University of Wrocław, ul. F. Joliot-Curie 14, 50-383 Wrocław, Poland
* Correspondence: aneta.jezierska@chem.uni.wroc.pl

Abstract: The hydrogen bonds properties of 2,6-difluorobenzamide, 5-hydroxyquinoline and 4-hydroxybenzoic acid were investigated by Car–Parrinello and path integral molecular dynamics (CPMD and PIMD), respectively. The computations were carried out in vacuo and in the crystalline phase. The studied complexes possess diverse networks of intermolecular hydrogen bonds (N-H...O, O-H...N and O-H...O). The time evolution of hydrogen bridges gave a deeper insight into bonds dynamics, showing that bridged protons are mostly localized on the donor side; however, the proton transfer phenomenon was registered as well. The vibrational features associated with O-H and N-H stretching were analyzed on the basis of the Fourier transform of the atomic velocity autocorrelation function. The spectroscopic effects of hydrogen bond formation were studied. The PIMD revealed quantum effects influencing the hydrogen bridges providing more accurate free energy sampling. It was found that the N...O or O...O interatomic distances decreased (reducing the length of the hydrogen bridge), while the O-H or N-H covalent bond was elongated, which led to the increase in the proton sharing. Furthermore, Quantum Theory of Atoms in Molecules (QTAIM) was used to give insight into electronic structure parameters. Finally, Symmetry-Adapted Perturbation Theory (SAPT) was employed to estimate the energy contributions to the interaction energy of the selected dimers.

Keywords: hydrogen bond; non-covalent interactions; spectroscopic signatures; CPMD; PIMD; QTAIM; SAPT

1. Introduction

Hydrogen bond (HB) can be considered to be the most important type of non-covalent interaction [1]. It is important to emphasize that hydrogen bonds are of great importance for the properties of water, the binding of drugs to receptors or the stability of macromolecules [2,3]. Furthermore, HBs are some of the most important factors that affect the packing in crystals [4]. A hydrogen bond is usually defined as X-H...Y, the interaction of a bridging hydrogen attached to an electronegative donor atom (denoted as X) with another electron-rich species (denoted as Y) [5]. A common feature of most types of hydrogen bonds is the elongation of the X-H covalent bond with the co-existing redshift in the X-H stretching vibrations and a decrease in H...Y distance. However, this classical definition does not encompass the whole diversity of hydrogen bonds. As shown, HBs can also be formed with carbon as a hydrogen donor or acceptor [1]. Theoretical studies of C-H proton donors with benzene and ethylene oxide species as acceptors have shown that redshift is not a definitive characteristic of all HBs—in the cited studies, the authors have demonstrated the existence of so-called anti-hydrogen or blue-shifting hydrogen bonds, in which hydrogen bond formation is accompanied by the C-H covalent bond contraction and an increase in its stretch (blue shift) [6,7]. A closely related group of non-covalent interactions, which are based on the same physical principles, is the so-called σ-hole bond family [8–13]. The σ-hole concept is related to the depletion of electron density on the bridge atom at the extension of its covalent bond, X-A (A denotes the bridge atom and X is any electronegative species).

Due to a local increase in molecular electrostatic potential (MEP), an atom can participate in a highly directional interaction with Lewis bases [14,15]. These electrostatically driven interactions are named after the family from which the bridge atom is derived [16]. Among the most well known are halogen, chalcogen and pnicogen bonds [17–23]. In general, the strength of these interactions increases with the increasing electronegativity, polarizability and basicity of the donor atom, bridge atom and Lewis base, respectively [24,25]. Redshift is usually a reliable measure of their strength, however, the change in their X-A stretching frequency is less correlated with the interaction strength compared to HBs [18,26]. Moreover, the heavier and more polarizable the bridge atom is, the smaller the co-occurring redshift [25].

Theoretical methods such as the quantum theory of atoms in molecules (QTAIM) [27], Reduced Density Gradient (RDG) [28], Electron Localization Function (ELF) [29] or various perturbational or variational energy-decomposition schemes [30,31] have proven to be invaluable in the detailed characterization of HBs and other secondary bonds [24,32]. From a theoretical perspective, there is also a way to characterize hydrogen bonding and it is usually performed using the QTAIM conceptual apparatus—the indicators of the presence of the HB in the examined system are: (i) the bond path between the proton donor and acceptor atoms with the Bond Critical Point (BCP) located on it; and (ii) relatively small electron density at the BCP between atoms (about an order to even two orders smaller than ρ_{BCP} for typical covalent bonds) with the Laplacian values close to zero. Considering the energy-decomposition of the hydrogen bonds it is usually described as mainly covalent in nature, especially in the case of strong HBs with an interaction energy above 24 kcal * mol^{-1} [33,34]. In this context, the term "covalent" is associated with a shortening of the H...Y distance and an increase in the importance of the induction and dispersion terms (in terms of the nomenclature used in the SAPT energy-partitioning scheme [30]). On the contrary, weak hydrogen bonds are described as interactions in which the electrostatics play the most important role [35].

However, it should not be forgotten that biological or chemical systems are inherently dynamic, hence the above-mentioned, so-called static approaches which, despite being very useful, can only provide information about one particular arrangement of atoms in the complex. Thus, Car–Parrinello molecular dynamics (CPMD) [36] and path integral molecular dynamics (PIMD) [37] are among the most often employed dynamical approaches to study H/D isotope effects [38], non-covalent interactions and hydrogen bridges in particular [39–41]. Furthermore, the quantization of nuclei using PIMD allows one to take into account the quantum nature of the examined system—the hydrogen atom is particularly sensitive to quantum effects, even at a standard temperature due to its small mass, and thus has a relatively large value of thermal de Broglie wavelength compared to other atoms. Noteworthy is the fact that the quantum-tunneling phenomena (which can be efficiently studied using path integration techniques [42,43]) are important in biochemistry and are associated with enzyme-mediated electron or hydrogen transfer [44,45]. Nowadays, attempts are being made to study the quantum effects at the ligand–receptor binding site. An example of this can be found in the work where binding affinities for histamine receptor ligands were studied [46]. Another important aspect is associated with the spectroscopic features' investigation on the basis of the CPMD method. The vibrational properties could be studied using standard approaches, that is, the Fourier transformation of the autocorrelation function of atomic velocity or dipole moment. However, it is also possible to apply a method, which enabled the a posteriori inclusion of quantum effects to the O-H, N-H stretching etc. The method was successfully used in studies where strongly anharmonic systems were investigated [47–50].

Due to the aforementioned reasons, we decided to take a hybrid approach in order to characterize in detail the HBs present in the studied compounds—the dynamical features as well as detailed static characteristics were taken into account. Interactions for dimers extracted from the crystal were quantified and assessed using the SAPT and QTAIM approaches. The metric parameters and spectroscopic signatures of the investigated compounds were obtained and thoroughly analyzed. Obtaining quantum statistics for the nuclear degrees of freedom via PIMD application allowed to estimate the importance of quantum effects in the description of intermolecular HBs, when compared to the classical-quantum CPMD approach.

In the current study, we investigated three aromatic compounds from the benzamide, quinoline and benzoic acid groups. The choice was dictated by the network of hydrogen bonds present in the crystal structures [51]. Benzamide is a derivative of benzoic acid. Some substituted benzamides are well-known commercial drugs, e.g., procainamide, imatinib and veralipride [52–54]. Benzamides are still an attractive group of compounds, especially in drug design, where very often their derivatives are taken into consideration as compounds that show a specific type of biological activity. Therefore, they are studied both experimentally and by molecular modeling methods, e.g., [55–59]. We chose 2,6-difluorobenzamide, which is a metabolite of pesticide diflubenzuron [60], to theoretically study its hydrogen bonding network. It was found that benzamide as well as 2,6-difluorobenzamide can form mutual intermolecular hydrogen bonds. However, the extended amide...amide dimer synthon in benzamide can form a network of HBs via NH_2 group. Concerning the 2,6-difluorobenzamide, the presence of fluorine atoms allows the formation of other intermolecular interactions [51]. The next compound chosen for our theoretical investigations is 5-hydroxyquinoline, where the O-H...N intermolecular hydrogen bond is present as the strongest intermolecular interaction, however, C-H...O interaction was noted in the crystalline phase [35]. Taking into account the fact that quinoline and its derivatives have diverse applications, e.g., in medicine as drugs, compounds exhibiting various biological activity, as dyes, and as solvents [61–67], it is of interest to investigate the properties of hydrogen bonds in the class of compounds. The last compound taken into account was 4-hydroxybenzoic acid, which is the simplest aromatic carboxylic acid. Benzoic acid occurs naturally in many plants [68]. Its salts are used as food and cosmetics preservatives [69]. Generally speaking, it is an important precursor for the industrial synthesis of many other organic substances [70,71]. 4-hydroxybenzoic acid is primarily known as the basis for the preparation of its esters (parabens), which are used as preservatives in, e.g., cosmetics [72,73]. In our case, it was interesting to explore the dynamical nature of intermolecular hydrogen bonds, wherein the carboxylic as well as hydroxy groups were involved [51].

We hope that our research contributes to the knowledge of hydrogen bond dynamics, and will help with the rational design of new derivatives with specific properties. Therefore, the main objective of this research is to perform multi-factor studies of non-covalent interactions in the examined compounds. In order to reproduce the dynamical nature of a hydrogen bonds network, we employed the Car–Parrinello molecular dynamics [36]. To be able to make comparisons the time evolution simulations were performed in vacuo and in the crystalline phase. The nuclear quantum effects (NQEs) was taken into account and path integral molecular dynamics (PIMD) simulations were performed for this purpose [37,74]. The quantum theory of atoms in molecules (QTAIM) [27] was applied for electron density topological studies enabling the estimation of the interaction strength and the detection of weaker interactions. An application of symmetry-adapted perturbation theory (SAPT) [30] method allowed the energy decomposition in the studied dimers.

2. Computational Methodology

2.1. Car–Parrinello Molecular Dynamics (CPMD)

The CPMD [36] computations were performed in the crystalline phase for three crystals taken from the Cambridge Crystallographic Data Centre (CCDC) [75]. Their CCDC codes are as follows: (A)—919101, (B)—908102 and (C)—908103 [51]. The molecular structures of the studied complexes are presented in Figures 1 and 2. Two different sets of structures were chosen to thoroughly describe the intermolecular hydrogen bonds present in these systems. The simulations of the monomers of the (A), (B) and (C) were conducted in the gas phase with the box edges set to: a = b = c = 15 Å. The Perdew–Burke–Ernzerhof functional (PBE) [76] and the norm-conserving Troullier-Martins pseudopotentials [77] were applied. The plane-wave kinetic energy cutoff was set to 100 Ry. The models for crystalline-phase simulations were constructed based on experimental data—the details of which are presented in Table 1. The periodic boundary conditions (PBCs) with real-space electrostatic summations for the eight nearest neighbors in each direction were employed during the crystalline phase computations. Additionally, simulations were performed for dimers in the gas phase as well. The sizes of the corresponding simulation boxes were benchmarked and adjusted to calculate the Hartree potential using the Hockney solver of the Poisson equation. The edges of the corresponding boxes for the dimers were equal to: a = 18 Å b = 21 Å c = 18 Å for the (D) dimer and a = 22 Å b = 18 Å c = 18 Å for the (E) dimer.

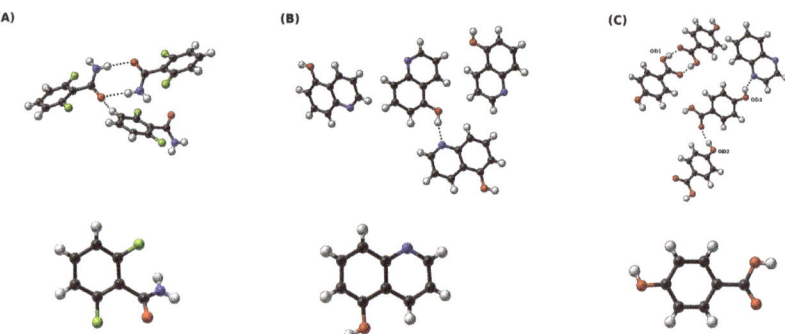

Figure 1. Crystalline (first row) and gaseous phases (second row) of (**A**) 2,6-difluorobenzamide, (**B**) 5-hydroxyquinoline, and (**C**) co-crystal of 4-hydroxybenzoic acid and quinoxaline. For clarity, the oxygen donor atoms for (**C**) are denoted as OD1, OD2 and OD3—this adopted nomenclature is used throughout the study. Dotted line indicates the intermolecular hydrogen bond. Color coding: white—hydrogen, grey—carbon, red—oxygen, blue—nitrogen, green—fluorine.

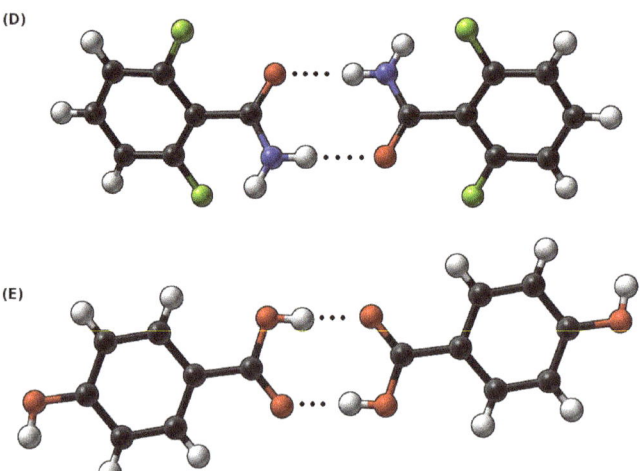

Figure 2. Dimers taken from the crystalline phase of (**D**) 2,6-difluorobenzamide and (**E**) 4-hydroxybenzoic acid to study intermolecular hydrogen bonds. Dotted line indicates the intermolecular hydrogen bond.

Table 1. CCDC code and unit cell data for the investigated compounds.

Designation	CCDC Code	Unit Cell Data
(A)	919101 [51]	Monoclinic a = 5.139 Å, b = 12.118 Å, c = 11.792 Å β = 112.482°, Z = 4
(B)	908102 [51]	Orthorhombic a = 3.835 Å, b = 12.718 Å, c = 14.067 Å, Z = 4
(C)	908103 [51]	Triclinic a = 6.910 Å, b = 12.289 Å, c = 12.647 Å α = 112.713°, β = 93.424°, γ = 103.103°, Z = 2

Subsequently, Car–Parrinello molecular dynamics (CPMD) [36] simulations were performed with the CPMD 4.3 suite of programs [78]. During the computations, the time step was set to 3 a.u., while the fictitious electron mass (EMASS) parameter was equal to 400 a.u. in both phases. The temperature was controlled by the Nosé–Hoover thermostat [79,80] and it was set to 297 K. The obtained CPMD trajectories were divided into equilibration (first 10,000 steps were excluded from the data analysis) and the production run. The CPMD production runs of the crystalline models were collected for ca. 72 ps and 42 ps for crystalline and gas-phase systems, respectively (monomers dynamics in the gas phase was simulated only to obtain the spectroscopic signatures of the O-H, C-H and N-H functional groups). Molecular dynamics corresponding to the models chosen to study dimer interactions were propagated for ca. 45 ps. The post-processing included the analysis of metric and spectroscopic properties. The vibrational features were studied using the Fourier transform of the atomic velocity autocorrelation function. This type of analysis enabled the decomposition of the computed IR spectra. The O-H and N-H stretching was obtained to give a deeper insight into spectroscopic signatures in the investigated hydrogen bridges. The metric parameters analysis was performed with the assistance of the VMD 1.9.3 program [81], while the Fourier transform power spectra of the atomic velocity were computed using home-made scripts. The graphs were obtained with the Gnuplot [82] program. The experimental unit cells were visualized and analyzed using the Mercury [83] program. The visualizations presented in Figures 1 and 2 were prepared in the SAMSON suite of programs [84].

2.2. Path Integral Molecular Dynamics (PIMD)

The quantum nature of the nuclear motion in the crystals was investigated using the path integral molecular dynamics (PIMD) approach [37,74]. The simulations were carried out in the gas and crystalline phases using the models prepared for the CPMD runs. The electronic structure setup was the same as described in the previous subsection. The computations were performed at 297 K temperature controlled by Nosé–Hoover thermostat [79]. The staging representation of the path integral propagator was used [37] and eight Trotter replicas (P = 8) were applied for imaginary time path integration. The initial 5000 steps of the simulation time were excluded from the analysis and treated as an equilibration phase. The trajectories of length 21 and 11 ps relative to the first and the second group of the structures under study, respectively, were collected and taken as production runs (11 ps in the case of (C) in the solid state). The data analysis was carried out with the assistance of home-made scripts. The PIMD simulations were performed with the CPMD version 4.3 program [78]. The Gnuplot program [82] was applied for the histogram preparation.

2.3. Quantum Theory of Atoms in Molecules (QTAIM)

Quantum theory of atoms in molecules (QTAIM) was applied to X-ray structures as well as to structures optimized at the ωB97XD/def2-TZVP level of theory [85,86] using the Gaussian 16 Rev. C.0.1 suite of programs [87]. The wavefunctions for further electronic structure analysis were obtained with the same computational setup. This part of the investigations was only performed for the gas phase models. The QTAIM analysis was performed with the assistance of the MultiWFN 3.8 program [88,89]. The graphical presentation of the obtained results was prepared using the VMD 1.9.3 program [81].

2.4. Symmetry-Adapted Perturbation Theory (SAPT)

Symmetry-adapted perturbation theory (SAPT) [30] was applied to gain an insight into the interaction energy between two studied molecules forming a dimer. The approximation of four-index integrals was carried out using the density-fitting technique (RI and JKI) with aug-cc-pVDZ (aDZ) [90] as auxiliary basis sets. The examined dimers were separated into two monomers in order to fulfill the conditions needed to eliminate the basis set superposition error (BSSE) [91]. The interaction energies were obtained at the SAPT2+/aDZ [92] level of theory. All calculations were performed using the Psi4 1.3.2 suite of programs [93].

3. Results and Discussion

3.1. Spectroscopic and Metric Parameters Associated with Intermolecular Hydrogen Bonds: Gas Phase vs. Crystalline Phase in the Light of Car–Parrinello Molecular Dynamics (CPMD) and Path Integral Molecular Dynamics (PIMD)

The gas phase as well as the crystalline phase simulations were carried out for 2,6-difluorobenzamide, 5-hydroxyquinoline and 4-hydroxybenzoic acid to investigate the intermolecular hydrogen bonds. Car–Parrinello molecular dynamics and Path Integral molecular dynamics allowed for the quantitative and qualitative description of the spectroscopic and geometric features of the studied compounds. In Figure 1, the models used for the crystalline phase (based on crystal structures (A) 919101, (B) 908102 and (C) 908103 [51]) and gas phase molecular dynamics simulations are presented. In order to obtain a full spectroscopic description of the aforementioned models, all simulations were carried out in the gas and in the crystalline phases. It is worth emphasizing that the results obtained in vacuo served as a reference for the O-H and N-H stretching discussion. On the basis of the Fourier transform of the atomic velocity autocorrelation function, the classical vibrational spectrum was obtained (see Figure 3). A major advantage of the methodology used is the ability to estimate the individual contributions of selected atoms to the entire spectrum—these are shown in the third column of the Figure 3. Due to the characteristics of the CPMD method (nuclei dynamics is inherently classical), the Fermi resonance, and thus the splitting of the bands of nearly identical energies and symmetries as well as tunneling phenomena, cannot be observed. Moreover, the classical amplitudes of motion at 297 K allow for sampling a narrower part of the potential energy surface than the true quantum particle with its nuclear wavefunction delocalization. This leads in many cases to the underestimation of the anharmonicity and is another factor of deviation between the CPMD-derived X-H stretching position and the experimental spectrum. It is therefore a good idea to concentrate not on the absolute wavenumbers, but on the shifts between the hydrogen-bonded protons and free protons. In the case of the results shown in Figure 3 (see the first and the second panels), the presence of the two regions characterized by the increased intensities can be seen: the region of the deformation vibrations, from ca. 500 to 1800 cm^{-1} in the solid and in the gas phase, as well as the region of stretching vibrations that extend from 2800 to 3700 cm^{-1}. The former can be attributed to the heavy atom oscillations, whereas the latter is the signature region for the protons, including those involved in the hydrogen bond formation. The third panel corresponds to the particular protons involved in the HBs in the crystalline phase. In addition, the gas phase results presented in the third panel show the characteristic, sharp stretching modes of O-H, N-H and C-H not involved in the hydrogen bonding. For (A), one may observe the blueshift of the C-H stretching vibration (from ca. 3050 to 3150 cm^{-1}) and thus CD-H...OA can be regarded as an anti-hydrogen bond [6,94]. In the case of ND-H...OA, it is visible that the N-H band for the solid phase is shifted towards lower wavenumber values (redshift). This is evidence for the ND-H weakening and the accompanying contraction and strengthening of the ND-H...OA bond, which can be interpreted as the charge transfer from the proton acceptor to the antibonding orbital of the ND-H and as a sign of the hydrogen bond formation. In the case of (B) and (C), similar observations can be made—the O-H stretching is redshifted in the solid phase, when compared to the gas phase. In summary, the spectroscopic features obtained allow us to conclude that the hydrogen bonds were formed in each of the compounds studied in the solid state.

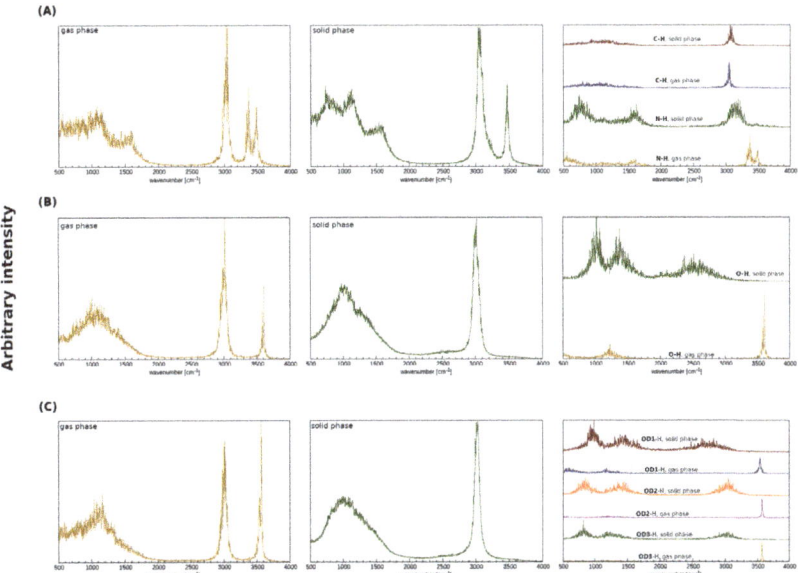

Figure 3. Atomic velocity power spectra obtained from the CPMD simulations. Left panel: the whole atomic spectra in the gas phase. Middle panel: the whole atomic spectra in the crystalline phase. Right panel: the contribution of the bridged protons in the crystalline and in the gaseous phases. (**A**) 2,6-difluorobenzamide, (**B**) 5-hydroxyquinoline and (**C**) 4-hydroxybenzoic acid.

The analysis of the Car–Parrinello (CPMD) and path integral molecular dynamics (PIMD) trajectories enables one to obtain a probability distribution of bridge proton positions in the studied compounds (see Figure 4). The characteristics of proton motion in all of hydrogen bridges highlighted in Figure 1 were obtained. Considering the impact of the NQEs on the proton behavior, one can observe that, in comparison with CPMD, the distance between the proton donor and its acceptor is insignificantly shortened for (A) (decrease of ca. 0.1 Å in each case). The same observation can be made for the ND-H and CD-H values—in these cases, the quantization of the nuclei does not change the proton dynamics. Indeed, both aforementioned hydrogen bonds cannot be regarded as strong hydrogen bonds—in the first case (A, ND), the proton donor is an amide group, with a behavior strongly altered by the presence of two fluorine substituents in -ortho positions. In the second case, the proton donor is the aromatic carbon atom and thus the HB can be considered weak (this is an example of the blue-shifting hydrogen bond, as we argued in the previous section). Different dynamical characteristics of the bridged proton were noticed for 5-hydroxyquinoline as well as the OD1 of the 4-hydroxybenzoic acid and quinoxaline co-crystal. For (B), one can observe that the proton mobility increased significantly, when the dynamics was performed in the PIMD scheme. On the contrary, the NQEs for the (C, OD1) provided further stabilization of the hydrogen atom at the donor side; in this case, the CPMD simulations indicated that the proton is more delocalized. Here, proton transfer, from the donor to the acceptor site in the hydrogen bond, has occurred, which can be easily discerned by the presence of the distinctive "tail" that extends towards larger r(OD-H) values. In the case of OD2-H...OA hydrogen bond (C, OD2), the only difference concerns the negligible shortening of OD2...OA distance in PIMD, when compared to the CPMD. For (C, OD3), one may observe that the NQEs induce minimal r(OD3-H) covalent bond elongation and the accompanying contraction of the r(OD3...NA) distance. Summarizing, it can be noted that the inclusion of the NQEs for the presented set of compounds results in changes in the quantitative as well as qualitative nature. The latter are especially pronounced for (B), where the free energy surface was sampled by the proton more efficiently and for (C, OD1) case, where the NQEs inclusion resulted in a more localized behaviour of the proton in the hydrogen bridge. In all studied cases, with the exception of the blue-shifting (A, CD) bond, a decrease in the donor–acceptor distance is accompanied by increased proton delocalization and sharing.

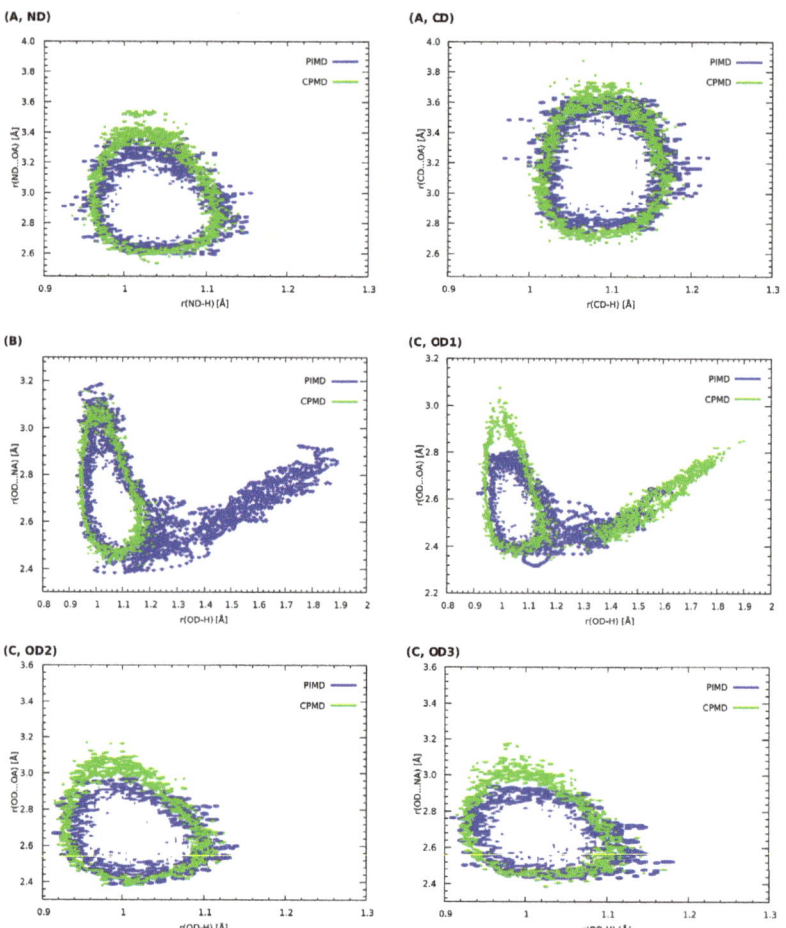

Figure 4. Histograms presenting the relationship between the length of the corresponding donor–proton covalent bonds as well as the distance between the proton and its acceptor. CPMD vs. PIMD in the crystalline phase. (**A**) 2,6-difluorobenzamide, (**B**) 5-hydroxyquinoline and (**C**) 4-hydroxybenzoic acid. ND, OD and CD indicate proton donors. Probability density isocontours drawn at the 1 Å^{-2} value.

The structures of the dimers discussed in this section were extracted from the 919,101 and 908,103 deposits in the CCDC database [51]. CPMD, as well as PIMD calculations, were performed to shed light on the dynamic features of the intermolecular hydrogen bridges present in these structures (see Figure 2). The time evolution of the metric parameters of the hydrogen bonds is presented in Figure 5. Let us start the discussion with the dimer (D). In this case, the amide nitrogen is a proton donor, whereas the oxygen from the amide group serves as the proton acceptor. The distances between the donor and the acceptor of the hydrogen atom varied between ca. 2.6 and 3.9 Å, whereas the H...OA distance changes were within 1.5–3.0 Å. Throughout the whole simulation time, the hydrogen atom is located at the proton donor side. In the case of (E), where the oxygen atoms belonging to the carboxylic groups are the proton donors and acceptors, the observations that can be made are strikingly different. Here, the bridged proton freely switches its donor from one to another; proton transfer actually happened 3 ps after the production run of the dynamics started. Moreover, the proton was on both the donor and acceptor sides for a similar amount of time during the MD run. This substantial difference in the behaviors of the hydrogen bridges between these two dimers can be attributed to the differences between the amide and carboxylic groups, the electronegativity of the donor atoms as well as the impact of fluorine substituents in the structure of (D). We will come back to this issue later, when the electronic structure of both dimers will be studied in a more detailed way via the static approaches. In particular, the apparent large difference in strength of the hydrogen bonds in (D) and (E), leading to such different dynamic characteristics, will be confirmed by the QTAIM descriptors.

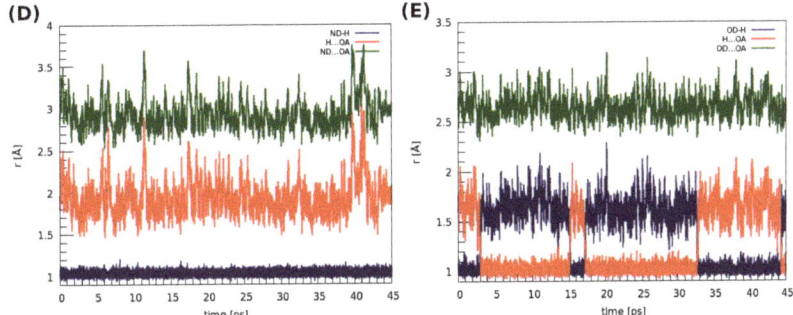

Figure 5. Time evolution of the metric parameters between the atoms involved in the hydrogen bond formation in the studied dimers. (**D**) 2,6-difluorobenzamide, (**E**) 4-hydroxybenzoic acid.

The estimation of the impact of the nuclear quantum effects (NQEs) on the proton was possible based on PIMD—the obtained results are presented in Figure 6. It can be seen that, in the case of dimer (D), using the quantum-classical isomorphism to impose the quantization on the nuclei does not significantly change the proton behavior. The PMF profile corresponding to the PIMD simulation resembles the classical harmonic one. The observations made for the dimer (E) show that the NQEs lower the barrier for the proton transfer (from ca. 3.1 kcal * mol^{-1} to 2.0 kcal * mol^{-1}) and predict two proton minima at the donor side, roughly at 1.55 Å and 1.70 Å of the H...OA distance. In the case of classical-quantum dynamics, we can observe two minima: one that occurs at ca. 1.05 Å, which is in accordance with the PIMD result, and the other one, which corresponds to 1.65 Å of the H...OA distance (proton at the donor side).

3.2. Electronic Structure Topological Analysis on the Basis of Quantum Theory of Atoms in Molecules (QTAIM)

Quantum theory of atoms in molecules (QTAIM) served as a method of choice to investigate the electronic properties of the studied molecules. The visualization and the results concerning the covalent and non-covalent interactions of the experimental and optimized structures of the studied dimers are presented in Table 2 and in Figure 7.

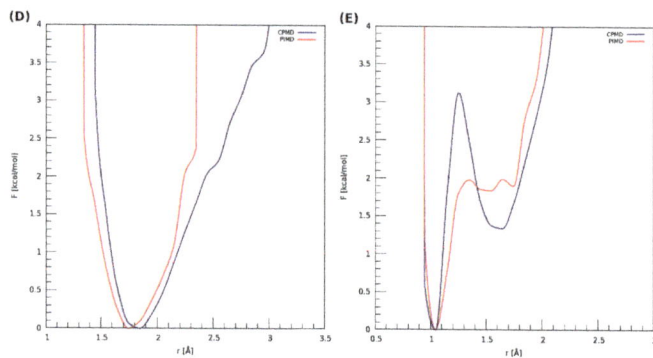

Figure 6. Potential of mean force (PMF) for the proton motion in the hydrogen bond of examined dimers with respect to the distance between the proton and its acceptor. (**D**) 2,6-difluorobenzamide and (**E**) 4-hydroxybenzoic acid.

Table 2. QTAIM-derived properties at BCPs for examined dimers from the structure deposited in the CCDC database and after geometry optimization at the ωB97XD/def2-TZVP level of theory. E1 is a bond energy based on the Espinosa model, given in kcal * mol^{-1}. Units of gathered quantities are as follows: electron density, ρ_{BCP}, is given in $e \cdot a_0^{-3}$ atomic units and the Laplacian of electron density, $\nabla^2 \rho_{BCP}$, is in $e \cdot a_0^{-5}$ units. V$_{BCP}$ stands for BCP potential energy density and H$_{BCP}$ denotes the energy density at the BCP.

System	BCP	ρ	V$_{BCP}$	$\nabla^2 \rho$	H$_{BCP}$	E1
		Experimental Structure				
(D1)	ND-H...OA	0.0183	−0.0124	0.0790	0.0037	3.9029
	F...N	0.0110	−0.0089	0.0601	0.0031	2.7829
	ND-H	0.4652	−1.0418	−3.5875	−0.9694	–
	C=OA	0.4046	−1.3240	−0.5213	−0.7272	–
(E1)	OD-H...OA	0.0384	−0.0381	0.1349	−0.0022	11.9557
	OD–H	0.4178	−0.9894	−3.3106	−0.9085	–
	C=OA	0.4069	−1.3567	−0.4383	−0.7331	–
		Optimized Structure				
(D2)	ND-H...OA	0.0302	−0.0254	0.1031	0.0002	7.9586
	ND-H	0.3229	−0.5649	−1.8740	−0.5167	–
	C=OA	0.4074	−1.3773	−0.4297	−0.7423	–
(E2)	OD-H...OA	0.0522	−0.0554	0.1237	−0.0122	17.3707
	OD–H	0.3092	−0.6476	−2.0757	−0.5833	–
	C=OA	0.4058	−1.3759	−0.4073	−0.7388	–

The use of the QTAIM method allowed a qualitative and quantitative description of the interactions between monomers. Bond energies were estimated on the basis of their linear dependence on the potential energy density at the BCP (V$_{BCP}$) via the Espinosa equation [95]. It can be observed that, in the case of both (D) and (E) dimers, the hydrogen bonds form quasi-rings, which provides structural stabilization. Their formation is indicated by the presence of the ring critical points (RCPs) (marked as small yellow spheres in Figure 7). For the dimer denoted as (D1), the properties at BCPs corresponding to two non-covalent interactions and two covalent bonds: ND-H...OA hydrogen bond and the intramolecular F...N contact as well as the ND-H and C=OA bonds were analyzed extensively. Both aforementioned non-covalent interactions can be considered weak, rather electrostatic in nature, since their estimated bond energies lie below 4 kcal * mol^{-1} and their Laplacian and energy density values are positive. Interestingly, when we inspect the data corresponding to the (D2) dimer, it can be seen that the relaxation of the structure causes the weakening of ND-H covalent bond, the intramolecular F...N interaction (which is not even detected by QTAIM) and the accompanying strengthening of the ND-H...OA hydrogen bond. In this case, the character of the ND-H...OA is noticeably more covalent, because the corresponding ρ and energy density values become larger and lower, respectively (H_{BCP} is indeed very close to zero) [34]. A similar analysis regarding the dimer (E) leads to the same conclusions: the optimization of the examined structures results in the strengthening of the hydrogen bond and the weakening of the corresponding covalent bond of the proton donor. More interesting is the characteristics of the hydrogen bond itself, namely OD-H...OA. In both the experimental and the optimized structures ((E1) and (E2)), the H_{BCP} values corresponding to the hydrogen bonds are negative and their bond energies are much larger compared to (D1) and (D2)—for this reason, one can say that they are more covalent in nature than their counterparts, ND-H...OA.

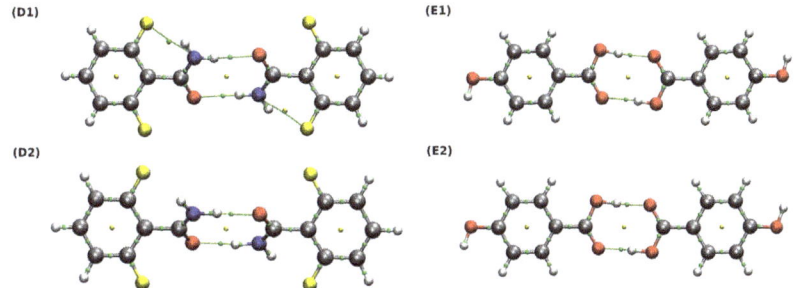

Figure 7. QTAIM molecular graphs of the studied dimers. Ball and stick model was used for visualization. Bond paths, BCPs and RCPs are presented as green lines and small green and yellow spheres, respectively. Color coding: white—hydrogen, grey—carbon, red—oxygen, blue—nitrogen, yellow—fluorine. (**D1,E1**) are the experimental structures, whereas (**D2,E2**) are the structures after the relaxation at the ωB97XD/def2-TZVP level of theory.

3.3. Decomposition of the Interaction Energy Using Symmetry-Adapted Perturbation Theory (SAPT)

Further insight into the nature of the interaction of the examined dimers was gained using the symmetry-adapted perturbation theory (SAPT) framework. SAPT is an invaluable method when it comes to the analysis of the interacting molecular fragments and their decomposition into physically grounded contributions. In the investigated dimers (see Figure 2), almost the whole interaction energy is constituted of two corresponding hydrogen bridges (results are presented in Table 3).

Table 3. Interaction energies between the examined dimers taken from the experimental structure [51] and after relaxation at the ωB97XD/def2TZVP level of theory. The calculations were performed at the SAPT2+/aug-cc-pVDZ level of theory, and the energies are given in kcal * mol^{-1}.

Complex	E_{elst}	E_{exch}	E_{ind}	E_{disp}	SAPT2+/aDZ
			Experimental Structure		
(D1)	−15.730	13.619	−4.656	−5.380	−12.146
(E1)	−27.448	33.520	−12.021	−8.788	−14.736
			Optimized Structures		
(D2)	−24.472	27.813	−10.886	−8.037	−15.582
(E2)	−35.921	46.756	−21.032	−10.774	−20.970

Let us start the discussion with the results obtained for the experimental structures of the analyzed dimers. Starting with (D1), it can be seen that the electrostatic term is almost counterbalanced by the exchange term—the sum of these contributions can be viewed as the electrostatic interaction of the symmetry-adapted reference state. Furthermore, for (D1), all of the contributions to the interaction energy are important; nonetheless, when we take into account their magnitudes, the dispersion turns out to be the most significant one. In the case of (E1), the sum of E_{elst} and E_{exch} is positive and thus repulsive. Due to this, the interaction energy between two monomers can be ascribed to the presence of the E_{ind} and E_{disp} terms.

The optimized structures ((D2) and (E2)) exhibit lower values of the total interaction energies. However, noteworthy is the fact that, for the optimized structures, the induction and the dispersion are, from a certain point of view, the most essential contributions to the interaction energies, respectively. It is evidently more visible for the (E2) structure and its corresponding HBs, which are more covalent in nature than their counterparts in (D2)—as such, the results obtained using the SAPT2+/aDZ level of theory are in accordance with the QTAIM method and agree with the electrostatic-covalent H-bond model proposed by Gilli [96]. For both experimental and optimized structures, the E_{ind} and E_{disp} terms play more significant roles for the dimer denoted as (E). Furthermore, both the Espinosa equation and the total energy obtained at the SAPT2+/aDZ level of the theory indicate that the OD-H...OA interaction strength is greater than its counterpart in (D), ND-H...OA. This difference is more pronounced in the results of the Espinosa equation, because it only covers the BCP of hydrogen bonding, while the SAPT method recognizes contributions from monomers, including the polarization of aromatic rings and other factors not directly related to hydrogen bonding.

4. Conclusions

The nature of intermolecular hydrogen bonds in exemplary compounds derived from benzamide, quinoline and benzoic acid groups was investigated. Quantum-chemical simulations were performed in vacuo and in the crystalline phase allowing a more in-depth analysis of the non-covalent interactions present in the chosen compounds for the current study: 2,6-difluorobenzamide, 5-hydroxyquinoline and 4-hydroxybenzoic acid. The computed power spectra of the atomic velocity reproduced the spectroscopic features of the investigated compounds indicating regions with O-H, N-H and C-H stretching. The comparison was made between the gas and crystalline phases, indicating the formation of the intermolecular hydrogen bonds. The metric parameter analysis based on the Car–Parrinello molecular dynamics results showed that the proton transfer phenomenon occurs for 4-hydroxybenzoic acid in the crystalline phase. Based on the results of the PIMD method, it was noted that the inclusion of quantum effects in the description of hydrogen bonds is important for strong interactions. Furthermore, theoretical investigations on the basis of static approaches (QTAIM and SAPT methods) revealed the strength of the non-covalent interactions as well as energy components. Hydrogen bonding energies estimated according to Espinosa's formula indicate that, in the studied dimers, the interaction is stronger in 4-hydroxybenzoic acid. Finally, the SAPT results provided a detailed look inside the energy components of the intermolecular interactions. It was shown that the dispersion and the induction contributions to the interaction energy are decisive factors in the intermolecular hydrogen bonds studied herein. The HBs strength is strongly correlated with their covalency.

Author Contributions: Conceptualization, K.W. and A.J.; methodology, K.W. and A.J.; validation, K.W.; formal analysis, K.W.; investigation, K.W. and A.J.; writing—original draft preparation, K.W.; writing—review and editing, K.W. and A.J.; visualization, K.W.; supervision, A.J.; project administration, K.W. and A.J. All authors have read and agreed to the published version of the manuscript.

Funding: This research received no external funding.

Institutional Review Board Statement: Not applicable.

Informed Consent Statement: Not applicable.

Data Availability Statement: Not applicable.

Acknowledgments: The authors thank Jarosław J. Panek for the scientific discussions. The authors gratefully acknowledge the Academic Computing Centre Cyfronet-Kraków (Prometheus supercomputer, part of the PL-Grid infrastructure) as well as the Poznań Supercomputing and Networking Center (PSNC) for generous grants of computer time and facilities. In addition, the ARCHER2 UK National Supercomputing Service (https://www.archer2.ac.uk) is acknowledged for generous CPU time and facilities in the framework of the DECI–17 access program.

Conflicts of Interest: The authors declare no conflict of interest.

Abbreviations

The following abbreviations are used in this manuscript:

BCP	Bond Critical Point
BSSE	Basis Set Superposition Error
CPMD	Car–Parrinello Molecular Dynamics
ELF	Electron Localization Function
HB	Hydrogen Bond
IR	Infrared Spectroscopy
MEP	Molecular Electrostatic Potential
NQE	Nuclear Quantum Effect
PBCs	Periodic Boundary Conditions
PIMD	Path Integral Molecular Dynamics
RCP	Ring Critical Point
RDG	Reduced Density Gradient
QTAIM	Quantum Theory of Atoms in Molecules
SAPT	Symmetry-Adapted Perturbation Theory

References

1. Hobza, P.; Havlas, Z. Blue-Shifting Hydrogen Bonds. *Chem. Rev.* **2000**, *100*, 4253–4264. [CrossRef] [PubMed]
2. Pauling, L.; Corey, R.B.; Branson, H.R. The structure of proteins: Two hydrogen-bonded helical configurations of the polypeptide chain. *Proc. Natl. Acad. Sci. USA* **1951**, *37*, 205–211. [CrossRef] [PubMed]
3. Pauling, L.; Corey, R.B. A Proposed Structure For The Nucleic Acids. *Proc. Natl. Acad. Sci. USA* **1953**, *39*, 84–97. [CrossRef] [PubMed]
4. Sobczyk, L.; Grabowski, S.J.; Krygowski, T.M. Interrelation between H-Bond and Pi-Electron Delocalization. *Chem. Rev.* **2005**, *105*, 3513–3560. [CrossRef] [PubMed]
5. Pauling, L. *The Nature of the Chemical Bond and the Structure of Molecules and Crystals; An Introduction to Modern Structural Chemistry*, 3rd ed.; Cornell University Press: Ithaca, NY, USA, 1960.
6. Hobza, P.; Špirko, V.; Selzle, H.L.; Schlag, E.W. Anti-Hydrogen Bond in the Benzene Dimer and Other Carbon Proton Donor Complexes. *J. Phys. Chem. A* **1998**, *102*, 2501–2504. [CrossRef]
7. Hobza, P.; Havlas, Z. The fluoroform···ethylene oxide complex exhibits a C–H···O anti-hydrogen bond. *Chem. Phys. Lett.* **1999**, *303*, 447–452. [CrossRef]
8. Murray, J.S.; Lane, P.; Politzer, P. A predicted new type of directional noncovalent interaction. *Int. J. Quantum Chem.* **2007**, *107*, 2286–2292. [CrossRef]
9. Murray, J.S.; Resnati, G.; Politzer, P. Close contacts and noncovalent interactions in crystals. *Faraday Discuss.* **2017**, *203*, 113–130. [CrossRef]
10. Politzer, P.; Riley, K.E.; Bulat, F.A.; Murray, J.S. Perspectives on halogen bonding and other σ-hole interactions: Lex parsimoniae (Occam's Razor). *Comput. Theor. Chem.* **2012**, *998*, 2–8. [CrossRef]
11. Politzer, P.; Murray, J. σ-Hole Interactions: Perspectives and Misconceptions. *Crystals* **2017**, *7*, 212. [CrossRef]
12. Politzer, P.; Murray, J.S.; Clark, T.; Resnati, G. The σ-hole revisited. *Phys. Chem. Chem. Phys.* **2017**, *19*, 32166–32178. [CrossRef] [PubMed]
13. Murray, J.S.; Lane, P.; Clark, T.; Riley, K.E.; Politzer, P. σ-Holes, π-holes and electrostatically-driven interactions. *J. Mol. Model.* **2011**, *18*, 541–548. [CrossRef] [PubMed]
14. Zierkiewicz, W.; Michalczyk, M.; Scheiner, S. Noncovalent Bonds through Sigma and Pi-Hole Located on the Same Molecule. Guiding Principles and Comparisons. *Molecules* **2021**, *26*, 1740. [CrossRef] [PubMed]
15. Scheiner, S. Principles Guiding the Square Bonding Motif Containing a Pair of Chalcogen Bonds between Chalcogenadiazoles. *J. Phys. Chem. A* **2022**, *126*, 1194–1203. [CrossRef] [PubMed]

16. Alkorta, I.; Elguero, J.; Frontera, A. Not Only Hydrogen Bonds: Other Noncovalent Interactions. *Crystals* **2020**, *10*, 180. [CrossRef]
17. Bauzá, A.; Frontera, A. Halogen and Chalcogen Bond Energies Evaluated Using Electron Density Properties. *ChemPhysChem* **2019**, *21*, 26–31. [CrossRef] [PubMed]
18. Lu, J.; Scheiner, S. Effects of Halogen, Chalcogen, Pnicogen, and Tetrel Bonds on IR and NMR Spectra. *Molecules* **2019**, *24*, 2822. [CrossRef]
19. Scheiner, S. Participation of S and Se in hydrogen and chalcogen bonds. *CrystEngComm* **2021**, *23*, 6821–6837. [CrossRef]
20. Auffinger, P.; Hays, F.A.; Westhof, E.; Ho, P.S. Halogen bonds in biological molecules. *Proc. Natl. Acad. Sci. USA* **2004**, *101*, 16789–16794. [CrossRef]
21. Scilabra, P.; Terraneo, G.; Resnati, G. The Chalcogen Bond in Crystalline Solids: A World Parallel to Halogen Bond. *Acc. Chem. Res.* **2019**, *52*, 1313–1324. [CrossRef]
22. Clark, T.; Hennemann, M.; Murray, J.S.; Politzer, P. Halogen bonding: the σ-hole. *J. Mol. Model.* **2006**, *13*, 291–296. [CrossRef] [PubMed]
23. Wang, W.; Ji, B.; Zhang, Y. Chalcogen Bond: A Sister Noncovalent Bond to Halogen Bond. *J. Phys. Chem. A* **2009**, *113*, 8132–8135. [CrossRef] [PubMed]
24. Kolář, M.H.; Hobza, P. Computer Modeling of Halogen Bonds and Other σ-Hole Interactions. *Chem. Rev.* **2016**, *116*, 5155–5187. [CrossRef] [PubMed]
25. Lu, J.; Scheiner, S. Relationships between Bond Strength and Spectroscopic Quantities in H-Bonds and Related Halogen, Chalcogen, and Pnicogen Bonds. *J. Phys. Chem. A* **2020**, *124*, 7716–7725. [CrossRef] [PubMed]
26. Michalczyk, M.; Zierkiewicz, W.; Wysokiński, R.; Scheiner, S. Theoretical Studies of IR and NMR Spectral Changes Induced by Sigma-Hole Hydrogen, Halogen, Chalcogen, Pnicogen, and Tetrel Bonds in a Model Protein Environment. *Molecules* **2019**, *24*, 3329. [CrossRef]
27. Bader, R. *Atoms in Molecules: A Quantum Theory*; International Series of Monographs on Chemistry; Clarendon Press: Oxford, UK, 1994.
28. Johnson, E.R.; Keinan, S.; Mori-Sánchez, P.; Contreras-García, J.; Cohen, A.J.; Yang, W. Revealing Noncovalent Interactions. *J. Am. Chem. Soc.* **2010**, *132*, 6498–6506. [CrossRef]
29. Savin, A.; Nesper, R.; Wengert, S.; Fässler, T.F. ELF: The Electron Localization Function. *Angew. Chem. Int. Ed.* **1997**, *36*, 1808–1832. [CrossRef]
30. Jeziorski, B.; Moszynski, R.; Szalewicz, K. Perturbation Theory Approach to Intermolecular Potential Energy Surfaces of van der Waals Complexes. *Chem. Rev.* **1994**, *94*, 1887–1930. [CrossRef]
31. Kitaura, K.; Morokuma, K. A new energy decomposition scheme for molecular interactions within the Hartree-Fock approximation. *Int. J. Quantum Chem.* **1976**, *10*, 325–340. [CrossRef]
32. Alcock, N. Secondary Bonding to Nonmetallic Elements. In *Advances in Inorganic Chemistry and Radiochemistry*; Elsevier: Amsterdam, The Netherlands, 1972; pp. 1–58. [CrossRef]
33. Rozas, I.; Alkorta, I.; Elguero, J. Behavior of Ylides Containing N, O, and C Atoms as Hydrogen Bond Acceptors. *J. Am. Chem. Soc.* **2000**, *122*, 11154–11161. [CrossRef]
34. Grabowski, S.J.; Sokalski, W.A.; Dyguda, E.; Leszczyński, J. Quantitative Classification of Covalent and Noncovalent H-Bonds. *J. Phys. Chem. B* **2006**, *110*, 6444–6446. [CrossRef]
35. Desiraju, G.R. Hydrogen Bridges in Crystal Engineering: Interactions without Borders. *Acc. Chem. Res.* **2002**, *35*, 565–573. [CrossRef] [PubMed]
36. Car, R.; Parrinello, M. Unified Approach for Molecular Dynamics and Density-Functional Theory. *Phys. Rev. Lett.* **1985**, *55*, 2471–2474. [CrossRef] [PubMed]
37. Tuckerman, M.E.; Marx, D.; Klein, M.L.; Parrinello, M. Efficient and general algorithms for path integral Car-Parrinello molecular dynamics. *J. Chem. Phys.* **1996**, *104*, 5579–5588. [CrossRef]
38. Brela, M.Z.; Wójcik, M.J.; Boczar, M.; Hashim, R. Car–Parrinello simulation of the vibrational spectra of strong hydrogen bonds with isotopic substitution effects: Application to oxalic acid dihydrate. *Chem. Phys. Lett.* **2013**, *558*, 88–92. [CrossRef]
39. Jezierska, A.; Błaziak, K.; Klahm, S.; Lüchow, A.; Panek, J.J. Non-Covalent Forces in Naphthazarin—Cooperativity or Competition in the Light of Theoretical Approaches. *Int. J. Mol. Sci.* **2021**, *22*, 8033. [CrossRef] [PubMed]
40. Tuckerman, M.E.; Marx, D. Heavy-Atom Skeleton Quantization and Proton Tunneling in "Intermediate-Barrier" Hydrogen Bonds. *Phys. Rev. Lett.* **2001**, *86*, 4946–4949. [CrossRef]
41. Tuckerman, M.E.; Marx, D.; Parrinello, M. The nature and transport mechanism of hydrated hydroxide ions in aqueous solution. *Nature* **2002**, *417*, 925–929. [CrossRef]
42. Warshel, A.; Bora, R.P. Perspective: Defining and quantifying the role of dynamics in enzyme catalysis. *J. Chem. Phys.* **2016**, *144*, 180901. [CrossRef] [PubMed]
43. Mavri, J.; Matute, R.A.; Chu, Z.T.; Vianello, R. Path Integral Simulation of the H/D Kinetic Isotope Effect in Monoamine Oxidase B Catalyzed Decomposition of Dopamine. *J. Phys. Chem. B* **2016**, *120*, 3488–3492. [CrossRef]
44. Sutcliffe, M.J.; Scrutton, N.S. Enzymology takes a quantum leap forward. *Philos. Trans. A Math. Phys. Eng. Sci.* **2000**, *358*, 367–386. [CrossRef]
45. Marcus, R.; Sutin, N. Electron transfers in chemistry and biology. *Biochim. Biophys. Acta Rev. Bioenerg.* **1985**, *811*, 265–322. [CrossRef]

46. Kržan, M.; Vianello, R.; Maršavelski, A.; Repič, M.; Zakšek, M.; Kotnik, K.; Fijan, E.; Mavri, J. The Quantum Nature of Drug-Receptor Interactions: Deuteration Changes Binding Affinities for Histamine Receptor Ligands. *PLoS ONE* **2016**, *11*, e0154002. [CrossRef] [PubMed]
47. Denisov, G.S.; Mavri, J.; Sobczyk, L. Potential energy shape for the proton motion in hydrogen bonds reflected in infrared and NMR spectra. In *Hydrogen Bonding—New Insights, (Challenges and Advances in Computational Chemistry and Physics, 3)*, 1st ed.; Grabowski, S.J., Ed.; Springer: Dordrecht, The Netherlands, 2006; pp. 377–416.
48. Jezierska, A.; Panek, J.J.; Koll, A.; Mavri, J. Car-Parrinello simulation of an O–H stretching envelope and potential of mean force of an intramolecular hydrogen bonded system: Application to a Mannich base in solid state and in vacuum. *J. Chem. Phys.* **2007**, *126*, 205101. [CrossRef]
49. Jezierska, A.; Panek, J.; Borštnik, U.; Mavri, J.; Janežič, D. Car-Parrinello Molecular Dynamics Study of Anharmonic Systems: A Mannich Base in Solution. *J. Phys. Chem. B* **2007**, *111*, 5243–5248. [CrossRef]
50. Jezierska, A.; Panek, J.J.; Koll, A. Spectroscopic Properties of a Strongly Anharmonic Mannich Base N-oxide. *ChemPhysChem* **2008**, *9*, 839–846. [CrossRef]
51. Mukherjee, A.; Tothadi, S.; Chakraborty, S.; Ganguly, S.; Desiraju, G.R. Synthon identification in co-crystals and polymorphs with IR spectroscopy. Primary amides as a case study. *CrystEngComm* **2013**, *15*, 4640. [CrossRef]
52. Osadchii, O.E. Procainamide and lidocaine produce dissimilar changes in ventricular repolarization and arrhythmogenicity in guinea-pig. *Fundam. Clin. Pharmacol.* **2013**, *28*, 382–393. [CrossRef] [PubMed]
53. Gambacorti-Passerini, C.; Antolini, L.; Mahon, F.X.; Guilhot, F.; Deininger, M.; Fava, C.; Nagler, A.; Casa, C.M.D.; Morra, E.; Abruzzese, E.; et al. Multicenter Independent Assessment of Outcomes in Chronic Myeloid Leukemia Patients Treated with Imatinib. *J. Natl. Cancer Inst.* **2011**, *103*, 553–561. [CrossRef]
54. Carranza-Lira, S. Actual status of veralipride use. *Clin. Interv. Aging* **2010**, *5*, 271–276. [CrossRef] [PubMed]
55. Yang, S.; Tian, X.Y.; Ma, T.Y.; Dai, L.; Ren, C.L.; Mei, J.C.; Liu, X.H.; Tan, C.X. Synthesis and Biological Activity of Benzamides Substituted with Pyridine-Linked 1, 2, 4-Oxadiazole. *Molecules* **2020**, *25*, 3500. [CrossRef] [PubMed]
56. Amanullah; Ali, U.; Ans, M.; Iqbal, J.; Iqbal, M.A.; Shoaib, M. Benchmark study of benzamide derivatives and four novel theoretically designed (L1, L2, L3, and L4) ligands and evaluation of their biological properties by DFT approaches. *J. Mol. Model.* **2019**, *25*, 223. [CrossRef] [PubMed]
57. Khan, G.S.; Pilkington, L.I.; Barker, D. Synthesis and biological activity of benzamide DNA minor groove binders. *Bioorganic Med. Chem. Lett.* **2016**, *26*, 804–808. [CrossRef] [PubMed]
58. Wang, W.; Wang, J.; Wu, F.; Zhou, H.; Xu, D.; Xu, G. Synthesis and Biological Activity of Novel Pyrazol-5-yl-benzamide Derivatives as Potential Succinate Dehydrogenase Inhibitors. *J. Agric. Food Chem.* **2021**, *69*, 5746–5754. [CrossRef]
59. Krasil'nikov, I.I.; Alferov, O.F.; Stepanov, A.V.; Tsikarishvili, G.V. Benzamide as a structural template for new drugs with diverse biological activity. *Pharm. Chem. J.* **1995**, *29*, 528–531. [CrossRef]
60. Rodriguez, E.; Gomez de Balugera, Z.; Goicolea, M.; Barrio, R. HPLC/diode-array method for the determination of the pesticide diflubenzuron and its major metabolites 2, 6-difluorobenzamide, 4-chlorophenylurea, and 4-chloroaniline in forestry matrices. *J. Liq. Chromatogr. Relat. Technol.* **1998**, *21*, 1857–1870. [CrossRef]
61. Shang, X.F.; Morris-Natschke, S.L.; Liu, Y.Q.; Guo, X.; Xu, X.S.; Goto, M.; Li, J.C.; Yang, G.Z.; Lee, K.H. Biologically active quinoline and quinazoline alkaloids part I. *Med. Res. Rev.* **2017**, *38*, 775–828. [CrossRef]
62. Shang, X.F.; Morris-Natschke, S.L.; Yang, G.Z.; Liu, Y.Q.; Guo, X.; Xu, X.S.; Goto, M.; Li, J.C.; Zhang, J.Y.; Lee, K.H. Biologically active quinoline and quinazoline alkaloids part II. *Med. Res. Rev.* **2018**, *38*, 1614–1660. [CrossRef]
63. Golden, E.B.; Cho, H.Y.; Hofman, F.M.; Louie, S.G.; Schönthal, A.H.; Chen, T.C. Quinoline-based antimalarial drugs: A novel class of autophagy inhibitors. *Neurosurg. Focus* **2015**, *38*, E12. [CrossRef]
64. Senerovic, L.; Opsenica, D.; Moric, I.; Aleksic, I.; Spasić, M.; Vasiljevic, B. Quinolines and Quinolones as Antibacterial, Antifungal, Anti-virulence, Antiviral and Anti-parasitic Agents. In *Advances in Microbiology, Infectious Diseases and Public Health: Volume 14*; Donelli, G., Ed.; Springer International Publishing: Cham, Switzerland, 2020; pp. 37–69. [CrossRef]
65. Tarnow, P.; Zordick, C.; Bottke, A.; Fischer, B.; Kühne, F.; Tralau, T.; Luch, A. Characterization of Quinoline Yellow Dyes as Transient Aryl Hydrocarbon Receptor Agonists. *Chem. Res. Toxicol.* **2020**, *33*, 742–750. [CrossRef]
66. Li, J.Y.; Chen, C.Y.; Ho, W.C.; Chen, S.H.; Wu, C.G. Unsymmetrical Squaraines Incorporating Quinoline for Near Infrared Responsive Dye-Sensitized Solar Cells. *Org. Lett.* **2012**, *14*, 5420–5423. [CrossRef] [PubMed]
67. Nasseri, M.A.; Zakerinasab, B.; Kamayestani, S. Proficient Procedure for Preparation of Quinoline Derivatives Catalyzed by NbCl$_5$ in Glycerol as Green Solvent. *J. Appl. Chem.* **2015**, *2015*, 743094. [CrossRef]
68. Widhalm, J.R.; Dudareva, N. A Familiar Ring to It: Biosynthesis of Plant Benzoic Acids. *Mol. Plant* **2015**, *8*, 83–97. [CrossRef] [PubMed]
69. Del Olmo, A.; Calzada, J.; Nuñez, M. Benzoic acid and its derivatives as naturally occurring compounds in foods and as additives: Uses, exposure, and controversy. *Crit. Rev. Food Sci. Nutr.* **2015**, *57*, 3084–3103. [CrossRef] [PubMed]
70. McCubbin, J.A.; Voth, S.; Krokhin, O.V. Mild and Tunable Benzoic Acid Catalysts for Rearrangement Reactions of Allylic Alcohols. *J. Org. Chem.* **2011**, *76*, 8537–8542. [CrossRef] [PubMed]
71. Ogata, Y.; Hojo, M.; Morikawa, M. Further Studies on the Preparation of Terephthalic Acid from Phthalic or Benzoic Acid. *J. Org. Chem.* **1960**, *25*, 2082–2087. [CrossRef]

72. Golden, R.; Gandy, J.; Vollmer, G. A Review of the Endocrine Activity of Parabens and Implications for Potential Risks to Human Health. *Crit. Rev. Toxicol.* **2005**, *35*, 435–458. [CrossRef]
73. Al-Halaseh, L.K.; Al-Adaileh, S.; Mbaideen, A.; Hajleh, M.N.A.; Al-Samydai, A.; Zakaraya, Z.Z.; Dayyih, W.A. Implication of parabens in cosmetics and cosmeceuticals: Advantages and limitations. *J. Cosmet. Dermatol.* **2022**, *21*, 3265–3271. [CrossRef]
74. Marx, D.; Parrinello, M. The Effect of Quantum and Thermal Fluctuations on the Structure of the Floppy Molecule $C_2H_3^+$. *Science* **1996**, *271*, 179–181. [CrossRef]
75. CCDC Structural Database. 2022. Available online: https://www.ccdc.cam.ac.uk/ (accessed on 9 August 2022).
76. Perdew, J.P.; Burke, K.; Ernzerhof, M. Generalized Gradient Approximation Made Simple [Phys. Rev. Lett. 77, 3865 (1996)]. *Phys. Rev. Lett.* **1997**, *78*, 1396. [CrossRef]
77. Troullier, N.; Martins, J.L. Efficient pseudopotentials for plane-wave calculations. *Phys. Rev. B* **1991**, *43*, 1993–2006. [CrossRef] [PubMed]
78. CPMD Version 4.3-4610, Copyright IBM Corp. (1990–2004) Copyright MPI für Festkoerperforschung Stuttgart (1997–2001). Available online: https://www.ibm.com/legal/copytrade (accessed on 1 October 2022).
79. Nosé, S. A unified formulation of the constant temperature molecular dynamics methods. *J. Chem. Phys.* **1984**, *81*, 511–519. [CrossRef]
80. Hoover, W.G. Canonical dynamics: Equilibrium phase-space distributions. *Phys. Rev. A* **1985**, *31*, 1695–1697. [CrossRef] [PubMed]
81. Humphrey, W.; Dalke, A.; Schulten, K. VMD–Visual Molecular Dynamics. *J. Mol. Graph.* **1996**, *14*, 33–38. [CrossRef]
82. Williams, T.; Kelley, C. Gnuplot 5.8.2: An interactive Plotting Program. 2019. Available online: http://www.gnuplot.info (accessed on 1 October 2022).
83. Mercury—Crystal Structure Visualisation. Available online: http://www.ccdc.cam.ac.uk/Solutions/CSDSystem/Pages/Mercury.aspx (accessed on 1 October 2022).
84. OneAngstrom. SAMSON 2022 R2. Available online: https://www.samson-connect.net/ (accessed on 1 October 2022).
85. Chai, J.D.; Head-Gordon, M. Long-range corrected hybrid density functionals with damped atom–atom dispersion corrections. *Phys. Chem. Chem. Phys.* **2008**, *10*, 6615–6620. [CrossRef]
86. Weigend, F.; Ahlrichs, R. Balanced basis sets of split valence, triple zeta valence and quadruple zeta valence quality for H to Rn: Design and assessment of accuracy. *Phys. Chem. Chem. Phys.* **2005**, *7*, 3297–3305. [CrossRef]
87. Frisch, M.J.; Trucks, G.W.; Schlegel, H.B.; Scuseria, G.E.; Robb, M.A.; Cheeseman, J.R.; Scalmani, G.; Barone, V.; Petersson, G.A.; Nakatsuji, H.; et al. *Gaussian~16 Revision C.01*; Gaussian Inc.: Wallingford, CT, USA, 2016.
88. Lu, T.; Chen, F. Multiwfn: A multifunctional wavefunction analyzer. *J. Comput. Chem.* **2012**, *33*, 580–592. [CrossRef]
89. Lu, T.; Chen, F. Quantitative analysis of molecular surface based on improved Marching Tetrahedra algorithm. *J. Molec. Graph. Model.* **2012**, *38*, 314–323. [CrossRef]
90. Dunning, T.H. Gaussian basis sets for use in correlated molecular calculations. I. The atoms boron through neon and hydrogen. *J. Chem. Phys.* **1989**, *90*, 1007–1023. [CrossRef]
91. Boys, S.; Bernardi, F. The calculation of small molecular interactions by the differences of separate total energies. Some procedures with reduced errors. *Mol. Phys.* **1970**, *19*, 553–566. [CrossRef]
92. Parker, T.M.; Burns, L.A.; Parrish, R.M.; Ryno, A.G.; Sherrill, C.D. Levels of symmetry adapted perturbation theory (SAPT). I. Efficiency and performance for interaction energies. *J. Chem. Phys.* **2014**, *140*, 094106. [CrossRef] [PubMed]
93. Smith, D.G.A.; Burns, L.A.; Simmonett, A.C.; Parrish, R.M.; Schieber, M.C.; Galvelis, R.; Kraus, P.; Kruse, H.; Di Remigio, R.; Alenaizan, A.; et al. PSI4 1.4: Open-source software for high-throughput quantum chemistry. *J. Chem. Phys.* **2020**, *152*, 184108. [CrossRef]
94. Starikov, E.B.; Steiner, T. Computational support for the suggested contribution of C—H...O=C interactions to the stability of nucleic acid base pairs. *Acta Crystallogr. D* **1997**, *53*, 345–347. [CrossRef] [PubMed]
95. Espinosa, E.; Molins, E.; Lecomte, C. Hydrogen bond strengths revealed by topological analyses of experimentally observed electron densities. *Chem. Phys. Lett.* **1998**, *285*, 170–173. [CrossRef]
96. Gilli, G.; Gilli, P. Towards an unified hydrogen-bond theory. *J. Mol. Struct.* **2000**, *552*, 1–15. [CrossRef]

Article

Hydrogen Bonding and Polymorphism of Amino Alcohol Salts with Quinaldinate: Structural Study

Nina Podjed and Barbara Modec *

Faculty of Chemistry and Chemical Technology, University of Ljubljana, Večna pot 113, 1000 Ljubljana, Slovenia; nina.podjed@fkkt.uni-lj.si
* Correspondence: barbara.modec@fkkt.uni-lj.si

Abstract: Three amino alcohols, 3-amino-1-propanol (abbreviated as 3a1pOH), 2-amino-1-butanol (2a1bOH), and 2-amino-2-methyl-1-propanol (2a2m1pOH), were reacted with quinoline-2-carboxylic acid, known as quinaldinic acid. This combination yielded three salts, (3a1pOHH)quin (**1**, 3a1pOHH$^+$ = protonated 3-amino-1-propanol, quin$^-$ = anion of quinaldinic acid), (2a1bOHH)quin (**2**, 2a1bOHH$^+$ = protonated 2-amino-1-butanol), and (2a2m1pOHH)quin (**3**, 2a2m1pOHH$^+$ = protonated 2-amino-2-methyl-1-propanol). The 2-amino-1-butanol and 2-amino-2-methyl-1-propanol systems produced two polymorphs each, labeled **2a/2b** and **3a/3b**, respectively. The compounds were characterized by X-ray structure analysis on single-crystal. The crystal structures of all consisted of protonated amino alcohols with NH$_3^+$ moiety and quinaldinate anions with carboxylate moiety. The used amino alcohols contained one OH and one NH$_2$ functional group, both prone to participate in hydrogen bonding. Therefore, similar connectivity patterns were expected. This proved to be true to some extent as all structures contained the NH$_3^+\cdots^-$OOC heterosynthon. Nevertheless, different hydrogen bonding and $\pi\cdots\pi$ stacking interactions were observed, leading to distinct connectivity motifs. The largest difference in hydrogen bonding occurred between polymorphs **3a** and **3b**, as they had only one heterosynton in common.

Keywords: hydrogen bond; synthon; crystal structure; polymorphism; amino alcohols; quinaldinic acid

Citation: Podjed, N.; Modec, B. Hydrogen Bonding and Polymorphism of Amino Alcohol Salts with Quinaldinate: Structural Study. *Molecules* **2022**, *27*, 996. https://doi.org/10.3390/molecules27030996

Academic Editor: Miroslaw Jablonski

Received: 21 December 2021
Accepted: 29 January 2022
Published: 1 February 2022

Publisher's Note: MDPI stays neutral with regard to jurisdictional claims in published maps and institutional affiliations.

Copyright: © 2022 by the authors. Licensee MDPI, Basel, Switzerland. This article is an open access article distributed under the terms and conditions of the Creative Commons Attribution (CC BY) license (https://creativecommons.org/licenses/by/4.0/).

1. Introduction

Crystal engineering, defined as preparation of new molecular solids with tailor-made properties by using intermolecular interactions [1], continues to draw the interest of a wide scientific community. A rational design of these solids is based on a thorough understanding of the supramolecular chemistry of functional groups, in particular those with a hydrogen bonding potential. Owing to their strength and directionality, hydrogen bonds are likely to dominate above all the other interactions. The extensive surveys of the Cambridge Structural Database (CSD) helped with the formulation of empirical guidelines concerning the design of molecular crystals [2]. A generally valid rule on hydrogen bonding states that all good proton donors and acceptors are normally engaged in interactions [3]. A new terminology has also emerged: a pair of complementary functional groups, linked via intermolecular interaction, such as a hydrogen bond, is known as a synthon [4]. A heterosynthon is composed of two different functional groups, whereas two identical groups make part of a homosynthon. A prominent example of a self-association motif is a well-known carboxylic acid dimer. Another rule concerns the synthon hierarchy: the heterosynthons are favored over the homosynthons. Recent reports agree that it is still impossible to predict the structure of the molecular solid [5,6]. In this context, a phenomenon of polymorphism is brought up. The term polymorphism describes the existence of the same compound in several crystal forms that differ in spatial arrangements of their components and some of their properties [7]. Polymorphs of the same compound generally differ in lattice energies by a few kJ/mol at most [8]. As claimed by McCrone [9], the number of forms known for a given compound is proportional to the time and money spent in research on that compound. A systematic study of crystal structures of a large number of molecular solids, fueled also by the pharmaceutical industry [10,11], has revealed that at least every other molecule exhibits polymorphism [12]. It has been shown that hydrogen bonding potential only slightly increases a likelihood for the molecule to be polymorphic, whereas chiral molecules are somewhat reluctant towards crystallization in more than one crystal form [13].

Herein, the solid-state structures of salts of three amino alcohols with quinaldinic acid are presented. The structural formulae of the acid and amino alcohols are depicted in Figure 1.

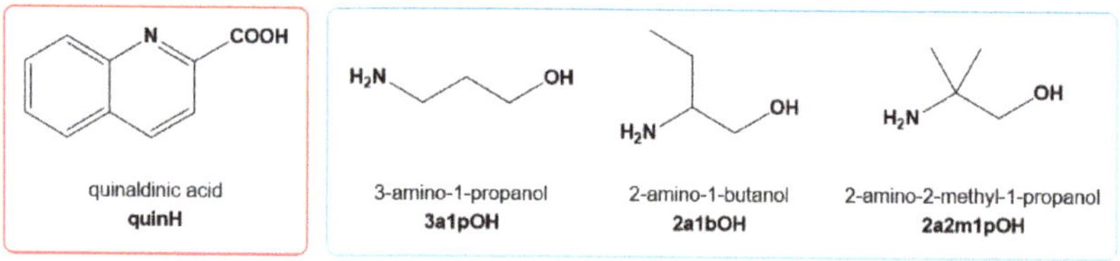

Figure 1. Structural formulae of quinaldinic acid, 3-amino-1-propanol, 2-amino-1-butanol, and 2-amino-2-methyl-1-propanol.

The salts contained protonated amino alcohols as cations and quinaldinate ions as counter-anions. Single crystals of all were obtained inadvertently as by-products of the $[Cu(quin)_2(H_2O)]$ reactions with the amino alcohol [14]. It has been observed previously that the amino alcohol OH group undergoes a spontaneous deprotonation in the presence of copper(II) complexes [15]. The resulting amino alcoholate ions coordinated to copper(II) in a chelating manner with the alkoxide oxygen serving as a bridge between two or among three metal ions. The amino alcoholate coordination probably assists in the deprotonation of amino alcohol. Some of our reaction systems provided a few more pieces of information concerning the formation of the amino alcoholate ions. The nature of the products, isolated from these reaction systems, strongly suggests a proton transfer from the OH group of the amino alcohol molecule to the NH_2 group of another molecule. In the reaction below, the $H_2N-(CH_2)_n-OH$ denotes amino alcohol in general.

$$2\ H_2N\text{-}(CH_2)_n\text{-}OH \leftrightarrow H_3N^+\text{-}(CH_2)_n\text{-}OH + H_2N\text{-}(CH_2)_n\text{-}O^-$$

The $H_2N-(CH_2)_n-O^-$ ions coordinated to copper(II), whereas the $H_3N^+-(CH_2)_n-OH$ ions crystallized as salts with quinaldinate. Later, a more straightforward synthesis of these salts was sought. A reaction of quinaldinic acid with the excess of amino alcohol in methanol with no copper(II) complex involved was met with success. Two of the salts were found to be polymorphic. A detailed account of the solid-state structures follows.

2. Results and Discussion

First, the common structural features of the title compounds are described. The crystal structures of all consist of NH_2-protonated amino alcohol molecules as counter-cations and quinaldinate anions with carboxylate moiety. In all, the C–O bond lengths of the carboxylate are the same within the experimental error. Interestingly, in some structures, the quinaldinate ions deviate from planarity. For convenience, we have described this deviation as a twist angle between the carboxylate plane and the quinoline plane. Depending upon the structure, the quinaldinates can stack one upon another. Geometric parameters of the $\pi\cdots\pi$ stacking interactions are conventionally given by the centroid$\cdots$centroid distance, dihedral angle, and shift distance [16]. Quinaldinate can participate in another interaction, a C–H$\cdots\pi$ interaction. All interactions involving π rings are given in Table 1. Both the cations and the anions possess groups that are hydrogen bond donors (NH_3^+ in protonated amino alcohol) or acceptors (carboxylate and quinaldinate nitrogen) or both (OH in protonated amino alcohol). With the first two being good hydrogen bond donors/acceptors, their participation in hydrogen bonding is likely to govern the connectivity patterns in solid state. A detailed list of hydrogen bonds is given in Table 2, whereas all possible heterosynthons and their actual occurrences in the structures of the title compounds are given in Table 3.

Table 1. $\pi\cdots\pi$ stacking and C–H$\cdots\pi$ interactions [Å, °] in title compounds.

1
$\pi\cdots\pi$ **Stacking Interactions**
Py$\cdots$Py [1−x, −y, 1−z], Cg$\cdots$Cg = 3.7402(15), dihedral angle = 0.02(11), shift distance = 1.552
Ph$\cdots$Py[1−x, −y, 1−z], Cg$\cdots$Cg = 3.6571(15), dihedral angle = 0.41(11), shift distance = 1.346
Ph$\cdots$Py[1−x, −1−y, 1−z], Cg$\cdots$Cg = 3.9681(16), dihedral angle = 0.41(11), shift distance = 1.656
Ph$\cdots$Ph[1−x, −1−y, 1−z], Cg$\cdots$Cg = 3.6899(16), dihedral angle = 0.00(11), shift distance = 0.821
2a
$\pi\cdots\pi$ **Stacking Interactions**
Py$\cdots$Py[−x, 1−y, 1−z], Cg$\cdots$Cg = 3.7080(10), dihedral angle = 0.02(6), shift distance = 1.635
Ph$\cdots$Py[−x, 1−y, 1−z], Cg$\cdots$Cg = 3.5163(9), dihedral angle = 0.39(7), shift distance = 1.123
C–H$\cdots\pi$ Interactions
C–H$\cdots$Ph[1−x, 1−y, 1−z], H$\cdots$Cg = 3.00, C–H$\cdots$Cg = 149, C$\cdots$Cg = 3.8620(17)
2b
C–H$\cdots\pi$ Interactions
C–H$\cdots$Ph[1+x, 1+y, z], H$\cdots$Cg = 2.79, C–H$\cdots$Cg = 131, C$\cdots$Cg = 3.495(3)
3a
C–H$\cdots\pi$ Interactions
C–H$\cdots$Py[2.5−x, 0.5+y, 0.5−z], H$\cdots$Cg = 2.95, C–H$\cdots$Cg = 150, C$\cdots$Cg = 3.7898(19)
C–H$\cdots$Ph[1.5−x, −0.5+y, 0.5−z], H$\cdots$Cg = 2.78, C–H$\cdots$Cg = 140, C$\cdots$Cg = 3.5438(18)
3b
$\pi\cdots\pi$ **Stacking Interactions**
Ph$\cdots$Py[−x, −y, 1−z], Cg$\cdots$Cg = 3.8194(7), dihedral angle = 2.51(6), shift distance = 1.462
C–H$\cdots\pi$ Interactions
C–H$\cdots$Py[1+x, 1+y, z], H$\cdots$Cg = 2.85, C–H$\cdots$Cg = 162, C$\cdots$Cg = 3.7689(15)
C–H$\cdots$Ph[1+x, 1+y, z], H$\cdots$Cg = 2.86, C–H$\cdots$Cg = 167, C$\cdots$Cg = 3.8079(14)

Table 2. Hydrogen bonds (Å) in title compounds.

Compound	Synthon	Details
1	$NH_3^+\cdots{}^-OOC$	N$\cdots$O[2−x, 1−y, 2−z] = 2.740(3)
	$NH_3^+\cdots{}^-OOC$	N$\cdots$O[x, 1+y, z] = 2.817(3)
	OH$\cdots{}^-$OOC	O$\cdots$O = 2.739(3)
2a	$NH_3^+\cdots{}^-OOC$	N$\cdots$O = 2.8147(14)
	$NH_3^+\cdots{}^-OOC$	N$\cdots$O[0.5+x, 0.5−y, 0.5+z] = 2.8216(16)
	$NH_3^+\cdots$OH	N$\cdots$O[1−x, −y, 1−z] = 2.9409(14)
	OH$\cdots{}^-$OOC	O$\cdots$O = 2.6178(12)
2b	$NH_3^+\cdots{}^-OOC$	N$\cdots$O = 2.772(3)
	$NH_3^+\cdots{}^-OOC$	N$\cdots$O[1+x, y, z] = 2.948(2)
	$NH_3^+\cdots{}^-OOC$	N$\cdots$O[1−x, 1−y, 1−z] = 3.008(2)
	$NH_3^+\cdots$N(quin$^-$)	N$\cdots$N[1+x, y, z] = 3.080(3)
	OH$\cdots{}^-$OOC	O$\cdots$O = 2.791(2)
3a	$NH_3^+\cdots{}^-OOC$	N$\cdots$O = 2.7404(18)
	$NH_3^+\cdots{}^-OOC$	N$\cdots$O[−1+x, y, z] = 2.7768(18)
	$NH_3^+\cdots{}^-OOC$	N$\cdots$O[1−x, 1−y, 1−z] = 2.7976(16)
	OH$\cdots$N(quin$^-$)	O$\cdots$N[−1+x, y, z] = 2.8187(17)
3b	$NH_3^+\cdots{}^-OOC$	N$\cdots$O[1−x, 2−y, 2−z] = 2.7307(13)
	$NH_3^+\cdots{}^-OOC$	N$\cdots$O[x, 1+y, z] = 2.8525(13)
	$NH_3^+\cdots$OH	N$\cdots$O[1−x, 2−y, 2−z] = 2.8148(12)
	OH$\cdots{}^-$OOC	O$\cdots$O = 2.6222(12)

Table 3. Heterosynthon occurrence in the structures of title compounds.

	1	2a	2b	3a	3b
$NH_3^+\cdots{}^-OOC$	✓	✓	✓	✓	✓
$OH\cdots{}^-OOC$	✓	✓	✓		✓
$NH_3^+\cdots N(quin^-)$	✓ [a]	✓ [a]	✓		✓ [a]
$NH_3^+\cdots OH$		✓			✓
$OH\cdots N(quin^-)$				✓	

[a] Weak interaction. The N⋯N contact is longer than the sum of the corresponding van der Waals radii, 3.1 Å [17].

The crystal structure of **1** consists of 3a1pOHH$^+$ cations and strictly planar quinaldinate ions. All hydrogen bond donors and acceptors participate in intermolecular interactions. The quinaldinate nitrogen interacts only weakly with the NH$_3^+$ group: the corresponding N⋯N distance amounts to 3.108(3) Å, the value that is almost the same as the sum of the van der Waals radii for nitrogen atoms, 3.1 Å [17]. The connectivity pattern consists of two types of hydrogen bonds: the OH⋯$^-$OOC and the NH$_3^+$⋯$^-$OOC hydrogen bonds. Each type occurs between the cation and the anion. The hydrogen bonding pattern produces infinite layers, which are coplanar with the *ab* plane and stack along the *c* crystallographic axis. Section of such a layer is depicted in Figure 2. The layers stack upon one another with significant π⋯π stacking interactions occurring between quinaldinates from adjacent layers (Figure S4). Parameters of the shortest π⋯π stacking interaction are Ph⋯Py type, Cg⋯Cg = 3.6571(15) Å, dihedral angle = 0.41(11)°, shift distance = 1.346 Å.

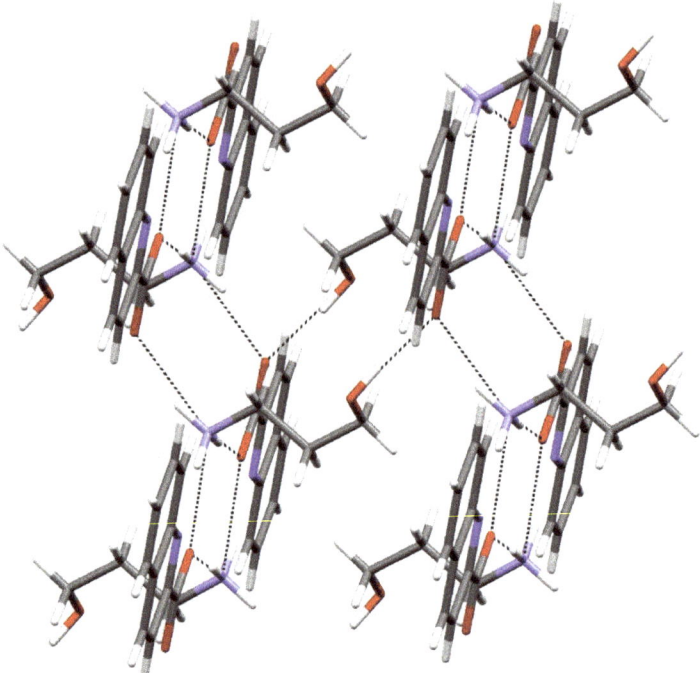

Figure 2. Perpendicular view to the section of a layer of hydrogen-bonded cations and anions in the structure of **1**.

The 2-amino-1-butanol salt was found in two polymorphic forms, **2a** and **2b**. Both crystallize in a monoclinic $P\,2_1/n$ unit cell. The quinaldinates of **2a** are non-planar with the twist angle of 11.4(2)°, whereas those of **2b** are nearly planar. The structures of both feature the OH···⁻OOC and the NH₃⁺···⁻OOC synthons. In **2a**, a weak interaction occurs between NH₃⁺ and OH groups. Once again, in neither of the two structures, the quinaldinate nitrogen is engaged in stronger intermolecular interactions. Its shortest contact occurs with the NH₃⁺ group with the corresponding N···N distance being 3.1669(16) Å (**2a**) or 3.080(3) Å (**2b**). Hydrogen bonds link cations and anions into layers (polymorph **2a**, Figure 3) or into chains (polymorph **2b**, Figure 4). In **2a**, significant $\pi\cdots\pi$ stacking interactions occur between quinaldinates from adjacent layers (Figure S5). Parameters of the shortest $\pi\cdots\pi$ stacking interaction are Ph···Py type, dihedral angle = 0.39(7)°, $Cg\cdots Cg$ = 3.5163(9) Å, and shift distance = 1.123 Å. The packing of chains in **2b** is such that no $\pi\cdots\pi$ stacking occurs.

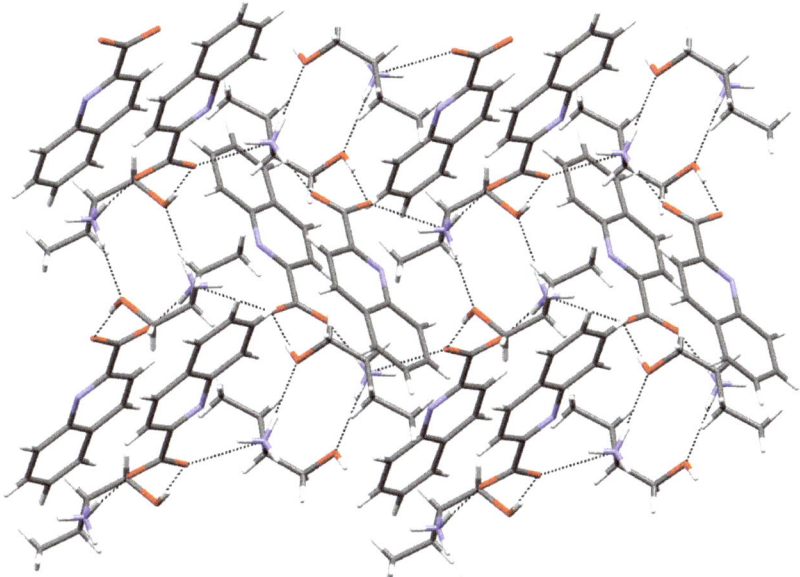

Figure 3. A perpendicular view to the layer in **2a**.

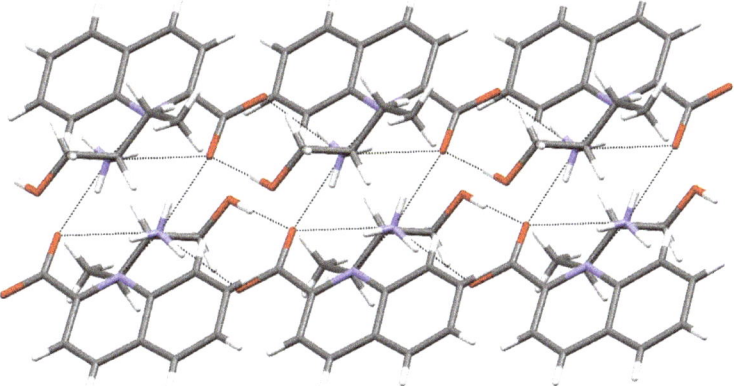

Figure 4. Section of a chain in **2b**.

The 2-amino-2-methyl-1-propanol salt also exists in two polymorphic forms. The one that crystallizes in a monoclinic $P\,2_1/n$ cell was labeled **3a**, and the one that crystallizes in a triclinic P−1 cell was labeled **3b**. The quinaldinates of the **3a** polymorph are non-planar with the twist angle of 25.48(10)°. Apart from the usual synthon, the $NH_3^+\cdots{}^-OOC$ hydrogen bond, there is a short contact between the hydroxyl group of the 2a2m1pOHH$^+$ cation and the quinaldinate nitrogen with the O⋯N distance being 2.8187(17) Å. The $NH_3^+\cdots{}^-OOC$ and the OH⋯N(quin$^-$) hydrogen bonds link ions into chains, which propagate along *a* crystallographic axis (Figure 5). The chains pack in a parallel fashion without any π⋯π stacking interactions.

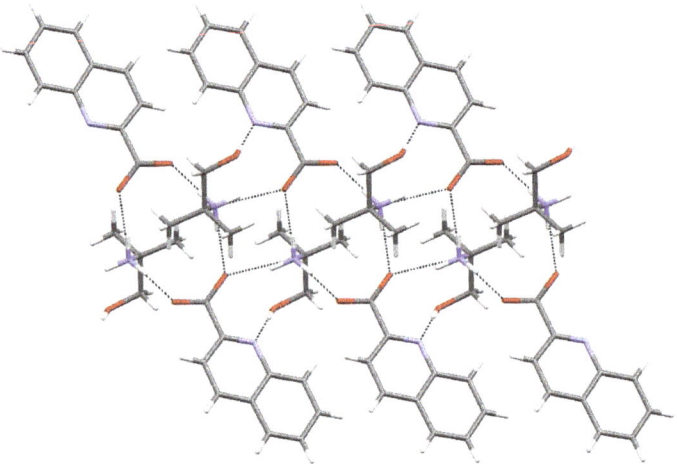

Figure 5. Section of a chain in **3a**.

The **3b** polymorph also consists of infinite chains. The chains propagate along *b* crystallographic axis. Yet, the hydrogen bonding motif markedly differs from that in **3a**. Firstly, the quinaldinate nitrogen is engaged in weak interaction with the adjacent NH_3^+ moiety. The corresponding N⋯N contact is 3.1037(14) Å. In the infinite chain, the following synthons may be recognized: in addition to the usual $NH_3^+\cdots{}^-OOC$ and OH⋯$^-$OOC hydrogen bonds, there is also the NH_3^+⋯OH hydrogen bond that links the cations (Figure 6). Of the two polymorphs, only **3b** displays hydrogen-bonding interactions between the cations. The packing of the chains is such that it allows π⋯π stacking interactions between neighboring chains (Figure S6). The quinaldinates are again non-planar with the 17.26(9)° twist angle.

Products obtained upon a direct reaction of a specific amino alcohol and quinaldinic acid may be classified as salts. The combinations involving amines and carboxylic acids do not always produce salts. The frequently employed ΔpK_a rule in predicting the nature of the product [18], ionic (a salt) or neutral (a co-crystal), can give indefinite answers. It has been stated that with the difference between the pK_a of the base and the pK_a of the acid in the −1 to 4 interval, the ionization of functional groups depends upon the whole crystal packing [18], and the product classification depends upon the position of the proton along a N⋯O hydrogen bond [19]. The combinations of amino alcohols, used in place of amines, and quinaldinic acid (quinoline-2-carboxylic acid) result in the ΔpK_a values that do not fall into the −1 to 4 domain. Although the hydroxyl group lowers the pK_a value relative to the "parent" alkylamine (For example, pK_a of 2-aminoethanol is by 1.15 unit lower than pK_a of ethylamine, 9.50 vs. 10.65 [20]), it is the quinaldinic acid that swings the balance in favor of the salt formation. The salt formation was further confirmed for all title compounds in the process of structure refinement by the location of proton in the electron difference maps.

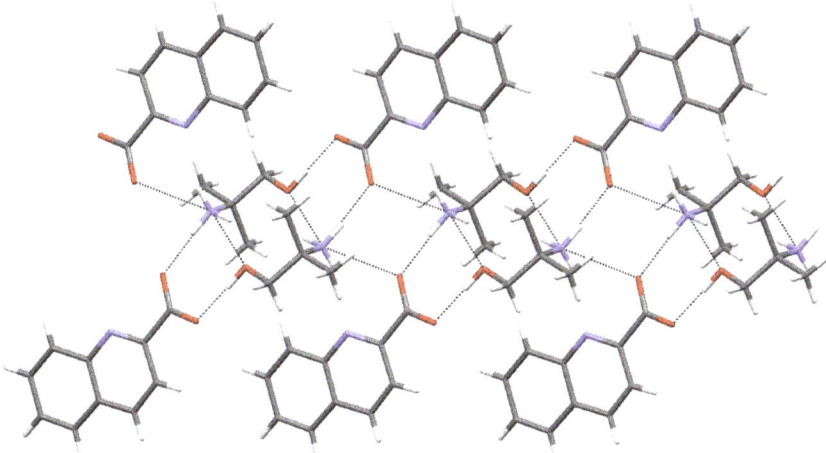

Figure 6. Section of a chain in **3b**.

The components of the three salts contain the same functional groups. Similar connectivity patterns are thus expected. The following discussion shows to what extent this expectation was realized. It is to be noted that three compounds present a very limited data set. The general validity of the conclusions is thus to be treated with caution. Firstly, in the solid-state structures of title compounds, all good proton donors and acceptors are used in the intermolecular connectivity. All five structures conform to the predicted synthon hierarchy [2]: only the heterosynthons may be displayed and no homomeric ones. As shown in Table 3, all our salts feature the $NH_3^+\cdots{}^-OOC$ synthon. The second one in the order of occurrence is the $OH\cdots{}^-OOC$ synthon, which is observed in all but **3a**. Interestingly, its formation is with no exception accompanied by a weak $NH_3^+\cdots N(quin^-)$ interaction. The **3a** salt, which lacks the $OH\cdots{}^-OOC$ interaction, also lacks the $NH_3^+\cdots N(quin^-)$ interaction. The absence of the $NH_3^+\cdots N(quin^-)$ interaction in **3a** is compensated by the $OH\cdots N(quin^-)$ hydrogen bond. The salt **3a** is the only compound that demonstrates this type of hydrogen bond; **3b**, the other (2a2m1pOHH)quin polymorph, also displays a specific feature, a $NH_3^+\cdots OH$ interaction. The latter is of interest because it occurs between ions of the same type, i.e., the 2a2m1pOHH$^+$ cations. The survey reveals that **1** and **2b** feature the same heterosynthons. The same observation pertains to the **2a/3b** pair. The **3a** polymorph differs from the other four structures. According to the literature, each pair of polymorphs, the **2a/2b** polymorphs and the **3a/3b** polymorphs, with differences in hydrogen bonding between their components may be thus classified as hydrogen bond isomers of the same solid [21]. The **2a/2b** polymorphs crystallized from the same reaction mixture, as opposed to the **3a/3b** polymorphs, which crystallized from different reaction mixtures. The **2a/2b** polymorphs are therefore concomitant polymorphs [22]. The structures of **2a** and **2b** reveal another important difference. Whereas **2a** features $\pi\cdots\pi$ stacking of quinaldinates, this type of interaction is lacking in **2b**. The same difference pertains to the **3a/3b** pair. On the other hand, the structures of all four share a common feature: the $C-H\cdots\pi$ type interactions.

The structures of **1–3b** have some structural features in common. The observed differences are a result of a complex interplay of short- and long-range intermolecular interactions that govern the supramolecular assembly during the crystallization procedure. Yet, each structure thus presents a specific situation and as such conforms with the current opinion in the field of crystal engineering that it is impossible to predict all molecular recognition events during the crystallization.

3. Materials and Methods

General. All reagents but acetonitrile were obtained from commercial sources (Aldrich and Fluorochem) and used as received. Acetonitrile was dried over molecular sieves [23]. In the case of the 2-amino-1-butanol reagent, a racemic mixture was used. The copper starting material, [Cu(quin)$_2$(H$_2$O)], was synthesized as previously reported [24]. Infrared (IR) spectra were recorded with the ATR module in the 4000–400 cm^{-1} spectral range on a Bruker Alpha II FT-IR spectrophotometer (Bruker, Manhattan, MA, USA). No corrections were made to the spectra. The spectra of all reveal strong bands in the 1560–1520 and 1370–1360 cm^{-1} spectral regions, which may be assigned as the ν_{as}(COO$^-$) and ν_s(COO$^-$) absorptions of the ionized quinaldinate. The engagement of the OH and NH$_3^+$ functional groups in hydrogen bonding prevents unambiguous identification of the stretching/deformation bands of these functional groups. ^{1}H nuclear magnetic resonance (NMR) spectra were recorded at 500 MHz on a Bruker Avance III 500 (Bruker BioSpin GmbH, Rheinstetten, Germany). The solvent was (CD$_3$)$_2$SO (DMSO-d_6) containing 0.03% tetramethylsilane (TMS), and all spectra were referenced to the central peak of the residual resonance for DMSO-d_6 at 2.50 ppm [25]. ^{1}H NMR spectra were processed using the MestReNova program [26]. Chemical shifts (δ) are given in ppm and coupling constants (J) in Hz. Multiplicities are labeled as follows: s = singlet, d = doublet, t = triplet, dd = doublet of doublet, and m = multiplet. Elemental analysis CHN was performed on a Perkin-Elmer 2400 II analyzer. Powder X-ray diffraction (PXRD) patterns were collected on a PANanlytical X'Pert PRO MD diffractometer (PANALYTICAL, Almelo, The Netherlands) using monochromatised Cu-K$_\alpha$ radiation (λ = 1.5406 Å). Thermogravimetric analyses were performed on a Mettler Toledo TG/DSC 1 instrument (Mettler Toledo, Schwerzenbach, Switzerland). Samples were placed into a 150 µL platinum crucible. Initial masses of samples were around 10 mg. Samples were heated from 25 to 450 °C with a heating rate of 10 °C min^{-1} and the furnace was purged with air at a flow rate of 50 mL min^{-1}. The baseline was subtracted. All three salts are stable up to about 120 °C and then the decomposition processes take place. No phase transitions were observed in the 25–120 °C temperature range.

(3a1pOHH)quin (1). Quinaldinic acid (100 mg, 0.58 mmol), methanol (10 mL), and 3-amino-1-propanol (88 µL) were added to an Erlenmeyer flask. The mixture was stirred until all the solid was consumed. The resulting solution was left to stand at ambient conditions. On the following day, it was concentrated under reduced pressure on a rotary evaporator. A glass vial with diethyl ether was carefully inserted into the Erlenmeyer flask with the concentrate. Colorless crystals of (3a1pOHH)quin were filtered off. Yield: 106 mg, 74%. Notes. The identity of the product was confirmed by PXRD (Figure S1). Single crystals of **1** were obtained as follows. A Teflon container was filled with CuO (50 mg, 0.63 mmol), quinaldinic acid (120 mg, 0.69 mmol), acetonitrile (7.5 mL), and 3-amino-1-propanol (150 mg). The container was closed and inserted into a steel autoclave, which was heated for 24 h at 105 °C. Afterwards, the reaction mixture was allowed to cool slowly to room temperature. Black solid was filtered off, and the resulting green filtrate was concentrated under reduced pressure on a rotary evaporator. The concentrate was stored at 4 °C. A mixture of colorless crystals of (3a1pOHH)quin (**1**) and blue needle-like crystals of *trans*-[Cu(quin)$_2$(3a1pOH)$_2$] was obtained. ^{1}H NMR (500 MHz, DMSO-d_6 with 0.03% v/v TMS): δ 8.30 (1H, d, J = 8.4 Hz, quin$^-$), 8.12 (1H, d, J = 8.4 Hz, quin$^-$), 8.04 (1H, d, J = 8.4 Hz, quin$^-$), 7.95 (1H, dd, J = 8.2, 1.1 Hz, quin$^-$), 7.74–7.71 (1H, m, quin$^-$), 7.60–7.57 (1H, m, quin$^-$), 3.50 (2H, t, J = 6.0 Hz, 3a1pOHH$^+$), 2.95 (2H, t, J = 7.3 Hz, 3a1pOHH$^+$), 1.80–1.74 (2H, m, 3a1pOHH$^+$) ppm. Elemental analysis calcd. for C$_{13}$H$_{16}$N$_2$O$_3$ (%): C, 62.89; H, 6.50; N, 11.28. Found (%): C, 62.80; H, 6.38; N, 11.35. IR (ATR, cm^{-1}): 3352m, 3061m, 2989m, 2947m, 2888m, 2745m, 2503w, 2090w, 1593s, 1559s, 1519s, 1502s, 1475m, 1462s, 1425s, 1384s, 1372vs, 1335s, 1298m, 1276m, 1214m, 1184m, 1168s, 1145w, 1129w, 1104m, 1068m, 1042m, 1023m, 1001m, 950w, 903s, 891s, 878m, 849m, 790vs, 776vvs, 746s, 629s, 592s, 541w, 530m, 520s, 500w, 477m.

(2a1bOHH)quin (2). Quinaldinic acid (100 mg, 0.58 mmol), methanol (10 mL), and 2-amino-1-butanol (109 µL) were added to an Erlenmeyer flask. The mixture was stirred until all the solid was consumed. The resulting solution was left to stand at ambient conditions. On the following day, it was concentrated under reduced pressure on a rotary evaporator. A glass vial with diethyl ether was carefully inserted into the Erlenmeyer flask with the concentrate. Colorless, needle-like crystals of (2a1bOHH)quin were filtered off. Yield: 116 mg, 77%. Notes. PXRD confirmed that the product is mostly **2b** polymorph (Figure S2). Single crystals of **2a** and **2b** polymorphs were obtained as follows. [Cu(quin)$_2$(H$_2$O)] (50 mg, 0.12 mmol), nitromethane (7.5 mL) and 2-amino-1-butanol (0.25 mL) were added to an Erlenmeyer flask. The mixture was stirred thoroughly until all the solid was consumed. After a few days, a mixture of crystals of **2a** and **2b** polymorphs was obtained. ^{1}H NMR (500 MHz, DMSO-d_6 with 0.03% v/v TMS): δ 8.30 (1H, d, J = 8.4 Hz, quin$^-$), 8.08 (1H, d, J = 8.5 Hz, quin$^-$), 8.00 (1H, d, J = 8.4 Hz, quin$^-$), 7.95 (1H, dd, J = 8.1, 1.4 Hz, quin$^-$), 7.76–7.72 (1H, m, quin$^-$), 7.60–7.57 (1H, m, quin$^-$), 3.64 (1H, dd, J = 11.7, 3.8 Hz, 2a1bOHH$^+$), 3.50 (1H, dd, J = 11.7, 6.2 Hz, 2a1bOHH$^+$), 3.04–2.99 (1H, m, 2a1bOHH$^+$), 1.62–1.53 (2H, m, 2a1bOHH$^+$), 0.91 (3H, t, J = 7.5 Hz, 2a1bOHH$^+$) ppm. Elemental analysis calcd. for C$_{14}$H$_{18}$N$_2$O$_3$ (%): C, 64.11; H, 6.92; N, 10.68. Found (%): C, 64.06; H, 6.68; N, 10.77. IR of **2a** polymorph (ATR, cm^{-1}): 3017w, 2965m, 2935m, 2873m, 2752m, 2635m, 2072w, 1594s, 1576s, 1554s, 1501s, 1462s, 1428m, 1371vvs, 1346s, 1306m, 1288w, 1272w, 1254m, 1219w, 1205m, 1171s, 1151m, 1133m, 1111w, 1066s, 1041s, 988s, 967w, 953m, 890m, 861s, 810s, 782vvs, 761s, 747s, 661m, 626s, 592s, 547w, 526w, 499m, 478m, 469m, 439m. IR of **2b** polymorph (ATR, cm^{-1}): 3232w, 3063m, 2963m, 2936m, 2868m, 1590m, 1553s, 1519s, 1503s, 1459s, 1427m, 1388s, 1367vs, 1340s, 1253w, 1219m, 1205m, 1170m, 1148m, 1068s, 1011w, 973w, 954w, 917w, 892m, 863s, 811s, 787vvs, 753m, 689m, 627s, 596s, 543w, 520m, 506m, 479w, 455w.

(2a2m1pOHH)quin (3). Quinaldinic acid (100 mg, 0.58 mmol), methanol (10 mL), and 2-amino-2-methyl-1-propanol (108 µL) were added to an Erlenmeyer flask. The mixture was stirred until all the solid was consumed. The resulting solution was left to stand at ambient conditions. On the following day, it was concentrated under reduced pressure on a rotary evaporator. A glass vial with diethyl ether was carefully inserted into the Erlenmeyer flask with the concentrate. Colorless crystals of (2a2m1pOHH)quin were filtered off. Yield: 111 mg, 73%. Notes. PXRD confirmed that the product is mostly **3b** polymorph (Figure S3). Single crystals of **3a** polymorph were obtained as follows. [Cu(quin)$_2$(H$_2$O)] (50 mg, 0.12 mmol), acetonitrile (7.5 mL), and 2-amino-2-methyl-1-propanol (0.5 mL) were added to an Erlenmeyer flask. The mixture was stirred thoroughly until all the solid was consumed. The resulting blue solution was left to stand at ambient conditions. On the following day, a mixture of colorless, needle-like crystals of **3a** polymorph and blue crystalline solid *syn*-[Cu$_2$(quin)$_2$(2a2m1pO)$_2$] was obtained. Single crystals of **3b** polymorph were obtained as follows. Teflon container was filled with [Cu(quin)$_2$(H$_2$O)] (50 mg, 0.12 mmol), acetonitrile (7.5 mL) and 2-amino-2-methyl-1-propanol (0.5 mL). The container was closed and inserted into a steel autoclave, which was heated for 24 h at 105 °C. Afterwards, the reaction mixture was allowed to cool slowly to room temperature. The resulting blue solution was left to stand at ambient conditions. After a few days, a mixture of colorless, needle-like crystals of **3b** polymorph and blue crystalline solid *syn*-[Cu$_2$(quin)$_2$(2a2m1pO)$_2$] was obtained. ^{1}H NMR (500 MHz, DMSO-d_6 with 0.03% v/v TMS): δ 8.30 (1H, d, J = 8.4 Hz, quin$^-$), 8.10 (1H, d, J = 8.5 Hz, quin$^-$), 8.01 (1H, d, J = 8.4 Hz, quin$^-$), 7.94 (1H, dd, J = 8.1, 1.4 Hz, quin$^-$), 7.75–7.72 (1H, m, quin$^-$), 7.60–7.57 (1H, m, quin$^-$), 3.43 (s, 2H, 2a2m1pOHH$^+$), 1.22 (s, 6H, 2a2m1pOHH$^+$) ppm. Elemental analysis calcd. for C$_{14}$H$_{18}$N$_2$O$_3$ (%): C, 64.11; H, 6.92; N, 10.68. Found (%): C, 63.97; H, 6.64; N, 10.71. IR of **3a** polymorph (ATR, cm^{-1}): 3185w, 2980m, 2894m, 2829s, 2724m, 2633m, 2593m, 2543m, 2168w, 1630s, 1578s, 1549vs, 1503m, 1482m, 1467s, 1426m, 1385vs, 1372vs, 1345s, 1327m, 1299m, 1264s, 1213w, 1173s, 1148m, 1114m, 1095m, 1067vs, 1009w, 980w, 958w, 946w, 912w, 893m, 873m, 853m, 804s, 778vvs, 752s, 737s, 697m, 651m, 630s, 592s, 551m, 523m, 480m, 459vvs, 421m. IR of **3b** polymorph (ATR, cm^{-1}): 3173w, 3010m, 2987m, 2975m, 2910m, 2831m, 2683m, 2583m, 2499m, 1619s, 1544vs, 1502s, 1474m, 1458s, 1423m, 1384vs, 1371vs, 1349s, 1307m, 1274s, 1251s, 1211m, 1192m, 1173m, 1144m, 1108m, 1093s, 1067s, 1017w, 999w, 988w, 972w, 952m, 919m, 888m, 873s, 834m, 803s, 775vvs, 739vs, 640w, 627s, 594s, 552m, 522m, 492m, 475m, 453s.

X-ray diffraction analysis. Agilent SuperNova diffractometer (Agilent Technologies XRD Products, Oxfordshire, UK) with molybdenum (Mo-K$_\alpha$, λ = 0.71073 Å) micro-focus sealed X-ray source was used to obtain X-ray diffraction data on single crystal at 150 K. The diffractometer was equipped with mirror optics and an Atlas detector. The crystals were placed on a glass fiber tip with silicon grease, which was mounted on the goniometer head. CrysAlis PRO [27] was used for data processing. Structures were solved with Olex^2 software [28] using intrinsic phasing in ShelXT [29] and refined with the least squares method in ShelXL [30]. Anisotropic displacement parameters were determined for all non-hydrogen atoms. With the exception of **2b**, NH$_3^+$ and OH hydrogen atoms of protonated amino alcohols were located from a difference Fourier map and refined with isotropic displacement parameters. Owing to the residual density in **2b**, the hydrogen atoms of NH$_3^+$ moiety were added in calculated positions. The residual density, i.e., a 2.30 e$^-$/Å^3 peak on a special position with too-short contacts to adjacent atoms, could not be interpreted. The data set, obtained from a crystal from a different batch, revealed the same problem. The remaining hydrogen atoms were placed in geometrically calculated positions in all structures and refined using riding models. Crystal structure analysis was performed with the program Platon [31], while the figures were made with Mercury [32]. The crystallographic data are summarized in Table 4. All crystal structures were deposited to the Cambridge Crystallographic Data Center (CCDC) and were assigned deposition numbers 2100261 (**1**), 2100262 (**2a**), 2100263 (**2b**), 2100264 (**3a**), and 2100265 (**3b**). These data can be obtained free of charge via http://www.ccdc.cam.ac.uk/conts/retrieving.html (accessed on 15 October 2021) (or from the CCDC, 12 Union Road, Cambridge CB2 1EZ, UK; Fax: +44 1223 336033; E-mail: deposit@ccdc.cam.ac.uk).

Table 4. Crystallographic data for **1–3b**.

	1	2a	2b	3a	3b
Empirical Formula	$C_{13}H_{16}N_2O_3$	$C_{14}H_{18}N_2O_3$	$C_{14}H_{18}N_2O_3$	$C_{14}H_{18}N_2O_3$	$C_{14}H_{18}N_2O_3$
Formula Weight	248.28	262.30	262.30	262.30	262.30
Crystal System	triclinic	monoclinic	monoclinic	monoclinic	triclinic
Space Group	$P-1$	$P\,2_1/n$	$P\,2_1/n$	$P\,2_1/n$	$P-1$
T (K)	150.00(10)	150.00(10)	150.00(10)	150.00(10)	150.00(10)
λ (Å)	0.71073	0.71073	0.71073	0.71073	0.71073
a (Å)	7.1378(16)	12.1437(11)	6.5579(4)	6.5428(4)	7.1342(4)
b (Å)	7.5269(7)	10.1451(5)	10.2309(6)	9.0723(4)	8.4346(3)
c (Å)	11.8314(14)	12.2312(15)	19.8329(13)	23.1232(10)	12.5059(7)
α (°)	99.172(9)	90	90	90	96.139(4)
β (°)	95.916(14)	119.527(14)	97.837(6)	93.835(5)	105.187(5)
γ (°)	90.647(13)	90	90	90	104.829(4)
V (Å^3)	623.93(17)	1311.2(3)	1318.22(14)	1369.48(12)	689.74(6)
Z	2	4	4	4	2
D_{calc} (g/cm^3)	1.322	1.329	1.322	1.272	1.263
μ (mm^{-1})	0.095	0.094	0.094	0.090	0.090
Collected Reflections	5349	11,880	6877	12,825	12,006
Unique Reflections	3186	3524	3413	3696	3689
Observed Reflections	1937	2780	2459	2528	2933
R_{int}	0.0587	0.0285	0.0233	0.0482	0.0224
R_1 ($I > 2\sigma(I)$)	0.0878	0.0413	0.0671	0.0513	0.0429
wR_2 (all data)	0.2559	0.1168	0.2000	0.1225	0.1274

4. Conclusions

Reactions of amino alcohols (3-amino-1-propanol, 2-amino-1-butanol, or 2-amino-2-methyl-1-propanol) and quinaldinic acid have produced salts, which consist of protonated amino alcohol and deprotonated quinaldinic acid. The obtained products obey the ΔpK_a rule. Of the three products, (3a1pOHH)quin (**1**), (2a1bOHH)quin (**2**), and (2a2m1pOHH)quin (**3**), the last two are polymorphic. A structural survey has revealed all five possible heterosynthons in their crystal structures. The supramolecular structures of all are built of the $NH_3^+\cdots{}^-OOC$ synthon in combination with one to up to three other heterosynthons. Interestingly, the $OH\cdots N(quin^-)$ synthon occurs only in one phase. The **2a/2b** and **3a/3b** polymorphic pairs differ both in the types of hydrogen bonds and in $\pi\cdots\pi$ stacking interactions. Due to the former, they are hydrogen bond isomers of the same compound. The presented series is yet another demonstration of polymorphism among molecular solids.

Supplementary Materials: The following supporting information can be downloaded, PXRD patterns (Figures S1–S3), packing diagrams (Figures S4–S6), IR spectra for 1–3b (Figures S7–S11), 1H NMR spectra for 1–3 (Figures S12–S14), and TG/DSC curves (Figures S15–S17).

Author Contributions: Conceptualization, B.M. and N.P.; validation, B.M. and N.P.; writing—review and editing, B.M. and N.P. All authors have read and agreed to the published version of the manuscript.

Funding: This research was funded by Slovenian Research Agency (Junior Researcher Grant for N.P. and the Program Grant P1-0134).

Institutional Review Board Statement: Not applicable.

Informed Consent Statement: Not applicable.

Acknowledgments: The authors thank Romana Cerc Korošec for running the TG/DSC experiments.

Conflicts of Interest: The authors declare no conflict of interest.

Sample Availability: Obtained from the authors upon request.

References

1. Desiraju, G.R.; Vittal, J.J.; Ramanan, A. *Crystal Engineering: A Textbook*; World Scientific Publishing Co.: Singapore, 2011; pp. 1–3.
2. Corpinot, M.K.; Bučar, D.-K. A practical guide to the design of molecular crystals. *Cryst. Growth Des.* **2019**, *19*, 1426–1453. [CrossRef]
3. Etter, M.C. Encoding and decoding hydrogen-bond patterns of organic compounds. *Acc. Chem. Res.* **1990**, *23*, 120–126. [CrossRef]
4. Desiraju, G.R. Supramolecular synthons in crystal engineering—A new organic synthesis. *Angew. Chem. Int. Ed. Engl.* **1995**, *34*, 2311–2327. [CrossRef]
5. Bučar, D.-K.; Lancaster, R.W.; Bernstein, J. Disappearing polymorphs revisited. *Angew. Chem. Int. Ed.* **2015**, *54*, 6972–6993. [CrossRef]
6. Corpinot, M.K.; Stratford, S.A.; Arhangelskis, M.; Anka-Lufford, J.; Halasz, I.; Judaš, N.; Jones, W.; Bučar, D.-K. On the predictability of supramolecular interactions in molecular cocrystals—The view from the bench. *CrysEngComm* **2016**, *18*, 5434–5439. [CrossRef]
7. Bernstein, J. *Polymorphism in Molecular Crystals*; Oxford University Press: Oxford, UK, 2002; pp. 1–9.
8. Gavezzotti, A.; Filippini, G. Polymorphic forms of organic crystals at room conditions: Thermodynamic and structural implications. *J. Am. Chem. Soc.* **1995**, *117*, 12299–12305. [CrossRef]
9. McCrone, W.C. *Physics and Chemistry of the Organic Solid State*; Fox, D., Labes, M.M., Weissberger, A., Eds.; Interscience: New York, NY, USA, 1965; Volume 2, pp. 725–767.
10. Morissette, S.L.; Almarsson, Ö.; Peterson, M.L.; Remenar, J.F.; Read, M.J.; Lemmo, A.V.; Ellis, S.; Cima, M.J.; Gardner, C.R. High-throughput crystallization: Polymorphs, salts, co-crystals and solvates of pharmaceutical solids. *Adv. Drug Delivery Rev.* **2004**, *56*, 275–300. [CrossRef] [PubMed]
11. David, W.I.F.; Shankland, K.; Pulham, C.R.; Blagden, N.; Davey, R.J.; Song, M. Polymorphism in benzamide. *Angew. Chem. Int. Ed.* **2005**, *44*, 7032–7035. [CrossRef]
12. Stahly, G.P. Diversity in single- and multiple-component crystals. The search for and prevalence of polymorphs and cocrystals. *Cryst. Growth Des.* **2007**, *7*, 1007–1026. [CrossRef]
13. Cruz-Cabeza, A.J.; Reutzel-Edens, S.M.; Bernstein, J. Facts and fictions about polymorphism. *Chem. Soc. Rev.* **2015**, *44*, 8619–8635. [CrossRef]
14. Podjed, N.; Modec, B.; Clérac, R.; Rouzières, M.; Alcaide, M.M.; López-Serrano, J. Structural diversity and magnetic properties of copper(II) quinaldinate compounds with amino alcohols. Manuscript in Preparation. University of Ljubljana, Faculty of Chemistry and Chemical Technology: Ljubljana, Slovenia, 2022.
15. Tudor, V.; Mocanu, T.; Tuna, F.; Madalan, A.M.; Maxim, C.; Shova, S.; Andruh, M. Mixed ligand binuclear alkoxo-bridged copper(II) complexes derived from aminoalcohols and nitrogen ligands. *J. Mol. Struct.* **2013**, *1046*, 164–170. [CrossRef]

16. Janiak, C. A critical account on π–π stacking in metal complexes with aromatic nitrogen-containing ligands. *J. Chem. Soc. Dalton Trans.* **2000**, 3885–3896. [CrossRef]
17. Bondi, A. van der Waals volumes and radii. *J. Phys. Chem.* **1964**, *68*, 441–451. [CrossRef]
18. Cruz-Cabeza, A.J. Acid–base crystalline complexes and the pK_a rule. *CrystEngComm* **2012**, *14*, 6362–6365. [CrossRef]
19. Childs, S.L.; Stahly, G.P.; Park, A. The salt−cocrystal continuum: The influence of crystal structure on ionization state. *Mol. Pharm.* **2007**, *4*, 323–338. [CrossRef]
20. Haynes, W.M.; Lide, D.R.; Bruno, T.J. *CRC Handbook of Chemistry and Physics*, 97th ed.; CRC Press: Boca Raton, FL, USA, 2016; pp. 5–91.
21. Braga, D.; Maini, L. Solid-state *versus* solution preparation of two crystal forms of [HN(CH$_2$CH$_2$)$_3$NH][OOC(CH$_2$)COOH]$_2$. Polymorphs or hydrogen bond isomers? *Chem. Commun.* **2004**, 976–977. [CrossRef]
22. Bernstein, J.; Davey, R.J.; Henck, J.-O. Concomitant polymorphs. *Angew. Chem. Int. Ed.* **1999**, *38*, 3440–3461. [CrossRef]
23. Williams, D.B.G.; Lawton, M. Drying of organic solvents: Quantitative evaluation of the efficiency of several desiccants. *J. Org. Chem.* **2010**, *75*, 8351–8354. [CrossRef]
24. Modec, B.; Podjed, N.; Lah, N. Beyond the simple copper(II) coordination chemistry with quinaldinate and secondary amines. *Molecules* **2020**, *25*, 1573. [CrossRef]
25. Gottlieb, H.E.; Kotlyar, V.; Nudelman, A. NMR chemical shifts of common laboratory solvents as trace impurities. *J. Org. Chem.* **1997**, *62*, 7512–7515. [CrossRef]
26. Willcott, M.R. MestRe Nova. *J. Am. Chem. Soc.* **2009**, *131*, 13180. [CrossRef]
27. Agilent. *CrysAlis PRO*; Agilent Technologies Ltd.: Yarnton, UK, 2014.
28. Dolomanov, O.V.; Bourhis, L.J.; Gildea, R.J.; Howard, J.A.K.; Puschmann, H. *Olex2*: A complete structure solution, refinement and analysis program. *J. Appl. Crystallogr.* **2009**, *42*, 339–341. [CrossRef]
29. Sheldrick, G.M. *SHELXT*—Integrated space-group and crystal-structure determination. *Acta Crystallogr. Sect. A* **2015**, *71*, 3–8. [CrossRef] [PubMed]
30. Sheldrick, G.M. Crystal structure refinement with *SHELXL*. *Acta Crystallogr. Sect. C* **2015**, *71*, 3–8. [CrossRef]
31. Spek, A.L. Structure validation in chemical crystallography. *Acta Crystallogr. Sect. D* **2009**, *65*, 148–155. [CrossRef]
32. Macrae, C.F.; Bruno, I.J.; Chisholm, J.A.; Edgington, P.R.; McCabe, P.; Pidcock, E.; Rodriguez-Monge, L.; Taylor, R.; van de Streek, J.; Wood, P.A. *Mercury CSD 2.0*—New features for the visualization and investigation of crystal structures. *J. Appl. Crystallogr.* **2008**, *41*, 466–470. [CrossRef]

MDPI
St. Alban-Anlage 66
4052 Basel
Switzerland
Tel. +41 61 683 77 34
Fax +41 61 302 89 18
www.mdpi.com

Molecules Editorial Office
E-mail: molecules@mdpi.com
www.mdpi.com/journal/molecules

www.ingramcontent.com/pod-product-compliance
Lightning Source LLC
LaVergne TN
LVHW070427100526
838202LV00014B/1543